高等学校计算机专业教材精选·算法与程序设计

Web框架技术
(Struts2+Hibernate+Spring3)
教程

张志锋 马军霞 范乃梅 徐洁 石东海 等 编著

清华大学出版社
北京

内容简介

本书旨在培养学生的 Java Web 框架技术实践和创新能力，为广大读者提供一本 Java Web 框架知识综合学习以及积累项目开发经验的书。

全书理论联系实践，引进以项目为驱动的教学模式，详细系统地讲解 Struts2、Hibernate 和 Spring3 框架技术，将项目开发贯穿整个知识体系。全书共分 11 章，内容包括 Struts2 框架技术入门、Struts2 核心组件详解、Struts2 的高级组件、基于 Struts2 的个人信息管理系统项目实训、Hibernate 框架技术入门、Hibernate 核心组件详解、Hibernate 的高级组件、基于 Struts2＋Hibernate 的教务管理系统项目实训、Spring3 框架技术入门、Spring3 的 AOP 框架、基于 Struts2＋Hibernate＋Spring3 的校园论坛 BBS 项目实训。通过 30 多个小项目、3 个大项目的开发实践，使读者能够掌握基本理论知识，并锻炼综合应用能力。

本书可作为普通高等院校的 Java Web 框架技术教材，也可作为 Java 工程师培训教材或作为 Java 工程师的参考书。

本书封面贴有清华大学出版社防伪标签，无标签者不得销售。
版权所有，侵权必究。侵权举报电话：010-62782989　13701121933

图书在版编目(CIP)数据

Web 框架技术(Struts2＋Hibernate＋Spring3)教程/张志锋等编著．--北京：清华大学出版社，2013
(2016.7 重印)
高等学校计算机专业教材精选·算法与程序设计
ISBN 978-7-302-31945-0

Ⅰ．①W… Ⅱ．①张… Ⅲ．①JAVA 语言—程序设计—高等学校—教材 Ⅳ．①TP312

中国版本图书馆 CIP 数据核字(2013)第 080429 号

责任编辑：白立军
封面设计：傅瑞学
责任校对：白　蕾
责任印制：刘海龙

出版发行：清华大学出版社
　　网　　址：http://www.tup.com.cn，http://www.wqbook.com
　　地　　址：北京清华大学学研大厦 A 座　　　邮　　编：100084
　　社 总 机：010-62770175　　　　　　　　　邮　　购：010-62786544
　　投稿与读者服务：010-62776969，c-service@tup.tsinghua.edu.cn
　　质 量 反 馈：010-62772015，zhiliang@tup.tsinghua.edu.cn
　　课 件 下 载：http://www.tup.com.cn，010-62795954

印 装 者：北京国马印刷厂
经　　销：全国新华书店
开　　本：185mm×260mm　　印　张：37.75　　字　数：919 千字
版　　次：2013 年 5 月第 1 版　　　　　　　　印　次：2016 年 7 月第 4 次印刷
印　　数：5501～7500
定　　价：59.00 元

产品编号：045226-01

前　言

目前，在 Java 工程师招聘时，95%的企业要求应聘人员必须具备 Java Web 框架技术（Struts、Spring、Hibernate）的应用能力，因此 Java Web 框架技术是 Java 工程师必备的从业技能。

本教材区别于其他传统教程，在"卓越工程师教育培养计划"思想的指导下，引进"以项目为驱动的教学模式"，旨在培养学生 Java Web 框架知识综合应用技能以及积累项目开发经验。

本书主要章节以及内容安排如下。

第1章 Struts2 框架技术入门，主要介绍 Struts2 框架的由来与发展、工作原理、核心组件、配置与使用以及应用实例。

第2章 Struts2 核心组件详解，主要介绍 Struts2 核心组件、OGNL 和标签库及其应用。

第3章 Struts2 的高级组件，主要介绍 Struts2 框架的国际化及其应用，拦截器及其应用，输入验证及其应用和文件上传、下载功能及其应用。

第4章基于 Struts2 的个人信息管理系统项目实训，通过该项目的练习实现整合前三章所学知识，同时培养项目实现能力，积累项目开发经验。

第5章 Hibernate 框架技术入门，主要介绍 Hibernate 框架的由来与发展、工作原理、核心组件、配置与使用及应用实例。

第6章 Hibernate 核心组件详解，主要介绍 Hibernate 框架核心组件的使用。

第7章 Hibernate 的高级组件，主要介绍 Hibernate 框架关联关系及其应用、数据查询及其应用、事务和 Cache 及其应用。

第8章基于 Struts2+Hibernate 的教务管理系统项目实训，通过该项目的练习整合前面章节所学知识，进一步积累项目开发经验。

第9章 Spring3 框架技术入门，主要介绍 Spring3 的由来与发展、Spring3 的下载与配置、Spring3 框架的体系结构、Spring3 IoC 的原理和主要组件及其应用。

第10章 Spring3 的 AOP 框架，主要介绍 AOP 框架、Spring3 的 AOP 框架主要术语、代理、创建通知、定义切入点和创建引入等。

第11章基于 Struts2+Hibernate+Spring3 的校园论坛 BBS 项目实训，综合运用 SSH 技术，积累项目开发经验。

参与本书编写的有郑州轻工业学院的张志锋、范乃梅、马军霞、徐洁、梁树军、赵晓君、刘育熙、黄艳、崔建涛、甘琤、程源、马欢、江楠、李红婵、冯媛、蔡增玉和刘书如。本书主编张志锋，副主编马军霞、范乃梅、徐洁。此外，石东海、孙雪津、张文志、任筱芳、何保锋参与了本书的审稿和编写工作。在本书的编写和出版过程中得到了郑州轻工业学院、河南经贸职业学院、河南理工大学、西亚斯国际学院、清华大学出版社的大力支持和帮助，在此表示感谢。

由于编写时间仓促，水平有限，书中难免有纰漏之处，敬请读者不吝赐教。

本书提供配套教学资源，包括实例及项目源代码、教学课件、教学日历、教学大纲、课后习题参考答案、期末试卷以及未收入教材的多个大项目。如有需要可在清华大学出版社网站(www.tup.com.cn)下载。

编 者

2013 年 3 月

目　录

第1章　Struts2框架技术入门 …………………………………………………………… 1
1.1　Struts2基础知识 ………………………………………………………………… 1
1.1.1　Struts2的由来与发展 …………………………………………………… 1
1.1.2　Struts2软件包的下载和配置 …………………………………………… 2
1.1.3　MVC设计模式 …………………………………………………………… 13
1.1.4　Struts2的工作原理 ……………………………………………………… 14
1.2　Struts2的核心组件 ……………………………………………………………… 15
1.2.1　Struts2的控制器组件 …………………………………………………… 15
1.2.2　Struts2的模型组件 ……………………………………………………… 16
1.2.3　Struts2的视图组件 ……………………………………………………… 18
1.3　基于Struts2的登录系统 ………………………………………………………… 19
1.3.1　使用NetBeans7开发项目 ……………………………………………… 19
1.3.2　使用MyEclipse 10开发项目 …………………………………………… 25
1.3.3　使用Eclipse4开发项目 ………………………………………………… 32
1.4　本章小结 ………………………………………………………………………… 32
1.5　习题 ……………………………………………………………………………… 32
1.5.1　选择题 …………………………………………………………………… 32
1.5.2　填空题 …………………………………………………………………… 33
1.5.3　简答题 …………………………………………………………………… 33
1.5.4　实训题 …………………………………………………………………… 33

第2章　Struts2核心组件详解 …………………………………………………………… 34
2.1　Struts2的配置文件struts.xml …………………………………………………… 34
2.1.1　struts.xml配置文件的结构 ……………………………………………… 34
2.1.2　Bean配置 ………………………………………………………………… 35
2.1.3　常量配置 ………………………………………………………………… 36
2.1.4　包含配置 ………………………………………………………………… 37
2.1.5　包配置 …………………………………………………………………… 38
2.1.6　命名空间配置 …………………………………………………………… 39
2.1.7　Action配置 ……………………………………………………………… 40
2.1.8　结果配置 ………………………………………………………………… 40
2.1.9　拦截器配置 ……………………………………………………………… 40
2.2　Struts2的核心控制器FilterDispatcher …………………………………………… 41
2.3　Struts2的业务控制器Action ……………………………………………………… 41

  2.3.1 Action 接口和 ActionSupport 类 ……………… 41
  2.3.2 Action 实现类 ……………… 45
  2.3.3 Action 访问 ActionContext ……………… 47
  2.3.4 Action 直接访问 Servlet ……………… 51
  2.3.5 Action 中的动态方法调用 ……………… 54
 2.4 Struts2 的 OGNL 表达式 ……………… 59
  2.4.1 Struts2 的 OGNL 表达式 ……………… 59
  2.4.2 Struts2 的 OGNL 集合 ……………… 62
 2.5 Struts2 的标签库 ……………… 63
  2.5.1 Struts2 的标签库概述 ……………… 63
  2.5.2 Struts2 的表单标签 ……………… 64
  2.5.3 Struts2 的非表单标签 ……………… 71
  2.5.4 Struts2 的数据标签 ……………… 74
  2.5.5 Struts2 的控制标签 ……………… 81
  2.5.6 Struts2 的 Ajax 标签 ……………… 89
 2.6 本章小结 ……………… 111
 2.7 习题 ……………… 112
  2.7.1 选择题 ……………… 112
  2.7.2 填空题 ……………… 112
  2.7.3 简答题 ……………… 112
  2.7.4 实训题 ……………… 113

第 3 章 Struts2 的高级组件 ……………… 114

 3.1 Struts2 的国际化 ……………… 114
  3.1.1 Struts2 实现国际化的流程 ……………… 114
  3.1.2 Struts2 国际化应用实例 ……………… 117
 3.2 Struts2 的拦截器 ……………… 122
  3.2.1 Struts2 拦截器的基础知识 ……………… 122
  3.2.2 Struts2 拦截器实现类 ……………… 123
  3.2.3 Struts2 拦截器应用实例 ……………… 124
 3.3 Struts2 的输入校验 ……………… 128
  3.3.1 Struts2 输入验证的基础知识 ……………… 128
  3.3.2 Struts2 的手工验证 ……………… 135
  3.3.3 Struts2 内置校验器的使用 ……………… 139
  3.3.4 Struts2 内置校验器应用实例 ……………… 145
 3.4 Struts2 的文件上传和下载 ……………… 151
  3.4.1 文件上传 ……………… 151
  3.4.2 文件下载 ……………… 157
 3.5 本章小结 ……………… 163

3.6	习题		163
	3.6.1	选择题	163
	3.6.2	填空题	163
	3.6.3	简答题	163
	3.6.4	实训题	164

第4章 基于Struts2的个人信息管理系统项目实训 ... 165

4.1	项目需求说明		165
4.2	项目系统分析		165
4.3	项目数据库设计		166
4.4	项目实现		168
	4.4.1	项目文件结构	168
	4.4.2	用户登录和注册功能的实现	169
	4.4.3	系统主页面功能的实现	202
	4.4.4	个人信息管理功能实现	206
	4.4.5	通讯录管理功能实现	220
	4.4.6	日程安排管理功能实现	236
	4.4.7	个人文件管理功能实现	254
4.5	本章小结		255
4.6	习题		255
	实训题		255

第5章 Hibernate框架技术入门 ... 256

5.1	Hibernate基础知识		256
	5.1.1	Hibernate的发展与特点	256
	5.1.2	Hibernate软件包的下载和配置	257
	5.1.3	Hibernate的工作原理	261
5.2	Hibernate的核心组件		261
5.3	基于Struts2和Hibernate的登录和注册系统		263
5.4	本章小结		276
5.5	习题		276
	5.5.1	选择题	276
	5.5.2	填空题	276
	5.5.3	简答题	276
	5.5.4	实训题	277

第6章 Hibernate核心组件详解 ... 278

6.1	Hibernate的配置文件		278
	6.1.1	hibernate.cfg.xml	278

- 6.1.2 hibernate.properties ……………………………………………… 280
- 6.2 Hibernate 的 PO 对象 ……………………………………………………… 282
  - 6.2.1 Hibernate 的 PO 对象基础知识 ……………………………………… 282
  - 6.2.2 Hibernate 的 PO 对象状态 …………………………………………… 284
- 6.3 Hibernate 的映射文件 …………………………………………………… 285
- 6.4 Hibernate 的 Configuration 类 ………………………………………… 288
- 6.5 Hibernate 的 SessionFactory 接口 ……………………………………… 289
- 6.6 Hibernate 的 Session 接口 ……………………………………………… 289
  - 6.6.1 Session 接口的基础知识 ……………………………………………… 289
  - 6.6.2 通过方法获取持久化对象 …………………………………………… 290
  - 6.6.3 操作持久化对象的常用方法 ………………………………………… 291
- 6.7 Hibernate 的 Transaction 接口 ………………………………………… 294
- 6.8 Hibernate 的 Query 接口 ………………………………………………… 296
  - 6.8.1 Query 接口的基本知识 ………………………………………………… 296
  - 6.8.2 Query 接口的常用方法 ………………………………………………… 296
- 6.9 基于 Struts2＋Hibernate 的学生信息管理系统 ……………………… 298
  - 6.9.1 项目介绍、主页面以及查看学生信息功能的实现 ………………… 299
  - 6.9.2 添加学生信息功能的实现 …………………………………………… 308
  - 6.9.3 修改学生信息功能的实现 …………………………………………… 312
  - 6.9.4 删除学生信息功能的实现 …………………………………………… 320
- 6.10 本章小结 ………………………………………………………………… 323
- 6.11 习题 ……………………………………………………………………… 323
  - 6.11.1 选择题 ………………………………………………………………… 323
  - 6.11.2 填空题 ………………………………………………………………… 324
  - 6.11.3 简答题 ………………………………………………………………… 324
  - 6.11.4 实训题 ………………………………………………………………… 324

第 7 章 Hibernate 的高级组件 ……………………………………………… 325
- 7.1 利用关联关系操纵对象 ………………………………………………… 325
  - 7.1.1 一对一关联关系 ……………………………………………………… 325
  - 7.1.2 一对一关联关系的应用实例 ………………………………………… 328
  - 7.1.3 一对多关联关系 ……………………………………………………… 341
  - 7.1.4 一对多关联关系的应用实例 ………………………………………… 343
  - 7.1.5 多对多关联关系 ……………………………………………………… 350
  - 7.1.6 多对多关联关系的应用实例 ………………………………………… 350
- 7.2 Hibernate 数据查询 ……………………………………………………… 359
  - 7.2.1 Hibernate Query Language …………………………………………… 359
  - 7.2.2 Criteria Query 方式 …………………………………………………… 362
  - 7.2.3 Native SQL 查询 ……………………………………………………… 362

    7.3　Hibernate 的事务管理 ·· 362
        7.3.1　事务的特性 ·· 362
        7.3.2　事务隔离 ·· 363
        7.3.3　在 Hibernate 配置文件中设置隔离级别 ····································· 364
        7.3.4　在 Hibernate 中使用 JDBC 事务 ··· 364
        7.3.5　在 Hibernate 中使用 JTA 事务 ··· 365
    7.4　Hibernate 的 Cache 管理 ··· 365
        7.4.1　一级 Cache ··· 366
        7.4.2　二级 Cache ··· 366
    7.5　本章小结 ·· 367
    7.6　习题 ·· 367
        7.6.1　选择题 ·· 367
        7.6.2　填空题 ·· 368
        7.6.3　简答题 ·· 368
        7.6.4　实训题 ·· 368

第 8 章　基于 Struts2＋Hibernate 的教务管理系统项目实训 ·································· 369
    8.1　项目需求说明 ··· 369
    8.2　项目系统分析 ··· 369
    8.3　项目数据库设计 ··· 371
    8.4　项目实现 ·· 373
        8.4.1　项目文件结构 ·· 373
        8.4.2　用户登录功能的实现 ··· 374
        8.4.3　学生管理功能的实现 ··· 400
        8.4.4　管理员管理功能的实现 ··· 413
        8.4.5　教师管理功能的实现 ··· 419
    8.5　本章小结 ·· 419
    8.6　习题 ·· 419
        实训题 ·· 419

第 9 章　Spring3 框架技术入门 ··· 420
    9.1　Spring3 基础知识 ··· 420
        9.1.1　Spring3 的由来与发展 ··· 420
        9.1.2　Spring3 的下载与配置 ··· 421
        9.1.3　Spring3 框架的体系结构 ··· 423
    9.2　Spring3 IoC 的原理和主要组件 ··· 426
        9.2.1　IoC 的基础知识以及原理 ··· 426
        9.2.2　IoC 的主要组件 ·· 432
        9.2.3　IoC 的应用实例 ·· 434

		9.2.4 注入的两种方式 ·· 442

- 9.3 基于Struts2＋Hibernate＋Spring3的登录系统 ······································ 446
  - 9.3.1 项目介绍 ·· 446
  - 9.3.2 在web.xml中配置Struts2和Spring3 ································ 447
  - 9.3.3 编写视图组件 ·· 448
  - 9.3.4 Action和JavaBean ·· 449
  - 9.3.5 Struts2、Spring3和Hibernate的配置文件 ···························· 453
  - 9.3.6 Struts2、Spring3和Hibernate整合中常见问题 ······················· 455
  - 9.3.7 项目部署和运行 ·· 456
- 9.4 本章小结 ·· 457
- 9.5 习题 ·· 457
  - 9.5.1 选择题 ·· 457
  - 9.5.2 填空题 ·· 458
  - 9.5.3 简答题 ·· 458
  - 9.5.4 实训题 ·· 458

第10章 Spring3的AOP框架 ·· 459

- 10.1 AOP框架基础知识 ·· 459
  - 10.1.1 AOP框架简介 ·· 459
  - 10.1.2 Spring3的AOP框架主要术语 ································ 461
- 10.2 代理 ·· 462
  - 10.2.1 静态代理 ·· 462
  - 10.2.2 动态代理 ·· 465
- 10.3 创建通知 ·· 467
  - 10.3.1 前置通知及应用实例 ································ 467
  - 10.3.2 后置通知及应用实例 ································ 470
  - 10.3.3 环绕通知及应用实例 ································ 472
  - 10.3.4 异常通知及应用实例 ································ 474
  - 10.3.5 引入通知 ·· 476
- 10.4 定义切入点 ·· 477
  - 10.4.1 静态切入点和动态切入点 ································ 477
  - 10.4.2 切入点的应用实例 ································ 478
- 10.5 创建引入 ·· 482
- 10.6 本章小结 ·· 482
- 10.7 习题 ·· 483
  - 10.7.1 选择题 ·· 483
  - 10.7.2 填空题 ·· 483
  - 10.7.3 简答题 ·· 483
  - 10.7.4 实训题 ·· 483

第 11 章 基于 Struts2＋Hibernate＋Spring3 的校园论坛 BBS 项目实训 …… 484
11.1 项目需求分析 …… 484
11.2 项目分析与设计 …… 485
11.3 项目的数据库设计 …… 486
11.4 项目实现 …… 488
11.4.1 项目的文件结构和主页面 …… 488
11.4.2 BBS 登录功能的实现 …… 530
11.4.3 BBS 板块管理功能的实现 …… 539
11.4.4 BBS 帖子管理功能的实现 …… 551
11.4.5 个人信息管理功能的实现 …… 580
11.5 本章小结 …… 590
11.6 习题 …… 590
实训题 …… 590

参考文献 …… 591

第 1 章　Struts2 框架技术入门

随着 Web 应用程序复杂性的不断提高,单纯依靠某种技术,很难达到快速开发、快速验证和快速部署的目的。必须整合 Web 相关技术形成完整的开发框架或应用模型,以满足各种复杂应用程序的需求,Struts2 框架就是解决这一问题的 Java Web 框架技术之一。

本章主要内容如下所示。
(1) Struts2 的发展及其配置。
(2) MVC 设计模式。
(3) Struts2 的工作原理。
(4) Struts2 的核心组件。
(5) 基于 Struts2 的登录系统。

1.1　Struts2 基础知识

Struts2 是 Java Web 项目开发中最经典的框架技术,受到许多软件开发人员的喜爱与追捧,是软件企业招聘 Java 软件人才时要求必备的技能之一。

1.1.1　Struts2 的由来与发展

Struts 是整合了当前动态网站开发中的 Servlet、JSP、JavaBean、JDBC、XML 等相关技术的一种主流 Web 开发框架,是一种基于经典 MVC 模式的框架。采用 Struts 可以简化 MVC 设计模式的 Web 应用开发工作,很好地实现代码重用,使开发人员从烦琐的工作中解脱出来,开发具有强扩展性的 Web 应用程序。

Struts 项目的创立者希望通过对该项目的研究,改进和提高 JSP、Servlet、标签库以及面向对象技术的水平。Struts 在英文中是支架、支撑的意思,体现其在 Web 应用程序开发中所起到的重要作用。如同建筑工程师使用支柱为建筑的每一层提供牢固的支撑一样,Java 工程师使用 Struts 为业务应用的每一层提供支持。它的目的是为了帮助程序开发人员减少运用 MVC 设计模型来开发 Web 应用所耗费的时间。

Struts 是 Apache 软件基金会下 Jakarta 项目(Apache 组织下的一套 Java 解决方案的开源软件的名称)的一部分。该基金会下除了 Struts 之外,还有其他优秀的开源产品,如 Tomcat。2000 年 Craig R. McClanahan(1960 年出生于丹麦,程序员,原 Sun 公司的高级员工,JSF 技术规范组负责人,Apache Struts Framework 创始人,Servlet 2.2、Servlet 2.3 和 JSP 1.1、JSP 1.2 规范的专家组成员之一,Tomcat4 的架构师)贡献了他编写的 JSP Model 2 架构之 Application Framework 原始程序代码给 Apache 基金会,成为 Apache Jakarta 计划 Struts Framework 的前身。从 2000 年 5 月开始开发 Struts,到 2001 年 6 月发布 Struts 1.0 版本。有 30 多个开发者参与进来,并有数千人参与到讨论组中。Struts 框架开始由一个志愿者团队来管理。到 2002 年,Struts 小组共有 9 个志愿者团队。Struts 框架的主要架构设计和开发者是 Craig R. McClanahan。

Struts 采取 MVC 模式，能够很好地帮助 Java 程序员利用 Java EE 开发 Java Web 应用项目。和其他的 Java 框架一样，Struts 也采用面向对象设计思想，将 MVC 模式的"分离显示逻辑和业务逻辑"能力发挥得淋漓尽致。

2001 年推出 Struts，2004 年开始升温，并逐渐成为 Java Web 应用开发最流行的框架技术之一。在目前的 Java 工程师招聘要求中，通常会强调 Struts 框架技术。精通 Struts 框架技术已经成为 Java 工程师必备的技能。

Struts 1.x 系列的版本一般称为 Struts1。经过 6 年多的发展，Struts1 已经成为了一个高度成熟的框架，无论是稳定性还是可靠性都得到了广泛的认可。市场占有率也很高，拥有丰富的开发人群，几乎已经成为了事实上的工业标准。但是随着时间的流逝，技术的进步，Struts1 的局限性也越来越多地暴露出来，并且制约了 Struts1 的继续发展。对于 Struts1 框架而言，由于与 JSP、Servlet 耦合非常紧密，因而导致了一些严重的问题。首先，Struts1 支持的表示层(V)技术单一。由于 Struts1 出现的年代比较早，那个时候没有 FreeMarker、Velocity 等技术，因此它不可能与这些视图层的模板技术进行整合。其次，Struts1 与 Servlet API 的紧耦合使应用程序难于测试。最后，Struts1 代码严重依赖于 Struts1 API，属于侵入性框架。从目前的技术层面上看，出现了许多与 Struts1 竞争的框架技术，例如，JSF、Spring MVC 等。这些框架技术由于出现的年代比较晚，应用了最新的设计理念，同时也从 Struts1 中吸取了经验，克服了很多不足。这些框架的出现也促进了 Struts 的发展。目前，Struts 已经分化成了两个框架：一个是在传统的 Struts1 基础上融合了另外一个优秀的 Web 框架 WebWork 的 Struts2；另一个就是 Struts1。Struts2 虽然是在 Struts1 的基础上发展起来的，但是实质上是以 WebWork 为核心的。

2007 年，Apache 发布 Struts 2.0，Struts2 是 Struts 的下一代产品，是在 Struts1 和 WebWork 框架基础上进行整合的全新的 Struts 框架。其全新的 Struts2 体系结构与 Struts1 体系结构差别巨大。Struts2 以 WebWork 为核心，采用拦截器机制来处理用户的请求，这样的设计也使得业务逻辑控制器能够与 Servlet 完全脱离开，所以 Struts2 可以理解为 WebWork 的更新产品。因此 Struts2 和 Struts1 有着很大的区别，但是相对于 WebWork 而言，Struts2 只有很小的变化。

1.1.2　Struts2 软件包的下载和配置

本书使用的是 Struts 2.3.4.1，它于 2012 年 8 月发布。

1. 软件包下载

Struts 的各版本可在 Apache 官方网站 http://struts.apache.org/download.cgi 下载。

要在 Apache 官方网站下载 Struts 2.3.4.1，可打开如图 1-1 所示的下载页面；单击图 1-1 左侧的 Struts2.3.4.1(GA)，出现如图 1-2 所示的下载页面；在图 1-2 中单击 Download Now 按钮，出现如图 1-3 所示的页面。在图 1-3 页面中选择下载 Full Distribution：struts-2.3.4.1-all.zip。

Struts2 下载提供以下选项：

图 1-1　Struts2 下载页面

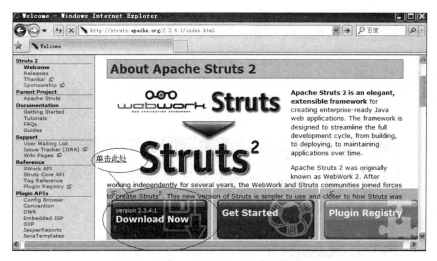

图 1-2　Struts 2.3.4.1 下载页面

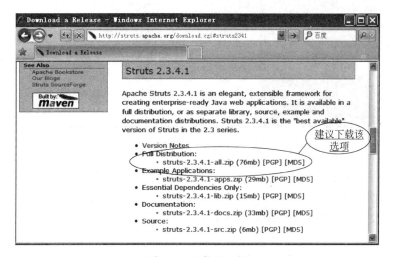

图 1-3　下载项选择

1) Full Distribution：struts-2.3.4.1-all.zip

这是 Struts2 的完整下载包，内容包括 Struts2 的核心类库、源代码、文档、实例等，建议选择该选项。

2) Example Applications：struts-2.3.4.1-apps.zip

该包只包含 Struts2 的实例，在完整版的 Struts2 下载包中已经包含了该选项中所有实例。

3) Essential Dependencies Only：struts-2.3.4.1-lib.zip

该包只包含 Struts2 的核心类库，在完整版的 Struts2 下载包中已经包含了该选项中所有类库。

4) Documentation：struts-2.3.4.1-docs.zip

该包只包含 Struts2 的相关文档，包括使用文档、参考手册和 API 等，在完整版的 Struts2 下载包中已经包含了该选项中所有文档。

5) Source：struts-2.3.4.1-src.zip

该包只包含 Struts2 的源代码，在完整版的 Struts2 下载包中已经包含了该选项中所有源代码。

2. Struts2 软件包

Struts2 下载完成后会得到一个 zip 文件，解压缩后，可得到如图 1-4 所示的文件夹结构。

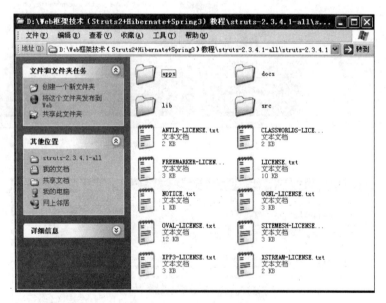

图 1-4　Struts2 文件夹结构

（1）apps 文件夹：该文件夹中存放基于 Struts2 的应用实例，这些实例对学习者来说是非常有用的资料。

（2）docs 文件夹：该文件夹中存放 Struts2 的相关文档，包括 Struts2 API、Struts2 快速入门等。

（3）lib 文件夹：该文件夹中存放 Struts2 框架的核心类库以及 Struts2 的第三方插件

类库。

(4) src 文件夹:该文件夹中存放 Struts2 框架的全部源代码。

3. Struts2 的配置

Struts 2.3.4.1 的 lib 文件夹中有 80 多个 JAR 文件。大多数情况下,使用 Struts2 开发 Java Web 应用程序并不需要使用 Struts2 的全部类库,因此没有必要把 lib 文件夹中的类库全部配置到项目中。一般只需配置 commons-fileupload-1.2.2.jar、commons-io-2.0.1.jar、freemarker-2.3.19.jar、javassist-3.11.0.GA.jar、ognl-3.0.5.jar、struts2-core-2.3.4.1.jar 和 xwork-core-2.3.4.1.jar 等文件。如果需要使用 Struts2 的更多特性,需要配置更多 lib 文件夹中的 JAR 文件到项目中。

1) 在 NetBeans 7.2 中安装 Struts2 插件

(1) NetBeans 7.2 中集成了 Struts 1.3.10,如果需要在 NetBeans 7.2 中使用 Struts2,可以安装 Struts2 插件,该插件下载地址是:www.netbeans.org,如图 1-5 所示。单击图 1-5 所示页面中的 Plugins,出现如图 1-6 所示的页面,在其中选择需要的 Struts2 插件。

图 1-5 Struts2 插件下载地址

(2) 在图 1-6 所示页面中单击 Struts2 Support Repack for NB 7.0+Xwork,出现如图 1-7 所示的页面,选择支持 NetBeans 7.2 的插件,单击该页面中的 Download 进行下载。

(3) 安装插件。下载的插件是名为 1345230225_nbstruts2-suite-1.3.4-for-7.2 的 zip 文件。首先解压缩该文件,然后单击 NetBeans 7.2 菜单栏中"工具"→"插件",弹出如图 1-8 所示的对话框。

(4) 在图 1-8 所示的对话框中单击"已下载"→"添加插件",弹出如图 1-9 所示的"添加插件"对话框,找到下载插件(需先解压缩插件文件)所在的位置,如图 1-9 所示,选定后单击"打开"按钮,在弹出的对话框中单击"安装"按钮,Struts2 插件即安装完成,最后重新启动 NetBeans 7.2。

2) 使用 NetBeans 7.2 新建 Struts2 项目

本书使用的工具是:JDK 7、NetBeans 7.2、MyEclipse 10.6、Eclipse 4.2、Tomcat 7。如需使用这些工具可在其官方网站下载。有关 JDK、NetBeans、MyEclipse、Eclipse、Tomcat 的下载、安装、配置和使用请参考相关资料或者参考作者编写的其他相关教材,在书中有介绍。

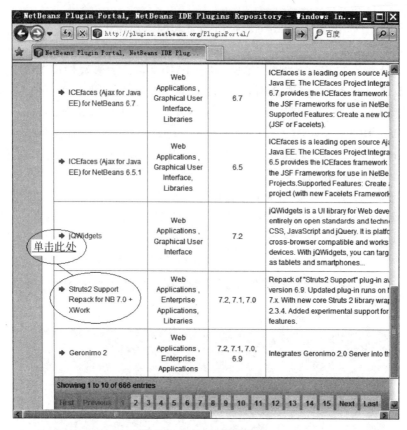

图 1-6　Struts2 插件地址

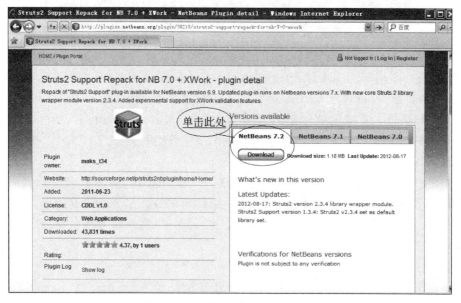

图 1-7　选择和 NetBeans 版本对应的 Struts2 插件进行下载

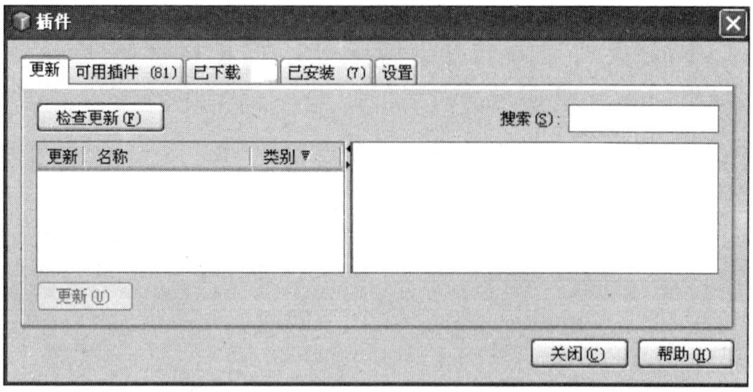

图 1-8 "插件"对话框

图 1-9 "添加插件"对话框

双击打开 NetBeans 7.2,出现如图 1-10 所示的 NetBeans 7.2 主界面。可以使用菜单项对 IDE 进行设置与使用。

图 1-10 NetBeans 7.2 主界面

(1)单击图1-10所示页面中的菜单"文件"→"新建项目"命令,弹出如图1-11所示的对话框,在"选择项目"中的"类别"框中选择Java Web,在"项目"框中选择Web应用程序,单击"下一步"按钮,弹出如图1-12所示的对话框。

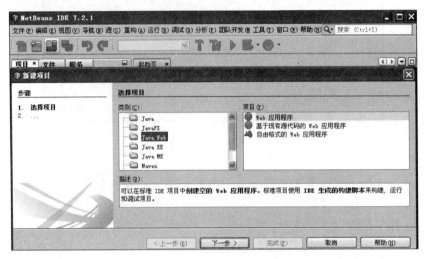

图1-11 "新建项目"对话框

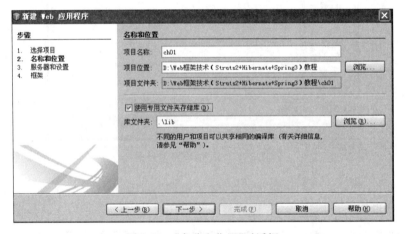

图1-12 "名称和位置"对话框

(2)在图1-12所示的对话框中,可以对项目的名称以及路径进行设置。在"项目名称"文本框中为Java Web项目命名,可以使用默认名称,也可以根据自己项目的需要命名;在"项目位置"文本框中对项目位置进行选择,可以使用默认路径,也可以根据编程需要进行选择;单击"下一步"按钮,弹出如图1-13所示的对话框。

(3)在图1-13所示对话框中"服务器和设置"部分的"服务器"框中,选择Web程序运行时使用的服务器。可以使用默认的服务器,也可以单击"添加"按钮选择其他服务器;在"Java EE版本"下拉框中,选择需要的Java EE版本;在"上下文路径"中设定项目路径。设置好后单击"下一步"按钮,弹出如图1-14所示的对话框,对"框架"进行选择,这里选择Struts2,单击"完成"按钮项目创建完成,将弹出如图1-15所示界面。

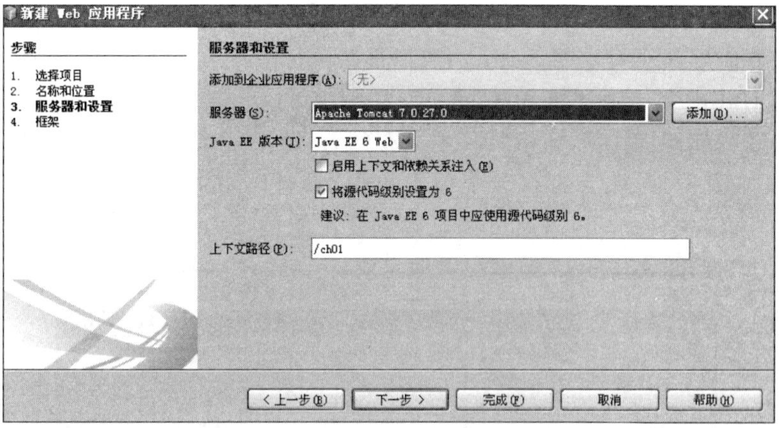

图 1-13 "服务器和设置"对话框

图 1-14 选择框架

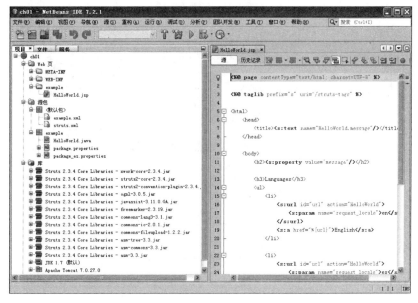

图 1-15 项目开发主界面

备注：本书使用的 Struts2 是最新版本 Struts 2.3.4，安装的插件中集成的也是 Struts 2.3.4。如果要使用其他版本的 Struts2，可以重新自行配置。

3）在 NetBeans 7.2 中配置 Struts2

方法1：首先删除图 1-15 所示"库"中原有的 Struts2 类库，然后在项目名称 ch01 上右击，接下来在如图 1-16 所示的右键菜单中单击"属性"，并在随后弹出的对话框中选择"库"→"添加 JAR/文件夹(F)"，通过浏览找到要配置的 Struts2 类库所在位置，如图 1-17 所示，最后单击"打开"，Struts2 类库配置即完成。

图 1-16　项目属性

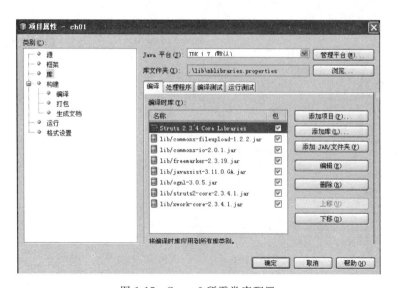

图 1-17　Struts2 所需类库配置

方法2：在项目的"库"上右击，在如图1-18所示的右键菜单中可以单击"添加JAR/文件夹"，在弹出的对话框中选择"库"→"添加JAR/文件夹(F)"，找到Struts2类库所在位置；也可以单击"添加库"，出现如图1-19所示对话框，单击其中的"创建"按钮会弹出如图1-20所示对话框，然后在"库名称"中为要添加的库命名为"Struts2.3.4"，再单击"确定"按钮，将弹出如图1-21所示的"定制库"对话框，最后单击"添加JAR/文件夹(F)"找到Struts2类库的所在位置。

图1-18　在"库"上右击

图1-19　"添加库"对话框

图1-20　"创建新库"对话框

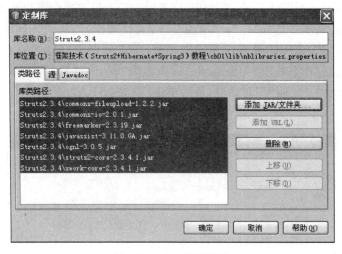

图 1-21 "定制库"配置

4）在 MyEclipse 中配置 Struts2

MyEclipse 10.6 中也已经集成了 Struts2 的插件，直接使用即可。要在 MyEclipse 10.6 中重新配置 Struts2，可在创建项目（如项目 ch01）后，单击菜单 MyEclipse → Project Capabilities→Add Struts Capabilities，如图 1-22 所示，会弹出如图 1-23 所示的对话框，选择 Struts2.1 后弹出另外一个对话框，在该对话框中单击 Next，弹出如图 1-24 所示对话框，在其中可以选择 Struts2 类库，完成选择后单击"打开"按钮，Struts2 类库在 MyEclipse 项目中的配置即完成。配置完成后，在项目 ch01 中将自动添加一个 Struts2 的包，即 Struts2 Core Libraries。如需使用 Struts2 的其他版本，可以导入到该包。

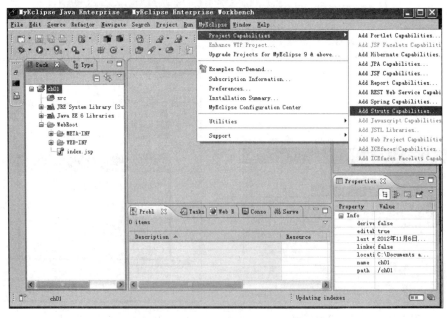

图 1-22 添加 Struts2

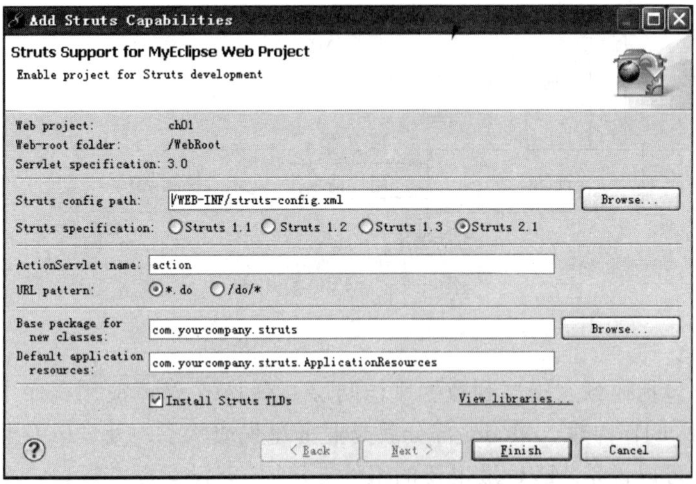

图 1-23 Struts 框架选定

图 1-24 添加 Struts2 类库

5）在 Eclipse 4.2 中配置 Struts2

Struts2 在 Eclipse 中的配置和在 MyEclipse 10.6 中的配置相似，这里不再赘述。

1.1.3　MVC 设计模式

　　MVC(Model、View 和 Controller)设计模式如图 1-25 所示，是一种目前广泛应用的软件设计模式。早在 20 世纪 70 年代，IBM 就进行了 MVC 设计模式的研究。近年来，随着 Java EE 的成熟，它成为在 Java EE 平台上最常见的一种设计模型，是广大 Java 开发者非常感兴趣的设计模型。随着网络应用的快速发展，MVC 模式对于 Web 应用的开发无疑是一种非常先进的设计思想，无论选择哪种语言(Java 或者 C#)，无论应用多复杂，它都能为理解应用模型提供最基本的分析方法，为构造产品提供清晰的设计框架，为软件工程提供规范的依据。

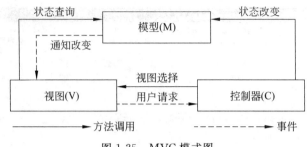

图 1-25　MVC 模式图

1. 模型

Model 部分包括业务逻辑层和数据库访问层。在 Java Web 应用程序中,业务逻辑层一般由 JavaBean 或 EJB 构建。Model 部分就是业务流程或状态的处理以及业务规则的制定。业务模型的设计可以说是 MVC 最主要的组件。MVC 并没有提供模型的设计方法,只是要求用户应该组织管理这些模型,以便于模型的重构和提高重用性。

2. 视图

在 Java Web 应用程序中,View 部分一般用 JSP、HTML 以及其他视图技术构建,也可以使用 XHTML、XML、Applet 或 JavaScript。客户在 View 部分提交请求,在业务逻辑层处理后,把处理结果又返回给 View 部分显示出来。因此,View 部分也是 Java Web 应程序的用户界面。一个 Web 项目可能有很多不同的视图,MVC 设计模式对于视图的处理仅限于视图上数据的采集和处理以及响应用户的请求,而不包括在视图上的业务流程的处理。业务流程的处理由模型负责。

3. 控制器

Controller 部分由 Servlet 组成。当用户请求从 V 部分传过来时,C 调用相应的 M 在控制器中进行业务的处理。C 再把处理结果转发给适当的 V 显示或者继续调用其他 M。因此,C 在视图层与业务逻辑层之间起到了桥梁作用,控制了两者之间的数据流向。

MVC 设计模式工作流程如下。

(1) 用户的请求(V)提交给控制器(C)。

(2) C 接收到用户请求后根据用户的具体需求,调用相应 M(JavaBean 或者 EJB)来处理用户的请求。

(3) C 调用 M 进行数据处理后,根据处理结果进行下一步的跳转,如跳转到另外一个 V 或者其他 C。

1.1.4　Struts2 的工作原理

Struts2 中使用拦截器来处理用户请求,从而允许用户的业务控制器 Action 与 Servlet 分离。Struts2 的工作原理如图 1-26 所示,用户请求提交后经过多个拦截器拦截后交给核心控制器 FilterDispatcher 处理。核心控制器读取配置文件 struts.xml,根据配置文件中的信息指定由某一个业务控制器 Action 来处理用户数据。业务控制器调用某些业务组件进行处理,在处理的过程中可以调用其他模型组件共同完成数据的处理。Action 处理完后会返回给核心控制器 FilterDispatcher 一个处理结果,核心控制器根据返回的处理结果读取配置文件 struts.xml,根据配置文件中的配置信息,决定下一步跳转到哪一个页面或者调用哪一个 Action。

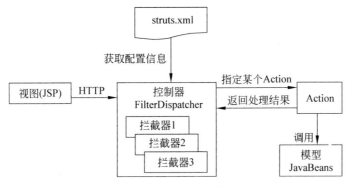

图 1-26 Struts2 的工作原理

一个客户请求在 Struts2 框架中处理的过程大概有以下几个步骤。

（1）客户提交请求到服务器。

（2）请求被提交到一系列的过滤器或者拦截器，最后到达 FilterDispatcher；FilterDispatcher 是核心控制器，是 Struts2 中 MVC 模式的控制器部分。

（3）FilterDispatcher 读取配置文件 struts.xml，根据配置信息调用某个 Action 来处理客户请求。

（4）Action 处理后，返回处理结果，FilterDispatcher 根据 struts.xml 的配置找到对应的页面跳转。

备注：Struts2 框架执行过程中，一个客户请求会获得 Struts2 类库中许多类的支持，由于在项目开发过程中编写程序时用不到这些类，为了简化对 Struts2 框架工作原理的理解，这里没有提及。

1.2 Struts2 的核心组件

Struts2 是基于 MVC 设计模式的 Java Web 框架技术之一，Struts2 框架按照 MVC 的设计思想把 Java Web 应用程序分为：控制器层，包括核心控制器 FilterDispatcher 和业务控制器 Action；模型层，包括业务逻辑组件和数据库访问组件；视图层，包括 HTML、JSP、Struts2 的标签等。

1.2.1 Struts2 的控制器组件

在基于 MVC 的应用程序开发中，控制器组件的主要功能是从客户端接收数据、调用模型(JavaBean)进行数据处理以及决定返回给客户某个视图。Struts2 的控制器主要有核心控制器 FilterDispatcher 和业务控制器 Action。

1. FilterDispatcher 控制器

基于 MVC 的 Java Web 框架需要在 Web 应用程序中加载一个核心控制器，Struts2 框架需要加载的是 FilterDispatcher。FilterDispatcher 是一个过滤器，是 Struts2 的核心控制器，控制着整个 Java Web 项目中数据的流向和操作。

FilterDispatcher 需在 web.xml 中进行配置。除了用 web.xml 配置文件配置核心控制器 FilterDispatcher 外，Struts2 控制数据的操作时，还需要用到 Struts2 的另一个配置文件 struts.xml。

2. struts.xml 配置文件

Struts2 的核心配置文件是 struts.xml。用户请求提交给核心控制器 FilterDispatcher 后，具体由哪个业务控制器 Action 来完成，是在 struts.xml 配置文件中事先配置好的，根据配置文件 struts.xml 中的数据，核心控制器 FilterDispatcher 调用某个具体的业务控制器 Action 来完成数据的处理，处理完数据后把处理结果通过其他对象返回给核心控制器 FilterDispatcher，核心控制器根据 struts.xml 配置文件中的配置，决定下一步的操作。

所以，Struts2 中的 struts.xml 文件是核心配置文件，在控制器操作中起到关键作用。

3. Action 控制器

Action 是 Struts2 的业务控制器，可以不实现任何接口或者不继承任何 Struts2 类。Action 类是一个基本的 Java 类，具有很高的可重用性。Action 中不实现任何业务逻辑，只负责组织调度业务模型组件。

Struts2 的 Action 类具有很多优势。

(1) Action 类完全是一个 POJO(Plain Old Java Objects，简单的 Java 对象)，实际上就是一个普通的 JavaBean，但为了避免和 EJB 混淆而另选了一个简称，所以 Action 具有良好的代码重用性。

(2) Action 类无须与 Servlet API 关联，降低了与 Servlet 的耦合度，所以应用和测试比较简单。

(3) Action 类的 execute()方法仅返回一个字符串作为处理结果，该处理结果可传到任何视图或者另外一个 Action。

1.2.2 Struts2 的模型组件

模型组件是可以实现业务逻辑的模块，如 JavaBean、POJO 或 EJB。在实际的项目开发中，对模型组件的区别和定义是比较模糊的，实际上也超出了 Struts2 框架的范围。Struts2 框架的业务控制器不会对用户请求进行实质的处理，用户请求最终由模型组件负责处理，业务控制器只是提供处理场合，是负责调度模型组件的调度器。

不同的开发者用不同的方式来编写模型组件，Struts2 框架的目的是使用 Action 来调用模型组件。例如，一个银行存款的模型组件代码如例 1-1 所示。

【例 1-1】 Bank 模型组件(Bank.java)。

```
package bankBean;

public class Bank {
    private String accounts;                              //账号
    private String money;                                 //资金
    public String getAccounts() {
        return accounts;
    }
    public void setAccounts(String accounts) {
```

```java
        this.accounts=accounts;
    }
    public String getMoney() {
        return money;
    }
    public void setMoney(String money) {
        this.money =money;
    }
    //模拟存款功能的方法
    public boolean saving(String accounts,String money){
        //调用其他类对数据库进行操作,下面省略的代码部分表示数据库查询的数据
         ⋮
        boolean bl;
        if(getAccounts().equals("accounts")&&getMoney().equals("money"))
        {
            bl=true;
        }
        else{
            bl=false;
        }
        return bl;
    }
}
```

例 1-1 中的代码是一个完成某一功能的业务逻辑模块,在执行 saving(String accounts, String money)方法时能够通过调用其他类(JavaBean)或者直接访问数据库完成存款功能。可以在业务控制器 Action 的 execute()方法中调用该业务逻辑组件,业务控制器的代码如例 1-2 所示。

【例 1-2】 业务控制器(BankSavingAction.java)。

```java
package bankAction;
import bankBean.Bank;

public class BankSavingAction {
    private String accounts;                          //账号
    private String money;                             //资金
    public String getAccounts() {
        return accounts;
    }
    public void setAccounts(String accounts) {
        this.accounts =accounts;
    }
    public String getMoney() {
        return money;
    }
```

```java
    public void setMoney(String money) {
        this.money =money;
    }
    public String execute(){
        Bank bk=new Bank();                        //调用模型组件 Bank 并实例化
        if(bk.saving(accounts, money)){
            return "success";
        }
        else{
            return "error";
        }
    }
}
```

业务控制器 BankSavingAction 通过创建模型组件实例的方式实现对银行存款业务组件的调用。当业务控制器需要获得业务逻辑组件实例时，通常并不会直接获取业务逻辑组件实例，而是通过工厂模式来获取业务逻辑组件的实例，或者使用其他 IoC 容器（如 Spring）来管理业务逻辑组件的实例。

1.2.3　Struts2 的视图组件

Struts1 视图组件的构成主要有 HTML、JSP 和 Struts1 标签。Struts2 视图组件除了有 HTML、JSP、Struts2 标签外，还可以采用模板技术作为视图技术，如 FreeMarker、Velocity 等视图技术。

1. HTML 和 JSP

HTML 和 JSP 是开发基于 Struts2 的视图组件的主要技术。

2. Struts2 标签

Struts2 框架提供了功能强大的标签库，使用 Struts2 的标签库开发视图，可以使页面更整洁，简化页面输出，支持更加复杂而丰富的功能且页面易维护，并能减少代码编写量和项目开发时间。

3. FreeMarker

FreeMarker 是一个"模板引擎"，是一个基于模板技术生成文本输出的通用工具。它是一个 Java 包，使用纯 Java 语言编写，是 Java 程序员可以使用的类库。本身并不是一个面向最终用户的应用程序。但是，程序员可以把它应用到他们的项目中。FreeMarker 被设计为可以生成 Java Web 页面（JSP）。它是基于 Servlet 且遵循 MVC 设计模式的应用，MVC 能够使网页设计人员和程序员的耦合减少。每个人都可以做他擅长的工作，网页设计人员可以改变网页的面貌，而并不需要程序员的重新编译，因为业务逻辑和页面的设计已经被分离开。模板是不能由复杂的程序片段组成的，即便网页设计人员和程序员是一个人，分离也是有必要的，它能使程序更加灵活和清晰。虽然 FreeMarker 能编程，但是它并不是一个编程语言，它是为程序显示数据而准备的。

FreeMarker 可利用模板加上数据生成文本页面。能生成任意格式的文本，如 HTML、XML、Java 源码等。Freemarker 并不是 Java Web 应用程序框架，可以说是一个 Java Web

应用框架的视图组件。FreeMarker 的下载地址为 http://www.freemarker.org/index.html。

4. Velocity

Velocity 是一个开放源码的"模板引擎",由 Apache 负责开发,现在最新的版本是 Velocity 1.7,可以到其官方网站 http://velocity.apache.org 了解 Velocity 的最新信息。

Velocity 是一个基于 Java 的模板引擎。它允许 Java Web 页面设计者引用 Java 代码预定义的方法。Java Web 设计者可以基于 MVC 模式和 Java 程序员并行工作,这意味着 Web 设计者可以单独专注于设计良好的站点,而程序员则可单独专注于编写底层代码。Velocity 将 Java 代码从 Web 页面中分离出来,使站点在长时间运行后仍然具有很好的可维护性,并提供了一个除 JSP 和 PHP 之外的可行的备选方案。

Velocity 可用来从模板产生 Web 页面、SQL 以及其他输出。它也可用于一个独立的程序以产生源代码和报告,或者作为其他系统的一个集成组件。项目完成后,Velocity 将为应用程序框架提供模板服务。

Velocity 的模板语言非常简单,它并没有复杂的数据类型和语法结构,即使没有编程经验的读者也可以轻松地掌握。

1.3 基于 Struts2 的登录系统

本节通过使用 NetBeans 7.2、MyEclipse 10.6 和 Eclipse 4.2 来开发一个简单的登录系统,来介绍如何使用它们开发基于 Struts2 的 Java Web 项目。

基于 Struts2 开发 Java Web 项目的主要包括以下步骤。

(1) 在 web.xml 中配置核心控制器 FilterDispatcher。

(2) 设计和编写视图组件(JSP 页面)。

(3) 编写用来实现视图组件的业务逻辑组件 JavaBean。

(4) 编写视图组件对应的业务控制器组件 Action,在该 Action 中调用业务逻辑组件 JavaBean,进行业务逻辑处理。

(5) 配置业务控制器 Action,即修改 struts.xml 配置文件。在 struts.xml 配置文件中配置逻辑视图与物理视图之间的跳转关系。Action 调用模型组件(业务逻辑组件)处理后返回处理结果(逻辑视图,即返回的字符串),根据处理结果进行下一步页面跳转。页面怎么跳转都是事先在 struts.xml 配置文件中配置好的。

1.3.1 使用 NetBeans7 开发项目

1. 项目介绍

该项目为登录系统,项目有一个登录页面(login.jsp),代码如例 1-4 所示。登录页面对应的业务逻辑组件 LoginBean 类代码如例 1-6 所示,对应的业务控制器 LoginAction 类代码如例 1-7 所示;如果登录成功(用户名、密码正确)跳转到 success.jsp 页面,代码如例 1-5 所示;如果登录失败(用户名、密码不正确)则重新回到登录页面(login.jsp)。此外还需要配置 web.xml,代码如例 1-3 所示;配置 struts.xml 文件的代码如例 1-8 所示。项目的文件结构如图 1-27 所示。

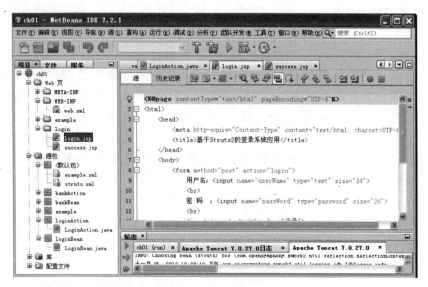

图 1-27　项目文件结构

2. 在 web.xml 中配置核心控制器 FilterDispatcher

在使用 NetBeans 7.2 开发基于 Struts2 的 Web 项目时，如果新建项目时在如图 1-14 所示步骤中选择了 Struts2，那么在新建项目的 WEB-INF 文件夹下，NetBeans 7.2 会自动创建一个 web.xml，所以使用 NetBeans 7.2 中的 Struts2 插件时，web.xml 中的核心控制器 FilterDispatcher 是自动配置好的。web.xml 的代码如例 1-3 所示。如果没有使用该插件新建项目，需要程序员自己配置 web.xml，代码如例 1-3 所示。

【例 1-3】 在 web.xml 中配置核心控制器（web.xml）。

```
<?xml version="1.0" encoding="UTF-8"?>
<web-app version="3.0" xmlns="http://java.sun.com/xml/ns/javaee"
    xmlns:xsi="http://www.w3.org/2001/XMLSchema-instance"
    xsi:schemaLocation="http://java.sun.com/xml/ns/javaee
    http://java.sun.com/xml/ns/javaee/web-app_3_0.xsd">
<filter>
    <!--配置 Struts2 核心控制器的名称-->
    <filter-name>struts2</filter-name>
    <!--配置 Struts2 核心控制器的类-->
    <filter-class>org.apache.struts2.dispatcher.FilterDispatcher</filter-class>
</filter>
<filter-mapping>
    <!--Struts2 控制器的名称-->
    <filter-name>struts2</filter-name>
    <!--拦截所有 URL 请求-->
    <url-pattern>/*</url-pattern>
</filter-mapping>
<!--指定默认的会话超时时间间隔,以分钟为单位-->
<session-config>
```

```xml
        <session-timeout>
            30
        </session-timeout>
    </session-config>
    <!--配置默认的访问界面-->
    <welcome-file-list>
        <welcome-file>example/HelloWorld.jsp</welcome-file>
    </welcome-file-list>
</web-app>
```

3. 编写视图组件（JSP 页面）

编写一个如图 1-28 所示的登录页面。

图 1-28 登录页面

该登录页面是一个 JSP 页面，代码如例 1-4 所示。

【**例 1-4**】 登录页面（login.jsp）。

```jsp
<%@page contentType="text/html" pageEncoding="UTF-8"%>
<html>
    <head>
        <meta http-equiv="Content-Type" content="text/html; charset=UTF-8">
        <title>基于Struts2的登录系统应用</title>
    </head>
    <body>
        <!--
            action属性中的login是连接的业务控制器指定的名字,login必须与struts.xml
            中的名字一致.
        -->
        <form method="post" action="login">
            用户名:<input name="userName" type="text" size="24">
            <br>
            密　码　:<input name="password" type="password" size="26">
            <br>
            <input type="submit" value="登录">
        </form>
        <hr>
```

```
        </body>
</html>
```

登录成功页面代码如例1-5所示。

【例1-5】 登录成功页面(success.jsp)。

```
<%@ page contentType="text/html" pageEncoding="UTF-8"%>
<html>
    <head>
        <meta http-equiv="Content-Type" content="text/html; charset=UTF-8">
        <title>登录成功页面</title>
    </head>
    <body>
        <h1>你登录成功,欢迎你! </h1>
    </body>
</html>
```

4. 编写登录页面对应的业务逻辑组件 JavaBean

登录页面使用的业务逻辑组件 LoginBean 类,代码如例1-6所示。

【例1-6】 登录页面的业务逻辑组件(LoginBean.java)。

```java
package loginBean;

public class LoginBean {
    private String userName;
    private String passWord;
    public String getUserName() {
        return userName;
    }
    public void setUserName(String userName) {
        this.userName = userName;
    }
    public String getPassWord() {
        return passWord;
    }
    public void setPassWord(String passWord) {
        this.passWord = passWord;
    }
    //处理用户登录的方法
    public boolean login(String userName, String passWord){
        boolean b= false;
        if(userName.equals("QQ")&&passWord.equals("123")){
            b=true;
            return b;
        }
```

```
        else{
            b=false;
            return b;
        }
    }
}
```

5. 编写业务控制器 Action

为了处理视图的业务逻辑,一般每个视图都会对应一个业务控制器 Action。login.jsp 对应的业务控制器如例 1-7 所示的 LoginAction 类,该类就是一个普通的 Java 类。

【例 1-7】 登录页面(login.jsp)对应的业务控制器(LoginAction.java)。

```
package loginAction;
import loginBean.LoginBean;

public class LoginAction {
    private String userName;
    private String passWord;
    public String getUserName() {
        return userName;
    }
    public void setUserName(String userName) {
        this.userName =userName;
    }
    public String getPassWord() {
        return passWord;
    }
    public void setPassWord(String passWord) {
        this.passWord =passWord;
    }
    public String execute() throws Exception{
        LoginBean lb=new LoginBean();
        if(lb.login(userName, passWord))
        {
            return "success";
        }
        else{
            return "error";
        }
    }
}
```

LoginAction 类就是业务控制器,首先该控制器保存数据,然后在该类的 execute()方法中可以调用模型组件 LoginBean,并在该方法中完成数据的处理,处理的结果是返回一个字符串,又称为逻辑视图。在 Struts2 配置文件 struts.xml 中配置了和返回结果对应的操作。

业务控制器 Action 默认的处理方法是 execute()，但也可以不使用该方法名而自行命名，一般建议使用该方法名。

6. 在 struts.xml 中配置 Action

业务控制器 LoginAction 需要在 struts.xml 中配置，只有这样核心控制器才能找到该业务控制器。另外，核心控制器根据业务控制器返回的值以及在 struts.xml 中的配置决定下一步跳转到哪个页面上去。struts.xml 的配置如例 1-8 中代码所示。

【例 1-8】 在 struts.xml 中配置 Action(struts.xml)。

```xml
<!DOCTYPE struts PUBLIC
    "-//Apache Software Foundation//DTD Struts Configuration 2.0//EN"
    "http://struts.apache.org/dtds/struts-2.0.dtd">
<!--根元素 -->
<struts>
    <!--
        导入一个配置文件,通过这种方式可以将Struts2的Action按模块配置到多个配置文
        件中.
    -->
    <include file="example.xml"/>
    <!--
        所有的Action配置都应该放在元素package下,name属性定义包名,extends属性定义
        继承的包空间struts-default.
    -->
    <package name="zzf" extends="struts-default">
        <!--
            Action配置可以有多对;name是对业务控制器命名,在表单中指定的名字需要与该
            名字一致;class指定Action类的位置.
        -->
        <action name="login" class="ch01Action.LoginAction">
            <!--
                定义两个逻辑视图和物理资源之间的映射,name的值是Action中返回的结果,
                即逻辑视图.
            -->
            <result name="error">/login/login.jsp</result>
            <result name="success">/login/success.jsp</result>
        </action>
    </package>
</struts>
```

7. 项目部署和运行

项目编写完成后在 Netbeans 7.2 中部署和运行 JSP 页面十分简单：在 login.jsp 上右击会弹出如图 1-29 所示快捷菜单，单击图 1-29 中的"运行文件"项，或者在 JSP 页面编辑区右击，出现如图 1-30 所示的快捷菜单，单击"运行文件"，项目将自动部署并运行。运行结果如图 1-28 所示，在该页面中输入用户名 QQ 和密码 123 后，单击"登录"按钮，业务控制器运行后页面跳转到成功页面。

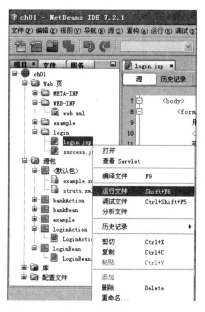
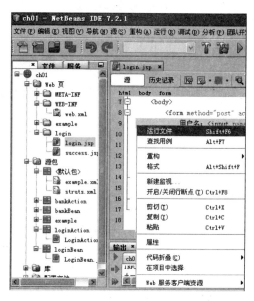

图 1-29 在 JSP 文件位置运行文件　　　　　图 1-30 在编辑区运行文件

1.3.2 使用 MyEclipse 10 开发项目

1. 项目介绍

该项目为登录系统,项目有一个登录页面(login.jsp),代码如例 1-10 所示;登录页面对应业务逻辑组件 LoginBean 类,代码如例 1-12 所示,对应的业务控制器是 LoginAction 类,代码如例 1-13 所示;如果登录成功(用户名、密码正确)跳转到 success.jsp 页面,代码如例 1-11 所示;如果登录失败(用户名、密码不正确)则重新回到登录页面(login.jsp)。此外,还需要配置 web.xml,代码如例 1-9 所示;配置 struts.xml,代码如例 1-14 所示。项目的文件结构如图 1-31 所示。

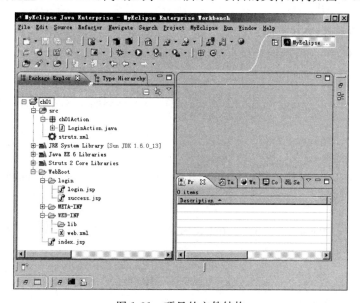

图 1-31 项目的文件结构

2. 在 web.xml 中配置核心控制器

在使用 MyEclipse 10.6 开发基于 Struts2 的 Java Web 项目时，新建项目后，如果要添加 Struts2，可参考 1.1.2 节中的图 1-22、图 1-23 和图 1-24，web.xml 中的核心控制器是 MyEclipse 10.6 自动配置好的。web.xml 的代码如例 1-9 所示。

【例 1-9】 在 web.xml 中配置核心控制器（web.xml）。

```xml
<?xml version="1.0" encoding="UTF-8"?>
<web-app version="3.0"
    xmlns="http://java.sun.com/xml/ns/javaee"
    xmlns:xsi="http://www.w3.org/2001/XMLSchema-instance"
    xsi:schemaLocation="http://java.sun.com/xml/ns/javaee
    http://java.sun.com/xml/ns/javaee/web-app_3_0.xsd">
    <display-name></display-name>
    <welcome-file-list>
        <welcome-file>index.jsp</welcome-file>
    </welcome-file-list>
    <filter>
        <filter-name>struts2</filter-name>
        <filter-class>
            org.apache.struts2.dispatcher.ng.filter.StrutsPrepareAndExecuteFilter
        </filter-class>
    </filter>
    <filter-mapping>
        <filter-name>struts2</filter-name>
        <url-pattern>*.action</url-pattern>
    </filter-mapping></web-app>
```

对比以上两个开发环境中的 web.xml 配置文件，可见每个开发平台对文件的组织格式并不完全相同。

3. 编写视图组件（JSP 页面）

编写一个如图 1-28 所示的登录页面，代码如例 1-10 所示。

【例 1-10】 登录页面（login.jsp）。

```jsp
<%@page language="java" import="java.util.*" pageEncoding="UTF-8"%>
<%
String path = request.getContextPath();
String basePath = 
request.getScheme()+"://"+request.getServerName()+":"+request.getServerPort()
+path+"/";
%>
<!DOCTYPE HTML PUBLIC "-//W3C//DTD HTML 4.01 Transitional//EN">
<html>
  <head>
    <base href="<%=basePath%>">
    <title>基于 Struts2 的登录系统应用</title>
```

```
    <meta http-equiv="pragma" content="no-cache">
    <meta http-equiv="cache-control" content="no-cache">
    <meta http-equiv="expires" content="0">
    <meta http-equiv="keywords" content="keyword1,keyword2,keyword3">
    <meta http-equiv="description" content="This is my page">
    <!--
    <link rel="stylesheet" type="text/css" href="styles.css">
    -->
  </head>
  <body>
      <!--表单的 action 属性值为 login.action ,NetBeas7.2 中为 login-->
      <form method="post" action="login.action">
          用户名:<input name="userName" type="text" size="24">
          <br>
          密码  :<input name="passWord" type="password" size="26">
          <br>
          <input type="submit" value="登录">     .
      </form>
      <hr>
  </body>
</html>
```

登录成功页面,代码如例 1-11 所示。

【例 1-11】 登录成功页面(success.jsp)。

```
<%@page language="java" import="java.util.*" pageEncoding=""%>
<%
String path =request.getContextPath();
String basePath =
request.getScheme()+"://"+request.getServerName()+":"+request.getServerPort()
+path+"/";
%>
<!DOCTYPE HTML PUBLIC "-//W3C//DTD HTML 4.01 Transitional//EN">
<html>
  <head>
    <base href="<%=basePath%>">
    <title>登录成功页面</title>
    <meta http-equiv="pragma" content="no-cache">
    <meta http-equiv="cache-control" content="no-cache">
    <meta http-equiv="expires" content="0">
    <meta http-equiv="keywords" content="keyword1,keyword2,keyword3">
    <meta http-equiv="description" content="This is my page">
    <!--
    <link rel="stylesheet" type="text/css" href="styles.css">
    -->
  </head>
```

```html
    <body>
        <hr>
        <h1>你登录成功,欢迎你!</h1>
        <hr>
    </body>
</html>
```

4. 编写登录页面对应的业务逻辑组件 JavaBean

登录页面对应的业务逻辑组件是 LoginBean 类,代码如例 1-12 所示。

【例 1-12】 登录页面对应的业务逻辑组件(LoginBean.java)。

```java
package loginBean;

public class LoginBean {
    private String userName;
    private String passWord;
    public String getUserName() {
        return userName;
    }
    public void setUserName(String userName) {
        this.userName = userName;
    }
    public String getPassWord() {
        return passWord;
    }
    public void setPassWord(String passWord) {
        this.passWord = passWord;
    }
    //处理用户登录的方法
    public boolean login(String userName,String passWord){
        boolean b=false;
        if(userName.equals("QQ")&&passWord.equals("123")){
            b=true;
            return b;
        }
        else{
            b=false;
            return b;
        }
    }
}
```

5. 编写业务控制器 Action

login.jsp 对应的业务控制器是如例 1-13 所示的 LoginAction 类。

【例 1-13】 登录页面对应的业务控制器(LoginAction.java)。

```java
package loginAction;
import loginBean.LoginBean;

public class LoginAction {
    private String userName;
    private String passWord;
    public String getUserName() {
        return userName;
    }
    public void setUserName(String userName) {
        this.userName =userName;
    }
    public String getPassWord() {
        return passWord;
    }
    public void setPassWord(String passWord) {
        this.passWord =passWord;
    }
    public String execute() throws Exception{
        LoginBean lb=new LoginBean();
        if(lb.login(userName, passWord))
        {
            return "success";
        }
        else{
            return "error";
        }
    }
}
```

6. 在 struts.xml 中配置 Action

业务控制器 LoginAction 需要在 struts.xml 中配置,代码如例 1-14 所示。

【例 1-14】 在 struts.xml 中配置 Action(struts.xml)。

```xml
<?xml version="1.0" encoding="UTF-8" ?>
<!DOCTYPE struts PUBLIC
    "-//Apache Software Foundation//DTD Struts Configuration 2.1//EN"
    "http://struts.apache.org/dtds/struts-2.1.dtd">
<struts>
    <package name="zzf" extends="struts-default">
        <action name="login" class="loginAction.LoginAction">
            <result name="error">/login/login.jsp</result>
            <result name="success">/login/success.jsp</result>
        </action>
```

```
        </package>
</struts>
```

7. 项目部署和运行

使用 MyEclipse 10.6 开发好项目后,要先发布项目(项目部署),然后启动服务器,最后运行页面。要发布项目可单击如图 1-32 所示的图标,弹出如图 1-33 所示的界面,在其中的 Project 后的下拉列表选择要发布的项目,然后单击 Add 按钮,弹出如图 1-34 所示的界面,选择需要使用的服务器后单击 Finish 按钮,再单击 OK 按钮,项目发布即完成。

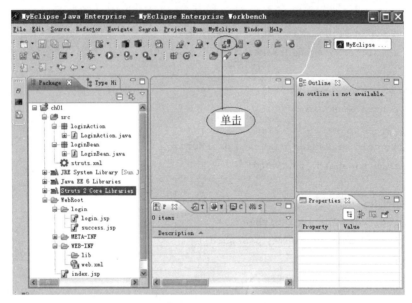

图 1-32　项目发布(部署)

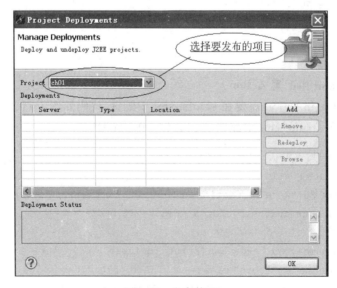

图 1-33　发布管理

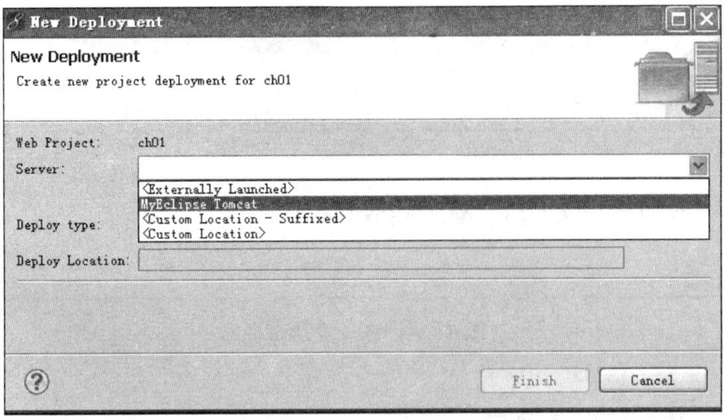

图 1-34　选择项目使用的服务器

项目发布后,单击如图 1-35 所示的 Start 命令,启动服务器。

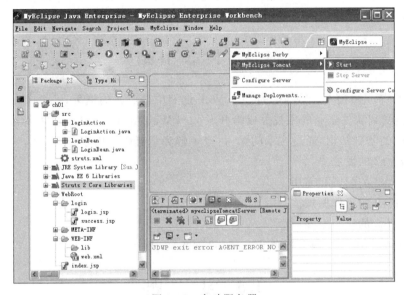

图 1-35　启动服务器

启动服务器后,在浏览器地址栏中输入 http：//localhost：8080/ch01/login/login.jsp,访问 JSP 页面,如图 1-36 所示。输入正确的用户名和密码后单击"登录"按钮,页面将跳转到如图 1-37 所示的登录成功页面。

图 1-36　使用 MyEclipse 开发的登录页面

图 1-37 登录成功页面

1.3.3 使用 Eclipse4 开发项目

MyEclipse 是在 Eclipse 平台上的一个插件,所以 MyEclipse 和 Eclipse 操作界面几乎一样,配置也几乎一样,限于篇幅,这里不再赘述。下载 Eclipse 以及相关插件的地址为 www.eclipse.org。

1.4 本章小结

Struts 框架是目前基于 MVC 的 Java Web 框架中最经典的框架,已经成为 Java 工程师必备的技能,受到软件企业的重视并成为开发 Java Web 项目事实上的工业标准。本章主要介绍 Struts2 的基础知识,通过本章的学习,应了解和掌握以下内容。

(1) Struts2 的发展史。
(2) Struts2 的配置和使用。
(3) Struts2 的工作原理。
(4) Struts2 的核心组件。
(5) 使用常用开发工具开发基于 Struts2 简单的项目。

1.5 习 题

1.5.1 选择题

1. 目前最经典的基于 MVC 的 Java Web 框架技术是(　　)。
   A. JSF　　　　　　B. FreeMarker　　　C. Velocity　　　D. Struts2
2. Struts2 属于(　　)基金会。
   A. Apache　　　　B. IBM　　　　　　C. Microsoft　　　D. W3C
3. 在 MVC 设计模式中控制器部分是(　　)。
   A. JavaBean　　　B. JSP　　　　　　C. Servlet　　　　D. Action
4. Struts2 的业务控制器是(　　)。
   A. FilterDispatcher　B. Action　　　　　C. Servlet　　　　D. ActionMaping
5. Struts2 的核心配置文件是(　　)。
   A. web.xml　　　　B. struts.xml　　　　C. server.xml　　　D. context.xml

1.5.2 填空题

1. Struts2 是基于_____设计模式的框架。
2. Struts2 集成了_____和_____框架的优点。
3. MVC 设计模式是_____公司推出的。
4. Struts2 的核心控制器是_____。
5. Struts2 的视图组件有 HTML、_____、_____、FreeMarker、Velocity 等。

1.5.3 简答题

1. 简述什么是 MVC 设计模式。
2. 简述 Struts2 的工作原理。

1.5.4 实训题

1. 把 1.3 节的登录系统改为连接数据库的登录系统，根据数据库的记录判断输入的用户名和密码是否正确，若正确页面跳转到成功页面，不正确页面跳转到登录页面。
2. 使用你所熟悉的 DBMS，编程实现一个基于 Struts2 的简单注册系统。

第 2 章 Struts2 核心组件详解

本章详细介绍 Struts2 的配置文件、核心控制器 FilterDispatcher、业务控制器 Action、OGNL 以及标签库。

本章主要内容如下所示。

(1) Struts2 的配置文件 struts.xml。
(2) Struts2 的核心控制器 FilterDispatcher。
(3) Struts2 的业务控制器 Action。
(4) Struts2 的 OGNL 和标签库。

2.1 Struts2 的配置文件 struts.xml

Struts2 的核心配置文件是 struts.xml,其具有重要的作用。所有用户请求被 Struts2 核心控制器 FilterDispatcher 拦截,然后查询配置文件 struts.xml 决定由哪个 Action 处理。Struts2 框架有两种配置文件格式: struts.xml 和 struts.properties,一般建议使用 struts.xml。

2.1.1 struts.xml 配置文件的结构

第 1 章已经对配置文件 struts.xml 有了简单了解,下面看一下一个常用 struts.xml 配置文件的结构,代码如例 2-1 所示。

【例 2-1】 struts.xml 配置文件的基本结构(struts.xml)。

```
<?xml version="1.0" encoding="UTF-8" ?>
<!--
    表明解析本 XML 文件的 DTD(Document Type Definition)文档位置,DTD 是对文档类型的定
    义,XML 解析器使用 DTD 文档来检查 XML 文件的合法性.不同的 Struts 版本对应的配置文件中
    的 DTD 信息会有一定的差异.切记,使用与版本相对应的 DTD 信息.
-->
<!DOCTYPE struts PUBLIC
    "-//Apache Software Foundation//DTD Struts Configuration 2.0//EN"
    "http://struts.apache.org/dtds/struts-2.0.dtd">
<struts>
    <!--Bean 配置-->
    <bean name="Bean 的名字" class="自定义的组件类"/>
    <!--常量配置,指定 Struts2 国际化资源文件的名字为 messageResource.-->
    <constant name="struts.custom.i18n.resources" value="messageResource"/>
    <!--
        导入一个配置文件,通过这种方式可以将 Struts2 的 Action 按模块配置到多个配置文
        件中.
```

```
        -->
        <include file="example.xml"/>
        <!--
            所有的 Action 配置都应该放在 package 下,name 定义包名,extends 定义继承包空间
            struts-default.
        -->
        <package name="zzf" extends="struts-default">
            <!--
                Action 配置可以有多对;name 是对业务控制器命名,在表单中指定的名字需要与该
                名字一致;class 指定 Action 类的位置.
            -->
            <action name="login" class="ch02Action.LoginAction">
                <!--
                    定义两个逻辑视图和物理资源之间的映射,name 值是 Action 中返回的结果,即
                    逻辑视图.
                -->
                <result name="error">/login/login.jsp</result>
                <result name="success">/login/success.jsp</result>
            </action>
        </package>
</struts>
```

在 struts.xml 文件中,为了配置不同的内容,许多元素可以重复出现多次。如可以有多个<package>、<bean>、<constant>、<include>、<action>、<result>等。

2.1.2 Bean 配置

Struts2 框架是一个具有高度可扩展性的 Java Web 框架,其主要核心组件不是以直接编码的方式写在代码中,而是通过配置文件以 IoC(控制反转)容器来管理这些组件。Struts2 可以通过配置方式管理其核心组件,从而允许开发者很方便地扩展这些核心组件。当需要扩展这些组件的时候,把自定义的组件类在 Struts2 的 IoC 容器中配置即可。

Bean 在 struts.xml 中的配置格式如下:

`<bean name="Bean 的名字" class="自定义的组件类"/>`

<bean/>元素的常用属性如下所示。

(1) name:指定 Bean 实例化对象名字,对于相同类型的多个 Bean 实例,其对应的 name 属性值不能相同,该项是可选项。

(2) class:指定 Bean 实例的实现类,即对应的类,该项是必选项。

(3) type:指定 Bean 实例实现的 Struts2 规范,该规范通常是通过某个接口来体现的,因此该属性的值通常是一个 Struts2 接口。如果配置 Bean 作为框架的一个核心组件来使用,就应该指定该属性,该项是可选项。

(4) scope:指定 Bean 的作用域,该项是可选项。

(5) optional:指定该 Bean 是否是一个可选 Bean,该项是可选项。

(6) static:指定允许不创建 Bean 实例,而是让 Bean 接受框架常量,这时该属性值设为

true。但是当指定了 type 属性时,该属性不能设为 true。

2.1.3 常量配置

Struts2 加载常量的顺序是 struts.xml、struts.properties 和 web.xml,如果在这 3 个文件中对某个变量有重复配置,后一个文件中配置的常量值会覆盖前面文件中同名的常量值。所以常量配置可以在不同的文件中进行配置,一般习惯在 struts.xml 中配置 Struts2 的属性,而不是在 struts.properties 文件中配置;之所以保留使用 struts.properties 文件配置 Struts2 属性的方式,主要是为了保持与 WebWork 框架的向后兼容性。

常量在 struts.xml 中的配置格式如下:

```
<constant name="属性名" value="属性值"/>
```

<constant/>元素的常用属性如下所示:

(1) name:指定常量(属性)的名字。
(2) value:指定常量的值。

例如,在 struts.xml 文件中配置国际化资源文件名和字符集的编码方式为 gb2312 的代码如下:

```
<!--常量配置,指定Struts2国际化资源文件的名字为messageResource-->
<constant name="struts.custom.i18n.resources" value="messageResource"/>
<!--常量配置,指定国际化编码方式-->
<constant name="struts.custom.i18n.encoding" value="gb2312"/>
```

在 struts.properties 文件中的相应配置代码如下:

```
struts.custom.i18n.resources=messageResource
struts.custom.i18n.encoding=gb2312
```

在 web.xml 文件中的相应配置代码如下:

```
<filter>
    <filter-name>struts2</filter-name>
    <filter-class>org.apache.struts2.dispatcher.FilterDispatcher</filter-class>
    <init-param>
        <!--指定国际化资源文件常量-->
        <param-name>
            struts.custom.i18n.resources
        </param-name>
        <param-value>
            messageResource
        </param-value>
        <!--指定编码方式常量-->
        <param-name>
            struts.custom.i18n.encoding
        </param-name>
        <param-value>
```

```
            gb2312
        </param-value>
    </init-param>
</filter>
```

2.1.4 包含配置

在开发一个项目时,一般采用模块开发的方式,即一个项目由多个模块组成,每个模块由某个项目组或者某些程序员来开发,每个程序员都可以创建使用自己的配置文件,然后把各个模块集成在一起。Struts2 的配置文件 struts.xml 提供了<include>元素,该元素能够把其他程序员开发的配置文件包含过来,但是被包含的每个配置文件必须和 struts.xml 格式一样。<include>元素可以和<package>元素交替出现,Struts2 框架将按照顺序加载配置文件。

在 struts.xml 中包含文件的格式如下:

```
<include file="文件名"/>
```

<include/>元素的常用属性如下:

flie:指定文件名,必选项。

例如,下载的 Struts2 实例放在图 1-4 所示窗口(请参考 1.1.2 节)中的 apps 文件夹中,其中有一个 struts2-portlet.war 项目,解压该项目后找到 struts.xml,如图 2-1 所示,该 struts.xml 的代码如例 2-2 所示。

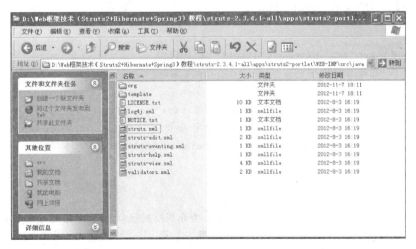

图 2-1 项目 struts2-portlet.war 的 struts.xml 位置

【例 2-2】 项目 struts2-portlet.war 的配置文件(struts.xml)。

```
<?xml version="1.0" encoding="UTF-8" ?>
<!DOCTYPE struts PUBLIC
    "-//Apache Software Foundation//DTD Struts Configuration 2.3//EN"
    "http://struts.apache.org/dtds/struts-2.3.dtd">
<struts>
```

```xml
    <include file="struts-view.xml"/>
    <include file="struts-edit.xml"/>
    <include file="struts-help.xml"/>
    <include file="struts-eventing.xml"/>
</struts>
```

包含配置能够避免开发复杂项目时配置的 struts.xml 过于庞大,导致读取配置文件速度较慢,同时有利于模块化开发。

2.1.5 包配置

在 Struts2 框架中,是通过包配置来管理 Action 和拦截器的。在包中可以配置多个 Action 和拦截器。在 struts.xml 配置文件中,包是通过<package>元素来配置的。

包配置在 struts.xml 中的格式如下:

`<package name="包名" extends="包名">…</package>`

<package>元素的常用属性如下所示。

(1) name:指定包名,是供其他包继承时使用的属性,必选项。
(2) extends:指定要继承的包名,可选项。
(3) namespace:定义包的名称空间,可选项。
(4) abstract:指定该包是否是一个抽象包,如果该包是抽象包,包中不能定义 Action。

包的配置代码如例 2-3 所示。

【例 2-3】 包的配置(struts.xml)。

```xml
…
<package name="zzf" extends="struts-default">
    <!--拦截器配置-->
    <interceptors>
        <interceptor-stack name="crudStack">
            <interceptor-ref name="params"/>
            <interceptor-ref name="defaultStack"/>
        </interceptor-stack>
    </interceptors>
    <!--
        对 Action 的配置可以有多对;name 为业务控制器命名,该名需要与在表单中指定的名字一致;class 指定 Action 类的位置.
    -->
    <action name="login" class="ch01Action.LoginAction">
        <!--
            定义两个逻辑视图和物理资源之间的映射,name 值是 Action 返回的结果,即逻辑视图.
        -->
        <result name="error">/login/login.jsp</result>
        <result name="success">/login/success.jsp</result>
    </action>
```

```
</package>
```
⋮

配置包时必须指定 name 属性,只有指定了这个属性后此包才可以被引用。extends 属性用来指定该包继承于其他包,其值必须是另外一个包的名字。通过继承,子包可以继承父包配置的 Action 和拦截器。上述代码定义的包名是 zzf,继承的包是 struts-default,struts-default 包是 Struts2 框架的默认包。

2.1.6 命名空间配置

在 Java 语言中为了避免同名 Java 类的冲突,可以使用包。例如,两个 login.java 文件存在同一个包中是不允许的,但是如果分别在两个包中是可以的。同样,Struts2 的配置中也存在同名的 Action 命名问题。

命名空间在 struts.xml 中的配置格式如下:

```
<package name="包名" extends="包名" namespace="/命名空间名">…</package>
```

例如,在项目的不同模块中都需要一个 LoginAction,如果用户在访问时不加以区分,项目就会出现问题。代码如例 2-4 所示。

【例 2-4】 名称空间配置(struts.xml)。

⋮
```xml
<package name="zzf" extends="struts-default">
    <action name="login" class=" ch02Action.zzf.LoginAction">
        <result name="error">/login/login.jsp</result>
        <result name="success">/login/success.jsp</result>
    </action>
</package>
<!--配置 zzf1 包,继承包 zzf,包名称空间/zzf1-->
<package name="zzf1" extends="zzf" namespace="/zzf1">
    <action name="login" class="ch02Action.zzf1.LoginAction">
        <result name="error">/manager/login.jsp</result>
        <result name="success">/manager/success.jsp</result>
    </action>
</package>
```
⋮

在例 2-4 中定义了两个包:zzf 和 zzf1,其中 zzf 继承了 struts-default 包,zzf1 继承了 zzf 包,两个包中都定义了 Action(LoginAction)。包 zzf 没有指定 namespace 属性,默认值是"",即为空。zzf1 指定名称空间为 namespace="/zzf1",说明用户请求访问该包下所有 Action 时,URL 应该是名称空间(namespace)+Action。例如,访问一个登录系统时使用的地址是 http://localhost:8084/ch02/login.action;访问另外一个时使用的是 http://localhost:8084/ch02/zzf1/login.action,即符合名称空间+Action。请求首先会在指定名称空间找对应的 Action,如果找不到再到默认的名称空间找 Action。

2.1.7 Action 配置

Struts2 中 Action 类的配置能够让 Struts2 的核心控制器知道 Action 的存在,并可以通过调用该 Action 来处理用户请求。Struts2 使用包来组织和管理 Action。

Action 在 struts.xml 中的配置格式如下:

`<action name="名称" class="Action 对应的类" >…</action>`

<action>元素的常用属性如下所示。

(1) name:指定客户端发送请求的地址名称,必选项。
(2) class:指定 Action 对应的实现类,可选项(参考 2.3.1 节)。
(3) method:指定 Action 类中处理方法名,如 get 方法或 post 方法等,可选项。
(4) converter:指定 Action 类型转换器的完整类名,可选项。

对 Action 的配置参考例 2-5。

【例 2-5】 Action 配置(struts.xml)。

```
  ⋮
<action name="login" class="ch02Action.LoginAction">
    <result name="error">/login/login.jsp</result>
    <result name="success">/login/success.jsp</result>
</action>
  ⋮
```

在例 2-5 中,<action>的 name 属性值在页面表单 action="***"或者传参数时使用。<action>的 class 属性值用来指定 Action 类所在位置,ch02Action 是包名,该包下有 LoginAction 类。

2.1.8 结果配置

<result>元素用来为 Action 的处理结果指定一个或者多个视图,配置 Struts2 中逻辑视图和物理视图之间的映射关系。

结果配置在 struts.xml 中的配置格式如下:

`<result name="字符串值" >…</result>`

<result>元素的常用属性如下所示。

(1) name:指定 Action 返回的逻辑视图,必选项。
(2) type:指定结果类型是定向到其他文件,该文件可以是 JSP 文件或者 Action 类,可选项。

代码参考例 2-5。在例 2-5 中 result name 的值是 Action 中 execute()方法返回的字符串值之一。/login/login.jsp 指定返回的页面是在 login 文件夹下的 login.jsp 文件。

2.1.9 拦截器配置

拦截器的作用就是执行 Action 处理用户请求之前或者之后进行某些拦截操作。例如,用户请求删除某些数据时,拦截器首先会判断用户是否有权删除,如果有权限,就通过

Action 删除，如果没有权限将不执行 Action 操作。

拦截器在 struts.xml 中的配置格式如下：

```
<!--拦截器配置-->
<interceptors>
        <!--定义拦截器-->
        <interceptor-ref name="拦截器名字" class="拦截器类"/>
          ⋮
    </interceptor-stack>
</interceptors>
```

<interceptor-ref>元素的常用属性如下所示。

(1) name：指定拦截器的名字，该名字用于在其他地方引用该拦截器。

(2) class：指定拦截器类。

2.2 Struts2 的核心控制器 FilterDispatcher

FilterDispatcher 是 Struts2 框架的核心控制器，该控制器作为一个过滤器运行在 Java Web 项目中，它负责拦截所有的用户请求，当用户请求到达时，该过滤器会过滤用户请求。

如果用户请求以 action 结尾，该请求将被转入 Struts2 框架处理。Struts2 框架获得了 *.action 请求后，将根据 *.action 请求的前面部分决定调用哪个业务控制器组件，例如，对于 login.action 请求，Struts2 调用名为 login 的 Action 来处理该请求。

Struts2 项目中的 Action 都被配置在 struts.xml 文件中。在该文件中配置 Action 时，要配置该 Action 的 name 属性和 class 属性，其中 name 属性决定了该 Action 处理哪个用户请求，而 class 属性指定了该 Action 的实现类。

Struts2 用于处理用户请求的 Action 实例，并不是用户实现的业务控制器，而是 Action 代理，因为用户实现的业务控制器并没有与 Servlet API 耦合，显然无法处理用户请求。而 Struts2 框架提供了一系列拦截器，该一系列拦截器负责将 HttpServletRequest 请求中的请求参数解析出来，传入到 Action 中，并通过 Action 的 execute()方法来处理用户请求。

2.3 Struts2 的业务控制器 Action

开发基于 Struts2 的 Java Web 应用项目时，Action 是数据处理的核心，需要编写大量的 Action 类，并在 struts.xml 文件中配置 Action。Action 类中包含了对用户请求的处理逻辑，因此也把 Action 称为 Action 业务控制器。

2.3.1 Action 接口和 ActionSupport 类

为了能够开发出更加规范的 Action 类，Struts2 提供了 Action 接口，该接口定义了 Struts2 中 Action 类应该遵循的规范。代码如例 2-6 所示。

【例 2-6】 Struts2 类库中的 Action 接口（Action.java）。

```
public interface Action {
```

```java
//常量声明
public static final String SUCCESS = "success";
public static final String NONE = "none";
public static final String ERROR = "error";
public static final String INPUT = "input";
public static final String LOGIN = "login";
//方法声明
public String execute() throws Exception;
}
```

在例 2-6 的 Action 接口声明中,定义了 5 个字符串常量,它们的作用是作为业务控制器中 execute()方法的返回值。Action 接口中声明一个 execute()方法,接口的规范规定了实现该接口的 Action 类应该实现该方法,该方法返回一个字符串。

另外,Struts2 为 Action 接口提供一个实现类 ActionSupport,该类的代码如例 2-7 所示。

【例 2-7】 ActionSupport 类(ActionSupport.java)。

```java
public class ActionSupport implements Action, Validateable, ValidationAware,
    TextProvider, LocaleProvider, Serializable {
    protected static Logger LOG = LoggerFactory.getLogger(ActionSupport.class);
    private final ValidationAwareSupport validationAware = new ValidationAwareSupport();
    private transient TextProvider textProvider;
    private Container container;
    public void setActionErrors(Collection<String> errorMessages) {
        validationAware.setActionErrors(errorMessages);
    }
    public Collection<String> getActionErrors() {
        return validationAware.getActionErrors();
    }
    public void setActionMessages(Collection<String> messages) {
        validationAware.setActionMessages(messages);
    }
    public Collection<String> getActionMessages() {
        return validationAware.getActionMessages();
    }
    public Collection<String> getErrorMessages() {
        return getActionErrors();
    }
    public Map<String, List<String>> getErrors() {
        return getFieldErrors();
    }
    public void setFieldErrors(Map<String, List<String>> errorMap) {
        validationAware.setFieldErrors(errorMap);
    }
    public Map<String, List<String>> getFieldErrors() {
```

```java
        return validationAware.getFieldErrors();
    }
    public Locale getLocale() {
        ActionContext ctx =ActionContext.getContext();
        if (ctx !=null) {
            return ctx.getLocale();
        } else {
            LOG.debug("Action context not initialized");
            return null;
        }
    }
    public boolean hasKey(String key) {
        return getTextProvider().hasKey(key);
    }
    public String getText(String aTextName) {
        return getTextProvider().getText(aTextName);
    }
    public String getText(String aTextName, String defaultValue) {
        return getTextProvider().getText(aTextName, defaultValue);
    }
    public String getText(String aTextName, String defaultValue, String obj) {
        return getTextProvider().getText(aTextName, defaultValue, obj);
    }
    public String getText(String aTextName, List<?>args) {
        return getTextProvider().getText(aTextName, args);
    }
    public String getText(String key, String[] args) {
        return getTextProvider().getText(key, args);
    }
    public String getText(String aTextName, String defaultValue, List<?>args) {
        return getTextProvider().getText(aTextName, defaultValue, args);
    }
    public String getText(String key, String defaultValue, String[] args) {
        return getTextProvider().getText(key, defaultValue, args);
    }
    public String getText (String key, String defaultValue, List <?> args, ValueStack stack) {
        return getTextProvider().getText(key, defaultValue, args, stack);
    }
    public String getText (String key, String defaultValue, String [ ] args, ValueStack stack) {
        return getTextProvider().getText(key, defaultValue, args, stack);
    }
    public ResourceBundle getTexts() {
        return getTextProvider().getTexts();
```

```java
    }
    public ResourceBundle getTexts(String aBundleName) {
        return getTextProvider().getTexts(aBundleName);
    }
    public void addActionError(String anErrorMessage) {
        validationAware.addActionError(anErrorMessage);
    }
    public void addActionMessage(String aMessage) {
        validationAware.addActionMessage(aMessage);
    }
    public void addFieldError(String fieldName, String errorMessage) {
        validationAware.addFieldError(fieldName, errorMessage);
    }
    public String input() throws Exception {
        return INPUT;
    }
    public String doDefault() throws Exception {
        return SUCCESS;
    }
    public String execute() throws Exception {
        return SUCCESS;
    }
    public boolean hasActionErrors() {
        return validationAware.hasActionErrors();
    }
    public boolean hasActionMessages() {
        return validationAware.hasActionMessages();
    }
    public boolean hasErrors() {
        return validationAware.hasErrors();
    }
    public boolean hasFieldErrors() {
        return validationAware.hasFieldErrors();
    }
    public void clearFieldErrors() {
        validationAware.clearFieldErrors();
    }
    public void clearActionErrors() {
        validationAware.clearActionErrors();
    }
    public void clearMessages() {
        validationAware.clearMessages();
    }
    public void clearErrors() {
        validationAware.clearErrors();
```

```
    }
    public void clearErrorsAndMessages() {
        validationAware.clearErrorsAndMessages();
    }
    public void validate() {
    }
    public Object clone() throws CloneNotSupportedException {
        return super.clone();
    }
    private TextProvider getTextProvider() {
        if (textProvider==null) {
            TextProviderFactory tpf=new TextProviderFactory();
            if (container!=null) {
                container.inject(tpf);
            }
            textProvider=tpf.createInstance(getClass(), this);
        }
        return textProvider;
    }
    public void setContainer(Container container) {
        this.container=container;
    }
}
```

ActionSupport 类是 Struts2 框架中默认的 Action 实现类,该类提供了许多默认的方法,如获取国际化信息的方法、数据验证的方法、默认处理用户请求的方法等。如果编写业务控制器类时继承了 ActionSupport 类,将会大大简化业务控制器类的开发。在开发 Java Web 项目时可以直接使用 ActionSupport 类作为业务控制器。在 struts.xml 中配置 Action 时,如果没有指定 class 属性(即没有提供用户的 Action 类),系统自动使用 ActionSupport 类作为业务控制器。

2.3.2 Action 实现类

Struts2 中的 Action 就是一个普通的 Java 类,该类不要求继承任何 Struts2 的父类,或者实现任何 Struts2 的接口,但是为了简化项目开发可以继承 ActionSupport 类。Action 类通常包含一个 execute()普通方法,该方法并没有任何参数,只是返回类型是字符串类型。Struts2 中的 Action 是如何获取用户 HTTP 请求中的参数值的?下面以例 2-8 来说明这个获取数据的过程,代码也可参考 1.3.1 节例 1-6,本例是在其基础上改进了一部分功能,即继承了 ActionSupport 类。

【例 2-8】 登录 Action(LoginAction.java)。

```
package ch02Action;
import com.opensymphony.xwork2.ActionContext;
import com.opensymphony.xwork2.ActionSupport;
```

```java
public class LoginAction extends ActionSupport{
    private String userName;
    private String password;
    public String getUserName()
    {
        return userName;
    }
    public void setUserName(String name)
    {
        this.userName=name;
    }
    public String getPassword()
    {
        return password;
    }
    public void setPassword(String password)
    {
        this.password =password;
    }
    public String execute() throws Exception
    {
        if(getUserName().equals("QQ")&&getPassword().equals("123"))
        {
            //将属性 userName 的值保存起来
            ActionContext.getContext().getSession().put("userName", getUserName());
            return SUCCESS;
        }
        else
        {
            return INPUT;
        }
    }
}
```

一般情况下，Struts2 中的 Action 会直接封装 HTTP 请求中的参数，所以通常情况下 Action 中会包含与 HTTP 请求参数对应的属性，并提供该属性的 getter 方法和 setter 方法。

在例 2-8 中 LoginAction 是 Action 的实现类，该类声明了 userName 和 password 两个属性，分别对应用户提交 form 表单中的两个参数，而且该类为每个属性声明了 getter 方法和 setter 方法。在 LoginAction 类中，execute()方法或者其他方法都可以使用该属性值获取 HTTP 请求中的参数。例如，在用户浏览器的 userName 文本框中输入 QQ，那么 LoginAction 类中的 userName 属性值就是 QQ，这个赋值过程由 Struts2 框架内部机制完成，即拦截器完成，该处理机制会调用 setUserName()方法设置相应属性的值。

Action 声明的属性名可以与用户 form 表单的属性名不同，但是在 Action 中一定要有与用户 form 表单参数对应的 getter 方法和 setter 方法。例如，在用户表单中有 userName

和 password 两个参数,那么在 Action 实现的类中必须有 getUserName()、setUserName()、getPassword()、setPassword()方法,在 Action 实现类中 userName 和 password 两个属性可以不要。

Action 不但可以设置和 HTTP 请求参数对应的属性,也可以设置 HTTP 请求参数中没有的属性,而且用户可以访问这些属性。对于 Struts2 框架来说,不会区分 Action 的属性是否为传入或者传出。Struts2 提供了类似"仓库"的机制,Action 可以使用 getter 方法和 setter 方法从"仓库"中取出或者存入属性的值,只要包含 HTTP 请求参数的 getter 方法和 setter 方法即可操作"仓库"中的数据。用户的 HTTP 也是通过 post 向"仓库"传入值或者从"仓库"中取出值。

2.3.3 Action 访问 ActionContext

Struts2 中的 Action 与 Servlet API 完全分离,体现了 Action 与 Servlet API 的非耦合性,也是对 Struts1 中的 Action 最大的改进。虽然 Struts2 框架中的 Action 已经与 Servlet 分离,但是在实现业务逻辑处理时,经常需要访问 Servlet 中的一些对象,如 request、session 和 application 等。Struts2 框架中提供 ActionContext 类,在 Action 中可以通过该类获取 Servlet 中的参数。

在 Java Web 应用程序中,需要访问的 Servlet 就是 HttpServlet、HttpSession 和 ServletContext,这 3 个类包含了 JSP 内置对象中所对应的 request、session 和 application 对象。

ActionContext 类是一个 Action 执行的上下文,Action 执行期间所用到的对象都保存在 ActionContext 中,如客户端提交的参数、session 会话信息等。而且 ActionContext 是一个局部线程,每个线程中的 ActionContext 内容都是唯一的,所以不用担心 Action 的线程安全问题。

创建 ActionContext 实例的方法如下:

```
ActionContext ac=ActionContext.getContext();
```

ActionContext 类中的常用方法如下所示。

(1) Object get(Object key):在 ActionContext 中查找 key 的值。
(2) Map getApplication():返回一个 application 级别的 Map 对象。
(3) static ActionContext getContext():获取当前线程的 ActionContext 对象。
(4) Map getParameter():返回一个 Map 类型的所有 HttpServletRequest 参数。
(5) Map getSession():返回 Map 类型的 HttpSession 值。
(6) void put(Object key,Object value):向当前 ActionContext 存入值。
(7) void setApplication(Map application):设置 application 对象的上下文。
(8) void setSession(Map session):设置 session 的值,参数为一个 Map 对象。

下面通过介绍登录系统,演示 Struts2 框架中 Action 是如何通过 ActionContext 类访问 Servlet API 的。本项目通过登录业务逻辑,分别将用户名保存在 application 和 session 中,最后通过 JSP 页面输出用户信息。下面是项目的开发步骤。

1. 项目介绍

该项目有登录页面(login1.jsp),代码如例 2-9 所示,其中使用了 Struts2 标签库中的标签,Struts2 标签库的用法将在后面章节中介绍;登录页面对应的业务控制器是 LoginAction1 类,代码如例 2-11 所示,该 Action 继承了 ActionSupport 类;如果登录成功(用户名、密码正确)跳转到 success1.jsp 页面,代码如例 2-10 所示,该页面中也使用了 Struts2 的标签库;如果登录失败(用户名或者密码不正确)则重新回到登录页面(login1.jsp)。此外还需要配置 web.xml,代码和 1.3.1 节中例 1-3 相同;配置 struts.xml 配置文件,代码如例 2-12 所示。登录系统的文件结构如图 2-2 所示。

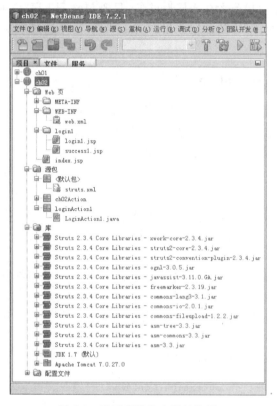

图 2-2 登录系统的文件结构图

2. 在 web.xml 中配置核心控制器 FilterDispatcher

参考 1.3.1 节中的例 1-3。

3. 编写视图组件(JSP 页面)

登录页面如图 2-3 所示,代码如例 2-9 所示。登录成功页面如图 2-4 所示,代码如例 2-10 所示。

图 2-3 登录页面

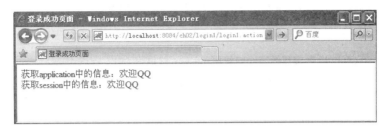

图 2-4　登录成功页面

【例 2-9】　登录页面（login1.jsp）。

```jsp
<%@page contentType="text/html" pageEncoding="UTF-8"%>
<!--通过 taglib 指令导入 Struts2 的标签库-->
<%@taglib prefix="s" uri="/struts-tags"%>
<html>
    <head>
        <meta http-equiv="Content-Type" content="text/html; charset=UTF-8">
        <!--Struts2 标签库的使用-->
        <title><s:text name="演示 Struts2 中 Action 通过 ActionContext 访问 Servlet API"/>
        </title>
    </head>
    <body>
        <hr/>
        <!--Struts2 标签库中表单标签的使用-->
        <s:form action="login1" method="post">
            <!--用户名输入框标签-->
            <s:textfield name="userName" label="用户名称" size="16"/>
            <br>
            <!--用户密码输入框标签-->
            <s:password name="password" label="用户密码" size="18"/>
            <br>
            <!--提交按钮标签-->
            <s:submit value="登录"/>
        </s:form>
        <hr/>
    </body>
</html>
```

【例 2-10】　登录成功页面（success1.jsp）。

```jsp
<%@page contentType="text/html" pageEncoding="UTF-8"%>
<%@taglib prefix="s" uri="/struts-tags"%>
<html>
    <head>
        <meta http-equiv="Content-Type" content="text/html; charset=UTF-8">
```

```
        <title><s:text name="登录成功页面"/></title>
    </head>
    <body>
        <!--OGNL 表达式和标签库的使用,value 的值为 OGNL 表达式-->
        获取 application 中的信息:欢迎<s:property value="#application.userName"/>
        <br/>
        获取 session 中的信息:欢迎<s:property value="#session.userName"/>
    </body>
</html>
```

4. 编写业务控制器 Action

业务控制器类 LoginAction1 是处理 login1.jsp 的业务逻辑器,代码如例 2-11 所示。

【例 2-11】 登录页面对应的业务控制器(LoginAction1.java)。

```java
package loginAction1;
import com.opensymphony.xwork2.ActionContext;
import com.opensymphony.xwork2.ActionSupport;

public class LoginAction1 extends ActionSupport{
    private String userName;
    private String password;
    public String getUserName()
    {
        return userName;
    }
    public void setUserName(String name)
    {
        this.userName=name;
    }

    public String getPassword()
    {
        return password;
    }
    public void setPassword(String password)
    {
        this.password =password;
    }
    public String execute() throws Exception
    {
        if(getUserName().equals("QQ")&&getPassword().equals("123"))
        {
            //获取 ActionContext
            ActionContext ac=ActionContext.getContext();
            //把登录名保存到 application 中
            ac.getApplication().put("userName", getUserName());
```

```
            //把登录名保存到 session 中
            ac.getSession().put("userName", getUserName());
            return SUCCESS;
        }
        else
        {
            return INPUT;
        }
    }
}
```

5. 修改 struts.xml 配置 Action

修改 struts.xml 配置文件,代码如例 2-12 所示。

【例 2-12】 在 struts.xml 中配置 Action(struts.xml)。

```
<!DOCTYPE struts PUBLIC
"-//Apache Software Foundation//DTD Struts Configuration 2.0//EN"
"http://struts.apache.org/dtds/struts-2.0.dtd">
<struts>
    <package name="login" extends="struts-default" >
        <action name="login1" class="loginAction1.LoginAction1">
            <!--在【例 2-11】中由于继承了 ActionSupport,所以返回的是 SUCCESS 和 INPUT,
            但是在这里应使用 success 和 input,即常量值-->
            <result name="input">/login/login1.jsp</result>
            <result name="success">/login/success1.jsp</result>
        </action>
    </package>
</struts>
```

6. 项目部署和运行

运行后效果如图 2-3 所示,输入用户名 QQ 和密码 123 后,单击"登录"按钮,转到如图 2-4 所示的登录成功页面。

2.3.4 Action 直接访问 Servlet

Struts2 中直接访问 Servlet 有 IoC 和非 IoC 两种方式。以 IoC 方式访问 Servlet 时 Action 实现类必须实现一些接口;以非 IoC 方式访问 Servlet 时可使用 Struts2 提供的辅助类来访问。Action 直接访问 Servlet 方式中提供的辅助类是 ServletActionContext。

IoC(Inversion of Control)将设计好的类交给系统去控制,而不是在类自己内部控制,称为控制反转。通俗的解释是"站着别动,我去找你"。在 Java 项目开发中,IoC 主要解决组件之间的依赖关系,降低模块之间的耦合度,也是 Spring 框架的核心。

1. IoC 方式

在 Struts2 框架中,可以通过 IoC 方式将 Servlet 对象注入到 Action 中,这需要在 Action 中实现以下接口。

(1) ServletRequestAware:实现该接口的 Action 可以直接访问 request 对象,该接口

中提供有 void setServletRequest(HttpServletRequest ruquest)。

（2）ServletResponseAware：实现该接口的 Action 可以直接访问 response 对象，该接口中提供有 void setServleResponse(HttpServletResponse response)。

（3）SessionAware：实现该接口的 Action 可以直接访问 session 对象，该接口中提供有 void setSession(Map map)方法。

IoC 方式访问 Servlet 的登录系统业务控制器代码如例 2-13 所示。

【例 2-13】 IoC 访问方式的 Action(IoCAction.java)。

```
package ch02Action;
import com.opensymphony.xwork2.ActionSupport;
import javax.servlet.http.HttpServletRequest;
import javax.servlet.http.HttpSession;
import org.apache.struts2.interceptor.ServletRequestAware;

public class IoCAction extends ActionSupport implements ServletRequestAware{
    private String userName;
    private String password;
    private HttpServletRequest request;
    public String getUserName()
    {
        return userName;
    }
    public void setUserName(String name)
    {
        this.userName=name;
    }
    public String getPassword()
    {
        return password;
    }
    public void setPassword(String password)
    {
        this.password =password;
    }
    //必须实现该方法,该方法是接口中的方法
    public void setServletRequest(HttpServletRequest hsr) {
        request=hsr;
    }
    public String execute() throws Exception
    {
        if(getUserName().equals("QQ")&&getPassword().equals("123"))
        {
            //通过 request 对象获取 session 对象
            HttpSession session=request.getSession();
            //把登录名传入 session 中
```

```
            session.setAttribute("userName", this.userName);
            return SUCCESS;
        }
        else
        {
            return INPUT;
        }
    }
}
```

例 2-13 中 Action 类继承了 ActionSupport 类,实现了 ServletRequestAware 接口,在 Action 类中实现了接口的 setServletRequest(HttpServletRequest hsr)方法,并获得了 request 对象;然后在 execute()方法中,通过 request 对象调用 getSession()方法获取 session 对象,实现了对 Servlet 的直接访问。

2. 非 IoC 方式

在非 IoC 方式中,Struts2 提供 ServletActionContext 类帮助获得 Servlet。该类中的常用方法如下所示。

(1) static getRequest():获取 Web 应用程序的 request 对象。

(2) static getResponse():获取 Web 应用程序的 response 对象。

非 IoC 方式访问 Servlet 的登录系统业务控制器代码如例 2-14 所示。

【例 2-14】 非 IoC 访问方式的 Action(NoIoCAction.java)。

```
package ch02Action;
import com.opensymphony.xwork2.ActionSupport;
import javax.servlet.http.HttpServletRequest;
import javax.servlet.http.HttpSession;
import org.apache.struts2.ServletActionContext;

public class NoIoCAction extends ActionSupport{
    private String userName;
    private String password;
    public String getUserName()
    {
        return userName;
    }
    public void setUserName(String name)
    {
        this.userName=name;
    }
    public String getPassword()
    {
        return password;
    }
    public void setPassword(String password)
```

```
        {
            this.password =password;
        }
        public String execute() throws Exception
        {
            if(getUserName().equals("QQ")&&getPassword().equals("123"))
            {
                //调用 ServletActionContext 的 getRequest()方法获取 request 对象
                HttpServletRequest request=ServletActionContext.getRequest();
                //调用 request 对象的 getSession()方法获取 session 对象
                HttpSession session=request.getSession();
                //调用 session 对象的方法设置数据
                session.setAttribute("userName", this.userName);
                session.setAttribute("password", this.password);
                return SUCCESS;
            }
            else{
                return INPUT;
            }
        }
    }
```

2.3.5 Action 中的动态方法调用

在实际项目开发中，需要一个 Action 中完成一组紧密相关的业务操作。例如，与一件商品相关的基本操作有增加商品、删除商品、修改商品和查看商品。

通过将增加商品、删除商品、修改商品和查看商品这些业务相关的操作合并到一个 Action 中，根据业务请求不同而动态地调用相应的方法，这就是 Action 的动态方法调用。该方法减少了 Struts2 框架中的 Action 数量，减少了重复编码，使应用更加便于维护。

Struts2 提供两种方式实现动态方法的调用：不指定 method 属性和指定 method 属性。

(1) 不指定 method 属性。Struts2 中的不指定 method 属性是指，表单元素的 action 属性并不是直接等于某个 Action 的名字，且 form 表单不需要指定 method 属性。

不指定 method 属性格式如下：

```
<form action="Action 名字!方法名字">
```

或者

```
<form action="Action 名字!方法名字.action">
```

如果在 JSP 页面中有多个提交按钮，每个提交按钮都将可以将请求提交到同一个 Action，但是每个业务对应着 Action 中不同的方法来处理不同的业务请求。在 struts.xml 中只需配置该 Action，而不必配置每个方法。

不指定 method 属性在 struts.xml 中的配置格式如下：

```xml
<action name="Action 名字" class="包名.Action 类名">
    <result name="***">/***.jsp</result>
    <result name="***">/***.jsp</result>
</action>
```

(2) 指定 method 属性。Struts2 中的指定 method 属性是指每个表单都有 method 属性，属性值指向在 Action 中定义的方法名。

指定 method 属性格式如下：

```xml
<form action="Action 名字" method="方法名字">
```

指定 method 属性需要在 struts.xml 中配置 Action 中的每个方法，而且每个 Action 配置中都要指定 method 属性，该属性值和表单属性值一致。

指定 method 属性在 struts.xml 中配置格式如下：

```xml
<action name="Action 名字" class="包名.Action 类名" method="方法名字">
    <result name="***">/***.jsp</result>
    <result name="***">/***.jsp</result>
</action>
```

对比 Struts2 提供的两种实现动态方法调用的方式，第一种方式只需在 struts.xml 中为 Action 配置一个＜action＞元素，使 struts.xml 文件比较简洁，但是逻辑结构不清楚。第二种方式是为 Action 中的每个业务逻辑方法在 struts.xml 中都配置一个＜action＞元素，业务逻辑结构清楚，但是增加了＜action＞元素配置的数量，使 struts.xml 文件过于庞大难以管理。在实际应用中，可以根据具体情况选择使用。

下面项目是使用不指定 method 方式在 Action 中实现登录和注册业务逻辑。项目的开发步骤如下。

1. 项目介绍

该项目有一个实现登录和注册功能的页面（loginRegister.jsp），代码如例 2-15 所示；登录和注册页面对应的业务控制器类是 LoginRegisterAction，代码如例 2-17 所示；登录和注册成功页面（success2.jsp）代码如例 2-16 所示。此外还需要配置 web.xml，代码和 1.3.1 节中例 1-3 相同；配置 struts.xml 配置文件，代码如例 2-18 所示。Action 中动态方法调用项目的文件结构如图 2-5 所示。

图 2-5 Action 动态方法调用应用项目的文件结构图

2. 在 web.xml 中配置核心控制器 FilterDispatcher

参考 1.3.1 节中的例 1-3。

3. 编写视图组件（JSP 页面）

登录注册页面如图 2-6 所示，其代码如例 2-15 所示。登录成功页面代码如例 2-16 所示。

【**例 2-15**】 登录注册页面（loginRegister.jsp）

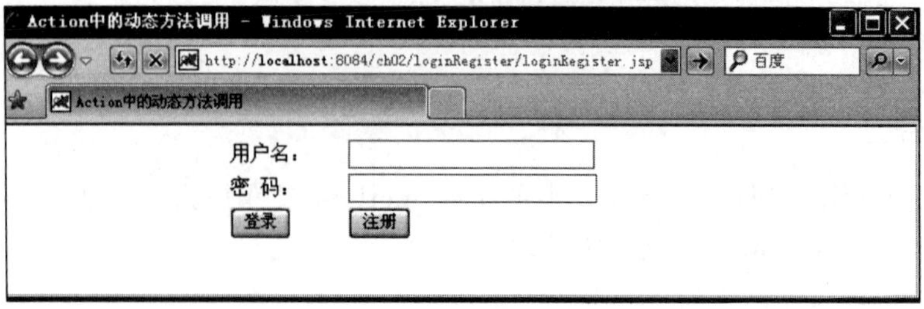

图 2-6 登录注册页面

```jsp
<%@ page contentType="text/html" pageEncoding="UTF-8"%>
<html>
    <head>
        <meta http-equiv="Content-Type" content="text/html; charset=UTF-8">
        <title>Action中的动态方法调用</title>
    </head>
    <body>
        <table width="360" align="center">
            <form   action="loginReg!execute">
            <tr>
                <td>用户名:</td>
                <td><input type="text" name="userName" size="26"/></td>
            </tr>
            <tr>
                <td>密   码:</td>
                <td><input type="password" name="password" size="28"/></td>
            </tr>
            <tr>
                <td><input type="submit" value="登录"/></td>
                <td><input type="submit" value="注册"
                    onclick="register();"/>
                </td>
            </tr>
            </form>
        <table>
        <script type="text/javascript">
            function register(){
                //获取页面的第一个表单
                targetForm = document.forms[0];
                //动态修改表单的 action 属性
                targetForm.action = "loginReg!regist";
            }
        </script>
    </body>
</html>
```

【例 2-16】 成功页面(success2.jsp)。

```jsp
<%@page contentType="text/html" pageEncoding="UTF-8"%>
<%@taglib prefix="s" uri="/struts-tags"%>
<html>
    <head>
        <meta http-equiv="Content-Type" content="text/html; charset=UTF-8">
        <title>成功页面</title>
    </head>
    <body>
        <s:property value="msg"/>
    </body>
</html>
```

4. 编写业务控制器 Action

业务控制器 LoginRegisterAction 是用于处理 loginRegister.jsp 页面的,代码如例 2-17 所示。

【例 2-17】 登录注册页面对应的业务控制器(LoginRegisterAction.java)。

```java
package loginRegisterAction;
import com.opensymphony.xwork2.ActionContext;
import com.opensymphony.xwork2.ActionSupport;
public class LoginRegisterAction extends ActionSupport{
    private String userName;
    private String password;
    //设置返回信息
    private String msg;
    public String getUserName()
    {
        return userName;
    }
    public void setUserName(String name)
    {
        this.userName=name;
    }
    public String getPassword()
    {
        return password;
    }
    public void setPassword(String password)
    {
        this.password =password;
    }
    public String getMsg() {
        return msg;
    }
    public void setMsg(String msg) {
```

```
        this.msg =msg;
    }
    //Action包含的注册控制逻辑
    public String regist() throws Exception{
        ActionContext.getContext().getSession().put("userName" , getUserName());
        setMsg("恭喜你,"+userName+",注册成功!");
        return SUCCESS;
    }
    //Action默认包含的控制逻辑
    public String execute() throws Exception{
        if(getUserName().equals("QQ")&&getPassword().equals("123"))
        {
            ActionContext.getContext().getSession().put("userName",getUserName());
            setMsg("你单击的是【登录】!"+"你的登录名为"+userName+",登录成功!");
            return SUCCESS;
        }
        else
        {
            return INPUT;
        }
    }
}
```

5. 修改 struts.xml 配置 Action

配置 struts.xml 配置文件,代码如例 2-18 所示。

【例 2-18】 在 struts.xml 中配置 Action(struts.xml)。

⋮
```
<!--不指定method属性的Action配置-->
<action name="loginReg" class="loginRegisterAction.LoginRegisterAction">
    <result name="input">/loginRegister/loginRegister.jsp</result>
    <result name="success">/loginRegister/success2.jsp</result>
</action>
```
⋮

6. 项目部署和运行

登录注册页面运行效果如图 2-6 所示,在其中输入用户名 QQ 和密码 123,单击"登录"按钮,运行效果如图 2-7 所示。若在图 2-6 所示页面中输入用户名"梦想"和密码 66,单击"注册"按钮,运行效果如图 2-8 所示。

图 2-7 登录成功

图 2-8 注册成功

2.4 Struts2 的 OGNL 表达式

对象图导航语言(Object-Graph Navigation Language,OGNL)是一种功能强大的表达式语言(Expression Language,EL),通过简单一致的表达式语法,可以存取对象的任意属性,调用对象的任意方法、遍历整个对象的数据以及自动实现字段类型转换等功能。

2.4.1 Struts2 的 OGNL 表达式

OGNL 有三个参数,即表达式、根对象和上下文环境。

表达式是 OGNL 的核心,所有的 OGNL 操作都是在解析表达式后进行的。表达式指出了 OGNL 操作要做的工作。例如,name、student.name 表达式表示取 name 的值或者 student 中 name 的值。

根对象是 OGNL 要操作的对象,在表达式规定了要完成的工作后,需要指定工作的操作对象。例如,<s:property value="♯request.name"/>中,request 就是对象,从这个对象中取出 name 属性的值。

上下文环境是 OGNL 要执行操作的地点。

如果使用 OGNL 要访问的不是根对象,则需要使用名称空间,用"♯"来表示;如果访问的是一个根元素,则不必使用名称空间,可以直接访问根对象的属性。

在 Struts2 中堆值就是 OGNL 的根对象。获取堆值的属性可以使用 $\{属性\}$,如 $\{name\}$ 可获取 name 的值。如果访问其他上下文路径中的对象,由于不是根对象,在访问时需要加"♯"前缀。

下面项目使用 OGNL 表达式实现对注册页面数据的提取及显示。项目的开发步骤如下。

1. 项目介绍

该项目有一个注册页面(register.jsp),代码如例 2-19 所示;注册页面对应的业务控制器是 OGNLAction 类,代码如例 2-21 所示,注册成功后页面跳转到注册成功页面 registerSuccess.jsp,代码如例 2-20 所示,该页面中使用了 OGNL 表达式。此外,还需要配置 web.xml,代码和 1.3.1 节中例 1-3 相同;配置 struts.xml 配置文件,代码如例 2-22 所示。该注册系统的文件结构如图 2-9 所示。

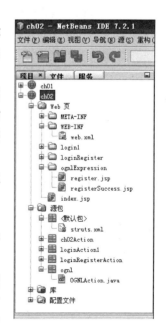

图 2-9 OGNL 表达式应用实例的文件结构图

2. 在 web.xml 中配置核心控制器 FilterDispatcher

参考 1.3.1 节中的例 1-3。

3. 编写视图组件(JSP 页面)

注册页面如图 2-10 所示。代码如例 2-19 所示。注册成功页面代码如例 2-20 所示。

图 2-10　注册页面

【例 2-19】　注册页面(register.jsp)。

```
<%@page contentType="text/html" pageEncoding="UTF-8"%>
<html>
    <head>
        <meta http-equiv="Content-Type" content="text/html; charset=UTF-8">
        <title>使用 OGNL 表达式获取数据</title>
    </head>
    <body>
        <form action="ognl">
            学号:<input type="text" name="no" ><br>
            姓名:<input type="text" name="name"><br>
            性别:<input type="text" name="sex"><br>
            年龄:<input type="text" name="age"><br>
            <input type="submit" value="注册"/><br>
        </form>
    </body>
</html>
```

【例 2-20】　注册成功页面(registerSuccess.jsp)。

```
<%@page contentType="text/html" pageEncoding="UTF-8"%>
<%@taglib prefix="s" uri="/struts-tags"%>
<html>
    <head>
        <meta http-equiv="Content-Type" content="text/html; charset=UTF-8">
        <title>使用 OGNL 表达式获取数据,注册成功</title>
    </head>
    <body>
        <h1>${name}</h1>
```

```
        <hr>
        获取 action 属性:<s:property value="name"/><br>
        获取 reqeust 属性:<s:property value="#request.name"/><br>
        获取 session 属性:<s:property value="#session.name"/><br>
        获取 application 属性:<s:property value="#application.name"/><br>
        <hr>
    </body>
</html>
```

4. 编写业务控制器 Action

注册页面(register.jsp)对应的业务控制器类是 LoginRegisterAction,代码如例 2-21 所示。

【例 2-21】 注册页面对应的业务控制器(OGNLAction.java)。

```
package ognl;
import com.opensymphony.xwork2.ActionContext;
import com.opensymphony.xwork2.ActionSupport;
import java.util.Map;
import javax.servlet.http.HttpServletRequest;
import org.apache.struts2.ServletActionContext;

public class OGNLAction extends ActionSupport{
    private String no;                              //学号
    private String name;                            //姓名
    private String sex;                             //性别
    private int age;                                //年龄
    public String getNo() {
        return no;
    }
    public void setNo(String no) {
        this.no =no;
    }
    public String getName() {
        return name;
    }
    public void setName(String name) {
        this.name =name;
    }
    public String getSex() {
        return sex;
    }
    public void setSex(String sex) {
        this.sex =sex;
    }
    public int getAge() {
        return age;
```

```java
    }
    public void setAge(int age) {
        this.age =age;
    }
    public String execute() throws Exception{
        //获取 request,并添加信息
        HttpServletRequest request=ServletActionContext.getRequest();
        request.setAttribute("name", getName());
        //获取 session,并添加信息
        Map session =ActionContext.getContext().getSession();
        session.put("name", getName());
        //获取 application,并添加信息
        Map application =ActionContext.getContext().getApplication();
        application.put("name", getName());
        return SUCCESS;
    }
}
```

5. 修改 struts.xml 配置 Action

配置 struts.xml 配置文件,代码如例 2-22 所示。

【例 2-22】 在 struts.xml 中配置 Action(struts.xml)。

⋮
```xml
<action name="ognl" class="ognl.OGNLAction">
<result name="success">/ognlExpression/registerSuccess.jsp</result>
</action>
```
⋮

6. 项目部署和运行

项目运行后出现如图 2-10 所示页面,输入数据后如图 2-11 所示,单击"注册"按钮,出现如图 2-12 所示页面。

图 2-11 在注册页面输入数据

图 2-12 注册成功页面

2.4.2 Struts2 的 OGNL 集合

OGNL 提供了对 Java 集合 API 非常好的支持,创建集合并对其操作是 OGNL 的一个

基本特性。如果需要一个集合元素时,如 List 对象或者 Map 对象,可以使用 OGNL 中与集合相关的表达式。

OGNL 中使用 List 对象的格式如下:

```
{e1,e2,e3}
```

该表达式会直接生成一个 List 对象,在生成的 List 对象中包含 3 个元素:e1、e2、e3。如果需要更多元素,可以继续添加。

OGNL 中使用 Map 对象的格式如下:

```
#{key1:value1, key2:value2, key3:value3,…}
```

该表达式会直接生成一个 Map 对象。

对于集合元素的判定,OGNL 表达式可以使用 in 和 not in 操作。in 表达式用来判断某个元素是否在指定的集合对象中;not in 用于判断某个元素是否不在指定的集合对象中。

例如:

```
<s:if test="'a' in {'a', 'b'}">
    ⋮
</s:if>
```

或者

```
<s:if test="'a' not in {'a', 'b'}">
    ⋮
</s:if>
```

除了 in 和 not in 之外,OGNL 还允许使用某些规则获取集合对象的子集,常用的相关操作如下所示。

(1) ?:用于获取符合逻辑的多个元素。
(2) ^:用于获取符合逻辑的第一个元素。
(3) $:用于获取符合逻辑的最后一个元素。

例如:

```
Student.sex{?#this.sex=='male'}        //获取 Student 的所有值为 male 的 sex 集合
```

2.5　Struts2 的标签库

Struts2 框架提供了丰富的标签库用于构建视图组件。Struts2 标签库大大简化了视图页面的开发,并且提高了视图组件的可维护性。

2.5.1　Struts2 的标签库概述

Struts2 标签库没有严格地对标签进行分类,而是把所有的标签整合到一个标签库中。但是按照标签库提供的功能可以将 Struts2 标签库分为三大类:UI 标签、非 UI 标签和 Ajax 标签。

(1) 用户界面标签(UI 标签)：主要用来生成 HTML 元素的标签。
(2) 非用户界面标签(非 UI 标签)：主要用来实现数据访问、逻辑控制。
(3) Ajax 标签：主要用来支持 Ajax 技术。

用户界面标签(UI 标签)又可以分为如下两大类。

① 表单标签：主要用于生成 HTML 中的表单信息。

② 非表单标签：主要包含一些常用的功能标签，如显示日期等。

非用户界面标签(非 UI 标签)又可以分为如下两大类：

① 控制标签：主要用来实现条件和循环流程控制。

② 数据标签：主要用来实现数据存储与处理。

Struts2 标签库的层次结构如图 2-13 所示。

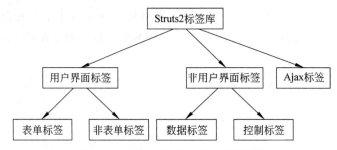

图 2-13　Struts2 标签库的层次结构图

2.5.2　Struts2 的表单标签

Struts2 的用户界面标签可以分为表单标签和非表单标签。HTML 表单元素和 Struts2 中大部分表单标签一一对应。

例如：

＜form action="login" method="post"＞

对应着

＜s:form action="login" method="post"＞

用户名

＜input type="text" name="userName"/＞

对应着

＜s:textfield name="userName" label="用户名"/＞

密码

＜input type=" password " name="userPassword"/＞

对应着

＜s:password name="password" label="密码"/＞

下面介绍 Struts2 中常用的表单标签。

1. ＜s：checkbox＞标签

checkbox 标签是复选框标签。常用属性如下所示。

(1) label：设置显示的字符串，可选项。

(2) name：设置表单元素的名字，表单元素的名字实际上封装着一个请求参数，而该请求参数被 Action 封装到其中，当该表单对应的 Action 需要使用参数的值且对应的属性有值时，该属性值就是表单元素 value 的值。name 属性是表单元素的通用属性，每个表单元素都会使用，必选项。

(3) value：该属性用于设置是否默认选定，可选项。

例如：

```
<s:checkbox label="学习" name="学习" value="true"/>
<s:checkbox label="电影" name="电影"/>
```

2. ＜s：checkboxlist＞标签

checkboxlist 标签可以一次创建多个复选框，在 HTML 中可以使用多行＜input type="checkbox"＞实现。常用属性是 list。

list：指定集合用于生成复选框的列表项，可以使用 List 集合或者 Map 对象，必选项。

例如：

```
<s:checkboxlist label="个人爱好" list="{'学习','看电影','编程序'}" name="love">
</s:checkboxlist>
```

3. ＜s：combobox＞标签

combobox 标签生成一个单行文本框和一个下拉列表框的组合，两个表单元素对应一个请求，单行文本框中的值对应请求参数，下拉列表框只是起到辅助作用。常用属性如下所示。

(1) list：指定集合生成下拉列表项，可以使用 List 集合或者 Map 对象，必选项。

(2) readonly：指定文本框是否可编辑，为 true 不可编辑，为 false 可编辑，默认为 false，可选项。

【例 2-23】 combobox 标签的使用(combobox.jsp)。

```
<%@page contentType="text/html" pageEncoding="UTF-8"%>
<%@taglib prefix="s" uri="/struts-tags"%>
<html>
    <head>
        <meta http-equiv="Content-Type" content="text/html; charset=UTF-8">
        <title>combobox 标签的使用</title>
    </head>
    <body>
        <s:form>
            <s:combobox label="颜色选择" name="colorName" readonly="false"
            headerValue="---请选择---" headerKey="1" list="{'红色','蓝色','黑色',
            '白色'}"/>
```

            </s:form>
        </body>
</html>

运行效果如图 2-14 所示。选择其中一个颜色后效果如图 2-15 所示。

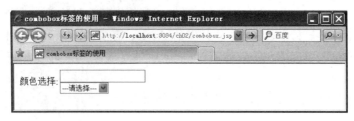

图 2-14　没有选择项前

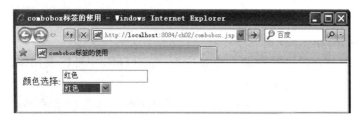

图 2-15　有选择项后

4.　<s：doubleselect>标签

doubleselect 标签生成一个相互关联的列表框，在第一个列表框中选择某一项后，第二个列表框中将自动为其选定相关信息。常用属性如下所示。

（1）headerValue：指定列表框默认值。

（2）headerKey：指定列表框默认项的值。

（3）doubleName：指定第二个下拉列表框的名字。

（4）list：指定第一个下拉列表框中选项的集合。

（5）doubleList：指定第二个下拉列表框中的选项集合。

（6）top：指定第一列表框。

【例 2-24】　doubleselect 标签的使用（doubleselect.jsp）。

```
<%@page contentType="text/html" pageEncoding="UTF-8"%>
<%@taglib prefix="s" uri="/struts-tags"%>
<html>
    <head>
        <meta http-equiv="Content-Type" content="text/html; charset=UTF-8">
        <title>doubleselect 标签的使用</title>
    </head>
    <body>
        <s:form>
            <s:doubleselect label="选择一项" headerValue="---请选择---"
                headerKey="1" doubleName="doublesel" list="{'颜色','水果'}"
                doubleList="top=='颜色'?{'红色','蓝色','黑色','白色'}:{'苹果','香蕉',
```

```
            '梨','葡萄'}" />
        </s:form>
    </body>
</html>
```

运行效果如图 2-16 所示。选择其中一项后效果如图 2-17 所示。

图 2-16 没有选择项前

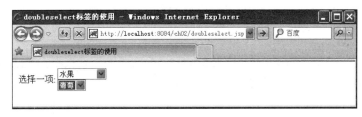

图 2-17 有选择项后

5．＜s：file＞标签

file 标签用于在页面上生成一个上传文件的元素。上传文件的具体实现请参考 3.4 节。

【**例 2-25**】 file 标签的使用(file.jsp)。

```
<%@page contentType="text/html" pageEncoding="UTF-8"%>
<%@taglib prefix="s" uri="/struts-tags"%>
<html>
    <head>
        <meta http-equiv="Content-Type" content="text/html; charset=UTF-8">
        <title>file 标签的使用</title>
        <s:head/>
    </head>
    <body>
        <s:form>
            <s:file name="UploadFileName" accept="text/*"/>
        </s:form>
    </body>
</html>
```

页面运行效果如图 2-18 所示，在其中单击"浏览…"按钮出现文件对话框，在文件对话框中可以选择要上传的文件。

6．＜s：select＞标签

select 标签用来生成一个下拉列表框，可通过指定 list 属性来指定下拉列表内容。常用

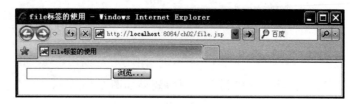

图 2-18 file 标签实例运行效果

属性如下所示。

(1) size：指定下拉列表框中可以显示的选择项个数，可选项。

(2) multiple：设置该列表框是否允许多选，默认值为 false，可选项。

【例 2-26】 select 标签的使用(select.jsp)。

```
<%@page contentType="text/html" pageEncoding="UTF-8"%>
<%@taglib prefix="s" uri="/struts-tags"%>
<html>
    <head>
        <meta http-equiv="Content-Type" content="text/html; charset=UTF-8">
        <title>select 标签的使用</title>
    </head>
    <body>
        <s:form>
            <s:select label="选择星期" headerValue="---请选择---" headerKey="3"
                list="{'星期一','星期二','星期三','星期四','星期五','星期六',
                '星期日'}" />
        </s:form>
    </body>
</html>
```

7. <s：radio>标签

radio 标签为一个单选框，用法和 checkboxlist 标签相似。

例如：

```
<s: radio label="性别" list="{'男','女'}" name="sex">
</s: radio>
```

8. <s：textarea>标签

textarea 标签用来生成一个文本区域，由行和列组成。

例如：

```
<s:textarea label="留言板" name="留言" cols="10" rows="10"/>
```

9. <s：token>标签

token 标签的使用目的是防止用户多次提交表单，避免恶意刷新页面。

例如：

```
<s:token/>
```

10. <s：optiontransferselect>标签

optiontransferselect 标签用来创建两个选项框以及转移项，该标签会自动生成两个下拉列表框，同时生成相关的按钮，这些按钮可以控制选项在两个下拉列表框之间的移动、排序。常用属性如下所示。

(1) addAllToLeftLabel：设置实现全部左移动功能按钮上的文本。
(2) addAllToRightLabel：设置实现全部右移动功能按钮上的文本。
(3) addToLeftLabel：设置实现左移动功能按钮上的文本。
(4) addToRightLabel：设置实现右移动功能按钮上的文本。
(5) allowAddAllToLeft：设置全部左移动功能的按钮。
(6) allowAddAllToRight：设置全部右移动功能的按钮。
(7) allowAddToLeft：设置左移动功能的按钮。
(8) allowAddToRight：设置右移动功能的按钮。
(9) leftTitle：设置左边列表框的标题。
(10) rightTitle：设置右边列表框的标题。
(11) allowSelectAll：设置全部选择功能的按钮。
(12) selectAllLabel：设置全部选择功能按钮上的文本。
(13) multiple：设置第一个列表框是否多选，默认是 true。
(14) doubleName：设置第二个列表框的名字。
(15) doubleList：设置第二个列表框的集合。
(16) doubleMultiple：设置第二个列表框是否允许多选，默认是 true。

【例 2-27】 optiontransferselect 标签的使用（optiontransferselect.jsp）。

```
<%@ page contentType="text/html" pageEncoding="UTF-8"%>
<%@ taglib prefix="s" uri="/struts-tags"%>
<html>
    <head>
        <meta http-equiv="Content-Type" content="text/html; charset=UTF-8">
        <title>optiontransferselect 标签的使用</title>
    </head>
    <body>
        <s:form>
            <s:optiontransferselect label="你喜欢的城市" name="left" leftTitle=
            "国内" rightTitle="国外" list="{'北京','上海','南京','深圳','海南','青岛'}"
            headerValue="---请选择---" headerKey="1" doubleName="right"
            doubleHeaderValue="---请选择---" doubleHeaderKey="1" doubleList=
            "{'东京','华盛顿','伦敦','芝加哥','温哥华','多伦多'}"/>
        </s:form>
    </body>
</html>
```

运行效果如图 2-19 所示。

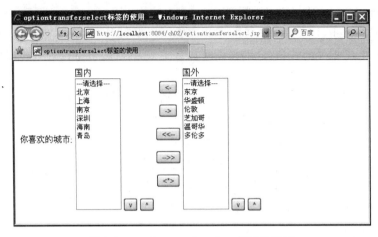

图 2-19　optiontransferselect 标签的使用

11. ＜s：updownselect＞标签

updownselect 标签用来在页面中生成一个下拉列表框，可以在选项内容中上下移动。常用属性如下所示。

　　(1) allowMoveUp：设置上移功能的按钮，默认值为 true，即显示该按钮。
　　(2) allowMoveDown：设置下移功能的按钮，默认值为 true，即显示该按钮。
　　(3) allowSelectAll：设置全选功能的按钮，默认值为 true，即显示该按钮。
　　(4) MoveUpLabel：设置上移功能按钮上的文本，默认值为 ∧。
　　(5) MoveDownLabel：设置下移功能按钮上的文本，默认值为 ∨。
　　(6) selectAllLabel：设置全选功能按钮上的文本，默认值为 *。

【例 2-28】 updownselect 标签的使用（updownselect.jsp）。

```
<%@page contentType="text/html" pageEncoding="UTF-8"%>
<%@taglib prefix="s" uri="/struts-tags"%>
<html>
    <head>
        <meta http-equiv="Content-Type" content="text/html; charset=UTF-8">
        <title>updownselec 标签的使用</title>
    </head>
    <body>
        <s:form>
            <s:updownselect label="你最喜欢的旅游城市" name="city" headerValue=
            "--------请选择城市--------" headerKey="1" list="{'北京','上海',
            '郑州','西安','杭州','苏州','青岛'}" emptyOption="true" selectAllLabel=
            "全选" moveUpLabel="上移" moveDownLabel="下移"/>
        </s:form>
    </body>
</html>
```

运行效果如图 2-20 所示。

图 2-20 updownselect 标签的使用

2.5.3 Struts2 的非表单标签

非表单标签主要用于在页面中生成非表单的可视化元素。下面对常用非表单标签进行介绍。

1. <s：a>标签

a 标签主要用于生成超链接。

例如：

```
<s:a href="register.action">注册</s:a>
```

2. <s：actionerror>和<s：actionmessage>标签

actionerror 标签和 actionmessage 标签的作用基本一样，都是在页面上输出 Action 方法中添加的信息。其中，actionerror 标签输出 Action 中 addActionErrors()方法添加的信息；而 actionmessage 标签输出的是 Action 中 AddActionMessage()方法添加的信息。

下面通过一个项目介绍两个标签的具体应用。首先编写一个 Action 类。类名为 ActionErrorActionMessage，代码如例 2-29 所示；在 struts.xml 中配置该 Action，代码如例 2-30 所示；最后写一个信息输出页面（showActionErrorMessage.jsp），代码如例 2-31 所示。项目的文件结构如图 2-21 所示。

图 2-21 项目文件结构图

1）编写 Action 类

【例 2-29】 在 Action 中封装信息（ActionErrorActionMessage.java）。

```
package actionerrorAndactionmessage;
import com.opensymphony.xwork2.ActionSupport;

public class ActionErrorActionMessage extends ActionSupport{
    public String execute(){
        //使用 addActionError()方法添加信息
        addActionError("使用 ActionError 添加错误信息!");
        addActionMessage("使用 ActionMessage 添加普通信息!");
        return SUCCESS;
```

        }
    }

2) 在 struts.xml 配置 Action

【例 2-30】 配置 Action(struts.xml)。

```
<!DOCTYPE struts PUBLIC
    "-//Apache Software Foundation//DTD Struts Configuration 2.0//EN"
    "http://struts.apache.org/dtds/struts-2.0.dtd">
<struts>
    <package name="notform" extends="struts-default" >
        <action name="em"
            class="actionerrorAndactionmessage.ActionErrorActionMessage">
            <result name="success">/ActionErrorAndActionMessage/
                    showActionErrorMessage.jsp
            </result>
        </action>
    </package>
<struts>
```

3) 编写页面显示信息

【例 2-31】 编写页面显示信息(showActionErrorMessage.jsp)。

```
<%@page contentType="text/html"  import="java.util.*" pageEncoding="UTF-8"%>
<%@taglib prefix="s" uri="/struts-tags"%>
<html>
    <head>
        <meta http-equiv="Content-Type" content="text/html; charset=UTF-8">
        <title>actionerror 标签和 actionmessage 标签的使用</title>
    </head>
    <body>
        <s:actionerror/>
        <br>
        <s:actionmessage/>
    </body>
</html>
```

4) 运行

项目部署后,在浏览器地址栏输入 http://localhost:8084/ch02/em.action,运行效果如图 2-22 所示,通过标签输出了 Action 封装的信息。

图 2-22 输出 Action 封装的信息

3. ＜s：component＞标签

使用 component 标签可以自定义组件,当需要多次使用某些代码段时,就可以自定义一个组件,在页面中使用 component 标签多次调用。该标签的主要属性如下所示。

(1) theme:该属性用来指定自定义组件所使用的主题,默认值为 xhtml。

(2) templateDir:该属性用来指定自定义组件使用的主题目录,默认值为 template。

(3) template:该属性用来指定自定义组件所使用的模板文件,自定义模板文件可以采用 JSP、FreeMarker 和 Velocity 这三种技术编写。

在 component 标签内还可以使用 param 标签,通过 param 标签可向模板标签中传递参数。

下面编写一个模板文件(myTemplate.jsp),代码如例 2-32 所示。

【例 2-32】 模板页面(myTemplate.jsp)。

```
<%@page contentType="text/html" pageEncoding="UTF-8"%>
<%@taglib prefix="s" uri="/struts-tags"%>
<html>
    <head>
        <meta http-equiv="Content-Type" content="text/html; charset=gb2312">
        <title></title>
    </head>
    <body>
        自定义模板
        <hr>
        <s:select label="你最喜欢的歌曲" list="parameters.songList"></s:select>
    </body>
</html>
```

上述代码使用了默认的主题(xhtml)、默认的主题目录(template)和 JSP 模板文件,该文件保存在 web/template/xhtml 文件夹下(NetBeans 开发环境)或者放在 WebRoot/template/xhtml 文件夹下(MyEclipse 和 Eclipse 开发环境),如图 2-23 所示。

编写完模板文件后,可通过 component 标签使用该模板文件。component 标签的使用实例代码如例 2-33 所示。

【例 2-33】 通过 component 标签使用模板(component.jsp)。

```
<%@ page contentType ="text/html" pageEncoding="UTF-8"%>
<%@taglib prefix="s" uri="/struts-tags"%>
<html>
    <head>
        <meta http-equiv="Content-Type" content=
        "text/html; charset=UTF-8">
        <title>component 标签的使用</title>
    </head>
```

图 2-23 模板页面位置

```
<body>
    <!--使用component标签调用模板-->
    <s:component template="myTemplate.jsp">
        <s:param  name="songList" value="{'中国人','真心英雄','青花瓷','传奇',
        '北京欢迎你'}" />
    </s:component>
</body>
</html>
```

最后运行 component.jsp 页面,页面效果如图 2-24 所示。

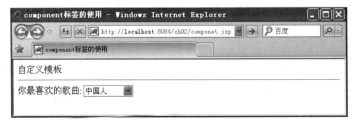

图 2-24 component 标签的使用

2.5.4 Struts2 的数据标签

Struts2 中的数据标签主要用于提供各种数据访问的相关功能,常用于显示 Action 中的属性以及国际化输出。下面介绍常用的 Struts2 数据标签。

1. <s：action>标签

action 标签用于在 JSP 页面中直接调用 Action。常用属性如下所示。

(1) id：指定被调用 Action 的引用 ID,可选项。

(2) name：指定被调用 Action 的名字,必选项。

(3) namespace：指定被调用 Action 所在的 namespace,可选项。

(4) executeResult：指定是否将 Action 处理结果包含到当前页面中,默认值为 false,即不包含,可选项。

(5) ignoreContextParams：指定当前页面的数据是否需要传给被调用的 Action,默认值为 false,即默认将页面中的参数传给被调用的 Action,可选项。

下面的项目演示 action 标签的使用。项目的文件结构如图 2-25 所示。

【例 2-34】 先编写一个 Action。

1) 编写 Action

【例 2-34】action 标签调用的 Action（ActionTagAction.java）。

图 2-25 项目的文件结构图

```java
package actionTagAction;
import org.apache.struts2.ServletActionContext;
import com.opensymphony.xwork2.ActionSupport;

public class ActionTagAction extends ActionSupport{
    private String name;
    public void setName(String name)    {
        this.name =name;
    }
    public String getName(){
        return name;
    }
    public String execute() throws Exception{
        return SUCCESS;
    }
    public String login() throws Exception{
        ServletActionContext.getRequest().setAttribute("name", getName());
        return SUCCESS;
    }
}
```

例 2-34 中包含两个方法：execute()和 login()，两个方法都是进行业务逻辑处理的,能够返回 SUCCESS，其中，在 login()方法中对 name 进行设置。

2) 在 struts.xml 中配置 Action(struts.xml)

【例 2-35】 在 struts.xml 中配置 Action(struts.xml)。

```xml
<!DOCTYPE struts PUBLIC
    "-//Apache Software Foundation//DTD Struts Configuration 2.0//EN"
    "http://struts.apache.org/dtds/struts-2.0.dtd">
<struts>
    <package name="actionTag" extends="struts-default" >
        <action name="tag1" class="actionTagAction.ActionTagAction">
            <result name="success">success3.jsp</result>
        </action>
        <action name="tag2" class="actionTagAction.ActionTagAction"
            method="login">
            <result name="success">loginSuccess3.jsp</result>
        </action>
    </package>
<struts>
```

由于在例 2-34 中有两个处理业务逻辑的方法，所以在例 2-35 中分别对它们进行配置。在返回结果上有两个视图，分别对应两个 JSP 页面，success3.jsp 页面的代码如例 2-36 所示，loginSuccess3.jsp 页面对应的代码如例 2-37 所示。

3）编写 struts.xml 配置 Action 中对应的页面

【例 2-36】 success3.jsp 页面（success3.jsp）。

```
<%@page contentType="text/html" pageEncoding="UTF-8"%>
<html>
    <head>
        <meta http-equiv="Content-Type" content="text/html; charset=UTF-8">
        <title>JSP Page</title>
    </head>
    <body>
        <h1>调用 Action!</h1>
    </body>
</html>
```

【例 2-37】 loginSuccess3.jsp 页面（loginSuccess3.jsp）。

```
<%@page contentType="text/html" pageEncoding="UTF-8"%>
<html>
    <head>
        <meta http-equiv="Content-Type" content="text/html; charset=UTF-8">
        <title>JSP Page</title>
    </head>
    <body>
        <s:property value="#request.name" /> 登录成功!
    </body>
</html>
```

4）编写调用 Action 的 JSP 页面

【例 2-38】 使用 action 标签调用 Action 的 JSP 页面（actionTag.jsp）。

```
<%@page contentType="text/html" pageEncoding="UTF-8"%>
<%@taglib prefix="s" uri="/struts-tags"%>
<html>
    <head>
        <meta http-equiv="Content-Type" content="text/html; charset=UTF-8">
        <title>action 标签库的使用</title>
    </head>
    <body>
        下面调用第一个 Action,并将结果包含到本页面中.
        <br>
        <!--使用 action 标签调用 Action 类,Action 处理返回 seccess3 页面,并将结果包含
        在页面中-->
        <s:action name="tag1" executeResult="true"/>
        <hr/>
        下面调用第二个 Action,并将结果包含到本页面中,阻止当前页面的参数传入 Action.
        <br>
        <s:action name="tag2" executeResult="true" ignoreContextParams="true"/>
```

```
        <hr/>
        下面调用第二个 Action,并不将结果包含到本页面中,但接受当前页面的参数传入 Action.
        <br>
        <s:action name="tag2" />
        当前页面传递的参数 name 的值：
        <s:property value="#request.name"/>
    </body>
</html>
```

5）运行

项目部署后,在浏览器输入 http://localhost:8084/ch02/actionTag.jsp?name=QQ,运行效果如图 2-26 所示。

图 2-26　action 标签库的使用

2. <s:bean>标签

bean 标签用于在 JSP 页面中创建 JavaBean 实例。在创建 JavaBean 实例时,可以使用 <s:param>标签为 JavaBean 实例传入参数。常用属性如下所示。

(1) name：指定实例化 JavaBean 的实现类,必选项。

(2) id：为实例化对象指定 id 名称,可选项。

例如,一个 Student 类是一个 JavaBean,代码如例 2-39 所示。

【例 2-39】　Student 类的 JavaBean,用于封装学生信息(Student.java)。

```
package beanTag;

public class Student {
    private String name;                            //姓名
    private String sex;                             //性别
    private int age;                                //年龄
    public String getName() {
        return name;
    }
    public void setName(String name) {
        this.name = name;
```

```
        }
        public String getSex() {
            return sex;
        }
        public void setSex(String sex) {
            this.sex = sex;
        }
        public int getAge() {
            return age;
        }
        public void setAge(int age) {
            this.age = age;
        }
    }
```

编写一个 JSP 页面使用 bean 标签访问 JavaBean(Student 类),代码如例 2-40 所示。

【例 2-40】 访问 bean 的页面(beanTag.jsp)。

```
<%@page contentType="text/html" pageEncoding="UTF-8"%>
<%@taglib prefix="s" uri="/struts-tags"%>
<html>
    <head>
        <meta http-equiv="Content-Type" content="text/html; charset=UTF-8">
        <title>bean 标签的使用</title>
    </head>
    <body>
        <s:bean name="beanTag.Student" id="s">
            <s:param name="name" value="'吴加一'"/>
            <s:param name="sex" value="'女'"/>
            <s:param name="age" value="18"/>
        </s:bean>
        姓名:<s:property value="#s.name"/>
        <br>
        性别:<s:property value="#s.sex"/>
        <br>
        年龄:<s:property value="#s.age"/>
    </body>
</html>
```

运行效果如图 2-27 所示。

3. <s:include>标签

include 标签用来在一个页面上包含另外一个 JSP 页面或者 Servlet 文件。

例如:

```
<s:include value="include-file.jsp"/>
```

或者

图 2-27　bean 标签的使用

```
<s:include value="include-file.jsp">
    <s:param name="user" value="'吴加一'"/>
</s:include >
```

4．<s：param>标签

param 标签用来为其他标签提供参数，如 include 标签、bean 标签等。

5．<s：set>标签

set 标签用来定义一个新的变量，并把一个已有的变量值赋给这个新变量，同时也可把新变量放到指定的范围内，如 session、application 范围内。常用属性如下所示。

（1）name：指定新变量的名字，必选项。

（2）scope：指定新变量的使用范围，如 action、page、request、response、session、application，可选项。

（3）value：为新变量赋值，可选项。

下面是 set 标签使用的实例，页面为 JSP 页面（setTag.jsp），代码如例 2-41 所示。

【例 2-41】 使用 set 标签设置新变量（setTag.jsp）。

```
<%@page contentType="text/html" pageEncoding="UTF-8"%>
<%@taglib prefix="s" uri="/struts-tags"%>
<html>
    <head>
        <meta http-equiv="Content-Type" content="text/html; charset=UTF-8">
        <title>set 标签的使用</title>
    </head>
    <body>
    <s:bean name="beanTag.Student" id="s">
            <s:param name="name" value="'吴加一'" />
    </s:bean>
    scope 属性值为 action 范围：
    <s:set value="#s" name="user" scope="action" />
        <s:property value="#attr.user.name" />
        <br>
        scope 属性值为 session 范围：
    <s:set value="#s" name="user" scope="session" />
        <s:property value="#session.user.name" />
    </body>
</html>
```

运行效果如图 2-28 所示。

图 2-28　set 标签的使用

6．<s：property>标签

property 标签用来输出 value 属性指定的值，该值可以使用 OGNL 表达式表示。

7．<s：url>标签

url 标签主要用来在页面中生成一个 URL 地址。常用属性如下所示。

（1）action：指定一个 Action 作为 URL 地址。

（2）method：指定使用 Action 的方法。

（3）value：用来指定生成 URL 的地址，如果不指定该属性，则使用 action 属性指定的 Action 作为 URL 地址。

（4）encode：指定编码方法。

（5）namespace：指定名称空间。

（6）includeContext：指定是否将当前上下文包含在 URL 地址中，默认值为 true。

（7）includeParams：指定是否包含请求参数，值有 none、get、all，默认值为 get。

8．<s：date>标签

date 标签用于格式化输出一个日期，还可以计算指定日期和当前时刻之间的时差。常用属性如下所示。

（1）format：使用日期格式化。

（2）nice：指定是否输出指定日期与当前时刻的时差，默认值为 false，即不输出时差。

（3）name：指定要格式化的日期值。

（4）var：指定格式化后的字符串将被放入 StaticContext 中，该属性可以用 id 属性代替。

【例 2-42】 date 标签的使用(dateTag.jsp)。

```
<%@page contentType="text/html" pageEncoding="UTF-8"%>
<%@taglib prefix="s" uri="/struts-tags"%>
<html>
    <head>
        <meta http-equiv="Content-Type" content="text/html; charset=UTF-8">
        <title>date 标签的使用</title>
    </head>
    <body>
        <s:bean id="d" name="java.util.Date"/>
        nice="false",且指定 format="dd/MM/yyyy"
        <br>
```

```
            <s:date name="#d" format="dd/MM/yyyy" nice="false"/>
            <hr>
            nice="true",且指定 format="dd/MM/yyyy"
            <br>
            <s:date name="#d" format="dd/MM/yyyy" nice="true"/>
            <hr>
            指定 nice="true"
            <br>
            <s:date name="#d" nice="true" />
            <hr>
            nice="false",且没有指定 format 属性
            <br>
            <s:date name="#d" nice="false"/>
            <hr>
            nice="false",没有指定 format 属性,指定了 var
            <br>
            <s:date name="#d" nice="false" var="abc"/>
            <hr>
            ${requestScope.abc} <s:property value="#abc"/>
    </body>
</html>
```

运行效果如图 2-29 所示。

图 2-29 date 标签的使用

2.5.5 Struts2 的控制标签

控制标签主要用来完成流程的控制,如条件分支、循环操作,也可以实现对集合的合并和排序。下面介绍常用的控制标签。

1. <s：if>标签、<s：elseif>标签和<s：else>标签

这 3 个标签是用来实现分支流程控制的,与 Java 语言中的 if、else if、else 语句相似。

【例2-43】 控制标签的使用(ifTag.jsp)。

```
<%@page contentType="text/html" pageEncoding="UTF-8"%>
<%@taglib prefix="s" uri="/struts-tags"%>
<html>
    <head>
        <meta http-equiv="Content-Type" content="text/html; charset=UTF-8">
        <title>控制标签的使用</title>
    </head>
    <body>
        <s:set name="score" value="86"/>
        <s:if test="#score>=90">优秀</s:if>
        <s:elseif test="#score>=80">良好</s:elseif>
        <s:elseif test="#score>=70">中等</s:elseif>
        <s:elseif test="#score>=60">及格</s:elseif>
        <s:else>不及格</s:else>
    </body>
</html>
```

其中,使用set标签进行传值,＜if＞和＜elseif＞语句中的test属性是必需的,是进行条件控制的逻辑表达式。

运行效果如图2-30所示。

图2-30 控制标签的使用

2. ＜s：iterator＞标签

iterator标签主要用于对集合进行迭代操作,集合可以是List、Map、Set和数组等。常用属性如下所示。

(1) id：指定集合元素的ID。

(2) value：指定迭代输出的集合,该集合可以是OGNL表达式,也可以是通过Action返回的一个集合。

(3) status：指定集合中元素的status属性。

【例2-44】 iterator标签的使用(iteratorTag.jsp)。

```
<%@page contentType="text/html" pageEncoding="UTF-8"%>
<%@taglib prefix="s" uri="/struts-tags"%>
<html>
    <head>
        <meta http-equiv="Content-Type" content="text/html; charset=UTF-8">
        <title>iterator标签的使用</title>
```

```
        </head>
        <body>
            <h2>iterator 标签的使用</h2>
            <hr>
            <s:iterator value="{'Java 程序设计与项目实训教程','JSP 程序设计技术教程',
            'JSP 程序设计与项目实训教程','Struts2+Hibernate 框架技术教程','Web 框架技术
            (Struts2+Hibernate+Spring3)教程'}"
                id="bookName">
                <s:property value="bookName"/><br>
            </s:iterator>
        </body>
</html>
```

运行效果如图 2-31 所示。

图 2-31　iterator 标签的使用

另外,iterator 标签的 status 属性还可以实现一些很有用的功能。指定 status 属性后,每次迭代都会产生一个 IteratorStatus 实例对象,该对象常用的方法如下所示。

(1) int getCount():判断当前迭代元素的个数。

(2) int getIndex():判断当前迭代元素的索引值。

(3) boolean isEven():判断当前迭代元素的索引值是否为偶数。

(4) boolean isOdd():判断当前迭代元素的索引值是否为奇数。

(5) boolean isFirst():判断当前迭代元素是否是第一个元素。

(6) boolean isLast():判断当前迭代元素是否是最后一个元素。

使用 iterator 标签的属性 status 时,其实例对象除了包含以上 6 个常用方法外,还包含了一些对应的属性,如#status.count、#status.even、#status.odd、#status.first 等。

【例 2-45】　iterator 标签 status 属性的使用(iteratorTag1.jsp)。

```
<%@page contentType="text/html" pageEncoding="UTF-8"%>
<%@taglib prefix="s" uri="/struts-tags"%>
<html>
    <head>
        <meta http-equiv="Content-Type" content="text/html; charset=UTF-8">
        <title>iterator 标签的使用</title>
```

```
        </head>
        <body>
            <h2>iterator 标签的使用</h2>
            <hr>
            <table border="1">
                <s:iterator value="{'Java 程序设计与项目实训教程','JSP 程序设计技术教程',
                'JSP 程序设计与项目实训教程','Struts2+Hibernate 框架技术教程','Web 框架
                技术(Struts2+Hibernate+Spring3)教程'}" id="bookName" status="st">
                    <tr <s:if test="#st.odd">style="background-color:red"</s:if>>
                        <td><s:property value="bookName"/><br></td>
                    </tr>
                </s:iterator>
            </table>
        </body>
</html>
```

该实例实现对奇数项颜色进行控制,运行效果如图 2-32 所示。

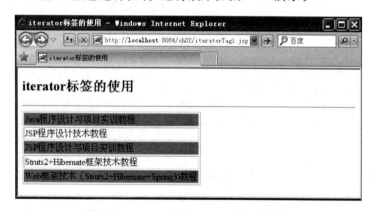

图 2-32　iterator 标签 status 属性的使用

3. <s：append>标签

append 标签用来将多个集合对象连接起来组成一个新的集合,并允许通过一个 iterator 标签完成对多个集合的迭代。常用属性如下所示。

id：指定连接生成的新集合的名字。

【例 2-46】　append 标签的使用(appendTag.jsp)。

```
<%@page contentType="text/html" pageEncoding="UTF-8"%>
<%@taglib prefix="s" uri="/struts-tags"%>
<html>
    <head>
        <meta http-equiv="Content-Type" content="text/html; charset=UTF-8">
        <title>append 标签的使用</title>
    </head>
    <body>
        <h2>append 标签的使用</h2>
```

```
        <hr>
        <s:append id="newList">
            <s:param value="{'Java 程序设计与项目实训教程','JSP 程序设计与项目实训教
程','Web 框架技术(Struts2+Hibernate+Spring3)教程'}"/>
            <s:param value="{'Java 程序设计','JSP 程序设计','SSH 技术'}"/>
        </s:append>
        <table border="1">
            <s:iterator value="#newList" status="st">
                <tr <s:if test="#st.odd">style="background-color:red"</s:if>>
                    <td><s:property /></td>
                </tr>
            </s:iterator>
        </table>
    </body>
</html>
```

运行效果如图 2-33 所示。

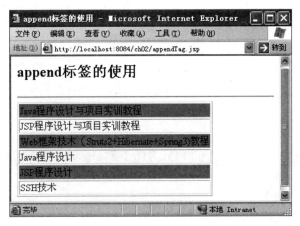

图 2-33 append 标签的使用

4. <s：merge>标签

merge 标签和 append 标签所实现的功能几乎一样,也是将多个集合连接成一个新集合,但是在生成新集合时这两个标签对元素的排序方式有所不同。

【例 2-47】 merge 标签和 append 标签比较(mergeTag.jsp)。

```
<%@page contentType="text/html" pageEncoding="UTF-8"%>
<%@taglib prefix="s" uri="/struts-tags"%>
<html>
    <head>
        <meta http-equiv="Content-Type" content="text/html; charset=UTF-8">
        <title>merge 标签的使用</title>
    </head>
    <body>
        <h3>merge 标签的使用</h3>
```

```
        <hr>
        <s:append id="newList_append">
            <s:param value="{'集合 1 中的元素 1','集合 1 中的元素 2','集合 1 中的元素 3'}"/>
            <s:param value="{'集合 2 中的元素 1','集合 2 中的元素 2'}"/>
        </s:append>
        <s:merge id="newList_merge">
            <s:param value="{'集合 1 中的元素 1','集合 1 中的元素 2','集合 1 中的元素 3'}" />
            <s:param value="{'集合 2 中的元素 1','集合 2 中的元素 2'}" />
        </s:merge>
        <br>
        迭代输出由 append 标签产生的新集合
        <s:iterator value="#newList_append" status="st">
            <br>
            <s:property />
        </s:iterator>
        <br>
        迭代输出由 merge 标签产生的新集合
        <s:iterator value="#newList_merge" status="st">
            <br>
            <s:property />
        </s:iterator>
    </body>
</html>
```

运行效果如图 2-34 所示。

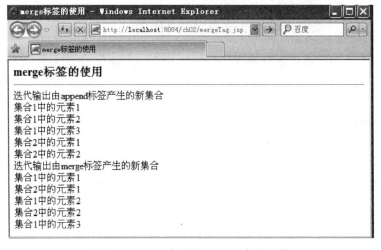

图 2-34　merge 标签和 append 标签比较

5．<s：generator>标签

generator 标签用来将一个字符串按指定的分隔符分割成多个子串，新生成的多个子字符串可以使用 iterator 标签进行迭代。常用属性如下所示。

（1）val：指定被解析的字符串，必选项。

（2）count：指定所生成集合中元素的总数。

（3）separator：用来指定分隔符，必选项。

（4）converter：指定一个转换器，该转换器将集合中的每个字符串转换成对象。

【例 2-48】 generator 标签的使用（generatorTag.jsp）。

```
<%@page contentType="text/html" pageEncoding="UTF-8"%>
<%@taglib prefix="s" uri="/struts-tags"%>
<html>
    <head>
        <meta http-equiv="Content-Type" content="text/html; charset=UTF-8">
        <title>generator 标签的使用</title>
    </head>
    <body>
        <h3>generator 标签的使用</h3>
        <hr>
        <s:generator val="'Java 程序设计与项目实训教程,JSP 程序设计与项目实训教程,
        Web 框架技术(Struts2+Hibernate+Spring3)教程'" separator=",">
            <s:iterator status="st">
                <br>
                <s:property />
            </s:iterator>
        </s:generator>
    </body>
</html>
```

运行效果如图 2-35 所示。

图 2-35　generator 标签的使用

6．<s：subset>标签

subset 标签用来从一个集合中截取一个子集。常用属性如下所示。

（1）source：指定源集合。

（2）count：指定子集合中元素的总数，默认值是源集合的元素总数。

（3）start：指定从源集合中第几个元素开始截取。

【例 2-49】 subset 标签的使用（subsetTag.jsp）。

```
<%@page contentType="text/html" pageEncoding="UTF-8"%>
<%@taglib prefix="s" uri="/struts-tags"%>
```

```
<html>
    <head>
        <meta http-equiv="Content-Type" content="text/html; charset=UTF-8">
        <title>subset 标签的使用</title>
    </head>
    <body>
        <h3>subset 标签的使用</h3>
        <hr>
        <s:subset source="{'Java 程序设计与项目实训教程','JSP 程序设计与项目实训教程','JSP 程序设计技术教程','Struts2+Hibernate 框架技术教程','Web 框架技术(Struts2+Hibernate+Spring3)教程'}" start="1" count="3">
            <s:iterator status="st">
                <br>
                <s:property />
            </s:iterator>
        </s:subset>
    </body>
</html>
```

例 2-49 中 start="1" count="3",表示从源集合中第二个元素开始,向后截取三个元素,由此生成一个新集合,并用 iterator 标签进行迭代。运行效果如图 2-36 所示。

图 2-36 subset 标签的使用

7. <s：sort>标签

sort 标签用来对指定集合进行排序,但是排序规则由开发者自己提供,即实现自己的 Comparator 实例。Comparator 是通过实现 Comparator 接口来实现的。常用属性如下所示。

(1) comparator:指定实现排序规则的 Comparator 实例,必选项。

(2) source:指定要排序的集合。

【例 2-50】 排序规则类(MyComparator.java)。

```
package sortTag;
import java.util.Comparator;

public class MyComparator implements Comparator{
    public int compare(Object element1, Object element2){
        return element1.toString().length()-element2.toString().length();
```

        }
    }

对应的 sort 标签页面(sortTag.jsp),代码如下:
```
<%@page contentType="text/html" pageEncoding="UTF-8"%>
<%@taglib prefix="s" uri="/struts-tags"%>
<html>
    <head>
        <meta http-equiv="Content-Type" content="text/html; charset=UTF-8">
        <title>sort 标签的使用</title>
    </head>
    <body>
        <h3>使用 sort 标签对集合进行排序</h3>
        <hr>
        <s:bean id="mc" name="sortTag.MyComparator" />
        <s:sort source="{'Java 程序设计与项目实训教程','JSP 程序设计与项目实训教程',
        'JSP 程序设计技术教程','Struts2+Hibernate 框架技术教程','Web 框架技术
        (Struts2+Hibernate+Spring3)教程'}" comparator="#mc">
            <s:iterator status="st">
                <br>
                <s:property />
            </s:iterator>
        </s:sort>
    </body>
</html>
```

运行效果如图 2-37 所示。

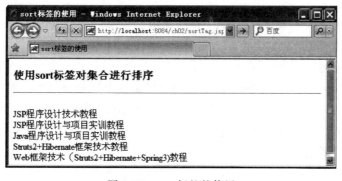

图 2-37 sort 标签的使用

2.5.6 Struts2 的 Ajax 标签

Struts2 提供了对 Ajax 的支持,主要包括 Ajax 插件和标签。Struts2 的 Ajax 功能主要依赖两个 Ajax 框架:Dojo 框架和 DWR 框架。Dojo 框架提供了丰富的组件库和页面效果,并使用大量的函数来简化 Ajax 过程。DWR 框架侧重于服务器端,能够在客户端页面通过 JavaScript 来调用远程的 Java 方法。关于 Ajax 技术以及框架请参考 Ajax 资料,这里不再介绍。

1. ajax 主题

Struts2 框架支持 Ajax 技术主要是通过内置的 ajax 主题来完成的，ajax 主题是对 xhtml 主题的扩展，它在 xhtml 的基础上加入了相关的 Ajax 特征。

Struts2 内置了对 Dojo 工具包的支持，它的 Ajax 标签基于 Dojo，能够在项目开发中提供快捷容易的 Ajax 使用方式。要使用 Ajax 标签，需要将标签的 theme 属性设置为 ajax，同时需要把 head 标签的 theme 设置为 ajax，以导入 ajax 头信息。对于一些更复杂的 Ajax 通信过程，可以通过使用 JSON 插件来实现。

例如：

<s:head theme="ajax"/>

Struts2 支持的 Ajax 常用标签如下所示。

(1) <s：div>：创建一个内容区域，可以通过 Ajax 来加载它的内容，以实现局部刷新，即可以动态从服务器获取数据。

(2) <s：a>：通过 Ajax 来更新某个或者多个元素的内容或提交表单。

(3) <s：submit>：通过 Ajax 来更新某个或者多个元素的内容或提交表单。

(4) <s：tabbedPanel>：创建一个标签面板，标签页的内容可以是静态的，也可以是动态的，由<s：div>标签来提供内容。

(5) <s：autocompleter>：用来生成一个带下拉按钮的单行文本框，页面加载时，生成下拉列表中的内容。

以上标签的通用属性包括如下。

(1) href：类型为 String，定义发送请求的地址（URL）。

(2) indicator：类型为 String，指定动态加载服务器端数据过程中的显示内容，一般指定为图标。

(3) listenTopics：类型为 String，它的值可以是一个逗号分隔的主题名称列表，当发布者发布该主题的事件时，标签将重新加载它的内容（div、autocompleter 标签）或者执行一个动作（a、submit 标签）。

(4) notifyTopics：类型为 String，它的值可以是一个逗号分隔的主题名称列表，这些主题的事件由标签来发布，并向事件处理函数传递 data、type 和 request 参数。

(5) showErrorTransportText：类型为 Boolean，设置是否要显示错误消息，默认为 true。

2. <s：div>标签

div 标签在页面上生成一个 div 元素，但这个 div 元素的内容不是静态内容，而是从服务器获取的内容，即 div 元素在获取并显示数据时采用的是异步通信方式，不需要刷新页面。

常用属性如下所示。

(1) afterLoading：指定获取内容后需要执行的 JavaScript 代码。

(2) afterNotifyTopics：指定在请求之后（请求成功）发表的话题清单，话题之间使用英文状态下的逗号分隔。

(3) autoStart：指定是否自动启动计时器。

(4) beforNotifyTopics：指定请求之前发表的话题清单，话题之间使用英文状态下的逗号分隔。

(5) closable：指定使用 div 标签作为选项卡的一个 Tab 页面时是否显示关闭按钮。

(6) delay：指定更新 div 内容的时间延迟，单位是 ms，如果没有指定 updateFreq 属性，则该属性没有意义。如果服务器包含了 JavaScript 代码，且希望在本页面内执行服务器响应的 JavaScript 代码，则可以为该 div 标签指定 executeScripts="true"。

(7) errorNotifyTopics：指定在请求之后（请求失败）发表的话题清单，话题之间使用英文状态下的逗号分隔。

(8) errorText：指定获取数据发生错误时的提示信息。

(9) executeScripts：指定是否在本页面执行服务器响应的 JavaScript 脚本代码，默认值为 false。

(10) formFilter：指定过滤表单字段的函数。

(11) fromId：请求参数的表单。

(12) handler：指定本页面的脚本函数作为处理函数。如果指定了该属性，将不会向服务器发送 Ajax 请求。

(13) highlightColor：指定突出显示颜色，对 targets 属性所指定的元素进行突出显示。

(14) highlightDuration：指定 targets 所指定元素突出显示的持续时间，单位为 ms。如果 highlightColor 属性无值，该属性无效。

(15) javascriptTooltip：指定是否使用 JavaScript 生成浮动提示框。

(16) loadingText：指定内容正在装载过程中的提示信息，主要用来提示用户正在装载的内容。

(17) openTemplate：打开 HTML 文件的显示模板。

(18) parseContent：指定是否分析返回的动态 Web 内容以查询组件。

(19) preload：指定是否加载页面的同时加载动态 Web 内容。

(20) refreshOnShow：指定是否需要在 div 元素变得可见时加载动态 Web 内容。该属性在 div 元素包含于 tabbedpanel 元素中时有效。

(21) separateScripts：指定是否需要为每个标签单独创建一个应用范围来运行脚本代码。

(22) showLoadingText：指定是否在封装内容时显示提示信息。

(23) startTimerListenTopics：指定一个监听的事件主题，当 Struts2 组件向该主题发布事件时 div 标签的计时器自动启动。

(24) stopTimerListenTopics：指定一个监听的事件主题，当 Struts2 组件向该主题发布事件时 div 标签的计时器自动停止。

(25) transport：指定传送请求参数的传输对象。

(26) updateFreq：指定内容的更新时间间隔，单位为 ms。如果不指定该属性则内容只有在页面加载时才会更新。

下面通过项目来介绍一下这些属性的用法。

项目 1：div 标签的 href 属性、updateFreq 属性和 delay 属性的使用。

1）项目介绍

项目有一个页面 divHUD.jsp，该页面能动态获取服务器端的数据，代码如例 2-51 所示，该页面对应的 Action 为 DivHUD，代码如例 2-53 所示。请求提交到 Action 处理后返回的视图页面为 showDate.jsp,代码如例 2-52 所示。另外还需要配置 web.xml,代码和 1.3.1 节中例 1-3 相同;配置 struts.xml 文件,代码如例 2-54 所示。该项目文件结构如图 2-38 所示。需要的 JAR 文件如图 2-39 所示。由于加载 Struts 2.3.4 类库时使用的是插件,该插件中没有 struts2-dojo-plugin-2.3.4.1.jar,所以需要在下载的 Struts 2.3.4.1 类库中加载该文件,请参考 1.1.2 节。否则运行 divHUD.jsp 后会出现如图 2-40 所示的异常。

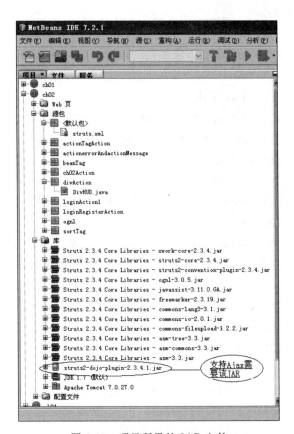

图 2-38　项目文件结构　　　　　　　图 2-39　项目所需的 JAR 文件

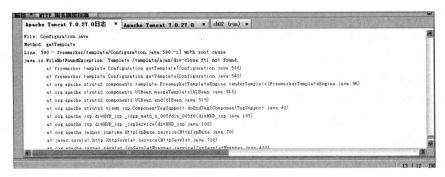

图 2-40　常见异常

解决该异常主要需要以下 4 个步骤。

第一步：在库中导入 struts2-dojo-plugin-2.3.4.1.jar。

第二步：在 divHUD.jsp 页面中导入 <%@taglib prefix="sx" uri="/struts-dojo-tags"%>。

第三步：把 divHUD.jsp 页面中的 <s:head theme="ajax"> 改为 <sx:head/>。

第四步：把 <s:div id="noUpdateFreq" theme="ajax" href="%{getDate}"></s:div> 改为 <sx:div id="noUpdateFreq" theme="ajax" href="%{getDate}"></sx:div>。

2）在 web.xml 中配置核心控制器 FilterDispatcher

参考 1.3.1 节中的例 1-3。

3）编写视图组件（JSP 页面）

【例 2-51】 divHUD.jsp。

```jsp
<%@page contentType="text/html" pageEncoding="UTF-8"%>
<%@taglib prefix="s" uri="/struts-tags"%>
<%@taglib prefix="sx" uri="/struts-dojo-tags"%>
<html>
    <head>
        <meta http-equiv="Content-Type" content="text/html; charset=UTF-8">
        <title>div 标签属性</title>
        <sx:head/>
    </head>
    <body>
        <s:url id="getDate" value="getDate.action" />
        <hr/>
        1. 只在页面加载时获取数据
        <sx:div id="noUpdateFreq" theme="ajax" href="%{getDate}"></sx:div>
        <hr/>
        2. 每 2s 更新一次数据
        <sx:div id="updateFreq" theme="ajax" href="%{getDate}"
            updateFreq="2000"></sx:div>
        <hr/>
        3. 每 6s 更新一次数据，但是延迟 2s
        <sx:div id="delay" theme="ajax" href="%{getDate}" updateFreq="6000"
            delay="2000"></sx:div>
        <hr/>
    </body>
</html>
```

【例 2-52】 showDate.jsp。

```jsp
<%@page contentType="text/html" pageEncoding="UTF-8"%>
<%@taglib prefix="s" uri="/struts-tags"%>
<html>
    <head>
        <meta http-equiv="Content-Type" content="text/html; charset=UTF-8">
```

```
        <title>JSP Page</title>
    </head>
    <body>
        <s:bean id="date" name="java.util.Date"/>
        服务器当前时间是:<s:date name="date" format="yyyy-MM-dd HH:mm:ss"/>
    </body>
</html>
```

4）编写业务控制器 Action

【例 2-53】 业务控制器类（DivHUD.java）。

```
package divAction;
import com.opensymphony.xwork2.ActionSupport;

public class DivHUD extends ActionSupport{
    public String exectue() throws Exception{
        return SUCCESS;
    }
}
```

5）修改 struts.xml 配置 Action

【例 2-54】 在 struts.xml 中配置 Action（struts.xml）。

```
<!DOCTYPE struts PUBLIC
"-//Apache Software Foundation//DTD Struts Configuration 2.0//EN"
"http://struts.apache.org/dtds/struts-2.0.dtd">
<struts>
    <package name="divTaglib" extends="struts-default" >
        <action name="getDate" class="divAction.DivHUD">
            <result name="success">/showDate.jsp</result>
        </action>
    </package>
</struts>
```

6）项目部署和运行

项目运行后效果如图 2-41 所示，第一个 div 元素与第二个 div 元素的时间差为 3s，而第三个 div 元素由于设置了 delay 属性，添加了时间延迟效果，所以此时还没有获取到服务器端数据，但等时间到后将显示获取到的时间，如图 2-42 所示。

图 2-41　第一次抓取页面效果

图 2-42　第二次抓取页面时的效果

项目 2：div 标签的 loadingText 属性和 indicator 属性的使用。

1）项目介绍

项目有一个页面 divLI.jsp，代码如例 2-55 所示，该页面对应的 Action 为 DivLI，代码如例 2-57 所示。请求提交到 Action 处理后返回的视图页面为 showNumber.jsp，代码如例 2-56 所示。另外还需要配置 web.xml，代码和 1.3.1 节中例 1-3 相同；配置 struts.xml 文件，代码如例 2-58 所示。该项目文件结构如图 2-43 所示。

2）在 web.xml 中配置核心控制器 FilterDispatcher
参考 1.3.1 节中的例 1-3。

3）编写视图组件（JSP 页面）

【例 2-55】 divLI.jsp。

图 2-43　项目文件结构

```
<%@page contentType="text/html" pageEncoding=
"UTF-8"%>
<%@taglib prefix="s" uri="/struts-tags"%>
<%@taglib prefix="sx" uri="/struts-dojo-tags"%>
<html>
    <head>
        <meta http-equiv="Content-Type" content
        ="text/html; charset=UTF-8">
        <title>div 标签属性</title>
        <sx:head/>
    </head>
    <body>
        <s:url id="getNumber" value="getNumber.
        action "/>
        <hr/>
        ① 获取数据时显示文本信息
        <sx:div id="loadingText" theme="ajax" href="%{getNumber}"
            loadingText="正在获取数据..."></sx:div>
        <hr/>
```

② 获取数据时显示图片

```
<img id="icon" src="image/girl.jpg"/>
<sx:div id="icon" theme="ajax" href="%{getNumber}"
    indicator="icon"></sx:div>
<hr/>
    </body>
</html>
```

【例2-56】 showNumber.jsp。

```
<%@page contentType="text/html" pageEncoding="UTF-8"%>
<%@taglib prefix="s" uri="/struts-tags"%>
<html>
    <head>
        <meta http-equiv="Content-Type" content="text/html; charset=UTF-8">
        <title>JSP Page</title>
    </head>
    <body>
        1加到666666666的和是:<s:property value="number"/>
    </body>
</html>
```

4）编写业务控制器Action

【例2-57】 业务控制器类（DivLI.java）。

```
package divAction;
import com.opensymphony.xwork2.ActionSupport;

public class DivLI extends ActionSupport{
    private double number;
    public double getNumber() {
        return number;
    }
    public void setNumber(double number) {
        this.number =number;
    }
    public String execute(){
        number=0;
        double i=1;
        //循环666666666次以便观察两个属性的效果
        while(i<666666666){
            number+=i;
            i++;
        }
        return SUCCESS;
    }
}
```

5）修改 struts.xml 配置 Action

【例 2-58】 在 struts.xml 中配置 Action(struts.xml)。

```
<!DOCTYPE struts PUBLIC
"-//Apache Software Foundation//DTD Struts Configuration 2.0//EN"
"http://struts.apache.org/dtds/struts-2.0.dtd">
<struts>
    <package name="divTaglib" extends="struts-default" >
        ⋮
        <action name="getNumber" class="divAction.DivLI">
            <result name="success">/showNumber.jsp</result>
        </action>
        ⋮
    </package>
</struts>
```

6）项目部署和运行

项目运行后效果如图 2-44 所示。

图 2-44 项目运行效果

项目 3：div 标签的 startTimerListen 属性和 stopTimerListen 属性的使用。

1）项目介绍

项目有一个页面 divSS.jsp，代码如例 2-59 所示。其他项目文件与 href 属性、updateFreq 属性和 delay 属性实例项目中的一样，即通过 Action 类 DivHUD 获取时间。返回文件以及 struts.xml 的配置也一样。

2）编写视图组件(JSP 页面)

【例 2-59】 divSS.jsp。

```
<%@page contentType="text/html" pageEncoding="UTF-8"%>
<%@taglib prefix="s" uri="/struts-tags"%>
<%@taglib prefix="sx" uri="/struts-dojo-tags"%>
<html>
```

```
<head>
    <meta http-equiv="Content-Type" content="text/html; charset=UTF-8">
    <title>div 标签属性</title>
    <sx:head/>
</head>
<body>
    <s:url id="getDate" value="getDate.action "/>
    <hr/>
    监听事件:
    <sx:div id="listenTopics" theme="ajax" href="%{getDate}"
        updateFreq="2000" startTimerListenTopics="/startTimerListenTopics"
        stopTimerListenTopics="/stopTimerListenTopics"></sx:div>
    <hr/>
    <input type="button" value="启动服务器" onclick="listener.startTimer()"/>
    <input type="button" value="停止服务器" onclick="listener.stopTimer()"/>
    <script title="text/javasrcipt">
        var listenter={
            startTimer : function(){},
            stopTimer : function(){}
        };
        dojo.event.topic.registerPublisher("/startTimerListenTopics",listener,
        "startTimer");
        dojo.event.topic.registerPublisher("/stopTimerListenTopics",listener,
        "stopTimer");
    </script>
</body>
</html>
```

3) 项目部署和运行

项目运行后效果如图 2-45 所示。

图 2-45 项目运行效果

3. <s:a>标签

a 和 submit 标签的作用几乎完全一样,除了外在的表现不一样(a 标签生成一个超链接,submit 标签生成一个提交按钮)。它们都用于向服务器发送异步请求,并将服务器响应加载在指定的 HTML 元素中。

常用属性如下所示。

(1) afterNotifyTopics：指定在请求之后(请求成功)发表的话题清单，话题之间使用英文状态下的逗号分隔。

(2) ajaxAfterValidation：指定如果验证成功是否发送一个异步请求，该属性只在 validation 属性值为 true 时有效。

(3) beforNotifyTopics：指定请求之前发表的话题清单，话题之间使用英文状态下的逗号分隔。

(4) errorNotifyTopics：指定在请求之后(请求失败)发表的话题清单，话题之间使用英文状态下的逗号分隔。

(5) errorText：指定获取数据发生异常时的提示信息。

(6) executeScripts：指定是否在本页面执行服务器响应的 JavaScript 脚本代码，默认值为 false。

(7) formFilter：指定过滤表单字段的函数。

(8) fromId：请求参数的表单。

(9) handler：指定本页面的脚本函数作为处理函数。如果指定了该属性，将不会向服务器发送 Ajax 请求。

(10) highlightColor：指定突出显示颜色，对 targets 属性所指定的元素进行突出显示。

(11) highlightDuration：指定 targets 所指定元素突出显示的持续时间，单位为 ms。如果 highlightColor 属性无值，该属性无效。

(12) javascriptTooltip：指定是否使用 JavaScript 生成浮动提示框。

(13) loadingText：指定内容正在装载过程中的提示信息，主要用来提示用户正在装载的内容。

(14) openTemplate：打开 HTML 文件的显示模板。

(15) parseContent：指定是否分析返回的动态 Web 内容以查询组件。

(16) separateScripts：指定是否需要为每个标签单独创建一个应用范围来运行脚本代码。

(17) showLoadingText：指定是否在封装内容时显示提示信息。

(18) targets：指定 HTML 元素的 ID，设置服务器响应时加载到该属性指定的几个 HTML 元素上。

(19) transport：指定传送请求参数的传输对象。

(20) validation：指定是否进行 Ajax 验证。

下面通过项目来介绍一下 a 标签的这些属性的使用。

项目 1：a 标签的 targets 属性的使用。

该项目与 div 标签的 href 属性、updateFreq 属性和 delay 属性实例类似，只需将例 2-51 中的 divHUD.jsp 页面修改为 aTargets.jsp，代码如例 2-60 所示，其他文件保持不变。

【**例 2-60**】 aTargets.jsp。

```
<%@page contentType="text/html" pageEncoding="UTF-8"%>
<%@taglib prefix="s" uri="/struts-tags"%>
<%@taglib prefix="sx" uri="/struts-dojo-tags"%>
```

```html
<html>
    <head>
        <meta http-equiv="Content-Type" content="text/html; charset=UTF-8">
        <title>a 标签属性的应用</title>
        <sx:head/>
    </head>
    <body>
        <s:url id="getDate" value="getDate.action "/>
        <hr/>
        更新指定页面元素的内容：
        <sx:div id="t1"></sx:div>
        <hr/>
        <!--a 标签中不使用 theme="ajax"-->
        <sx:a id="a1" href="%{getDate}" targets="t1">
            单击此处将更新 id 为 t1 的 div 元素内容</sx:a>
        <hr/>
    </body>
</html>
```

aTargets.jsp 页面的运行效果如图 2-46 所示。单击超链接后的页面效果如图 2-47 所示。

图 2-46　页面运行效果

图 2-47　单击超链接后的页面效果

项目 2：a 标签的 formId 属性的使用。

1）项目介绍

项目有一个页面 aFormId.jsp，代码如例 2-61 所示，该页面对应的 Action 为 AFormId.java，代码如例 2-63 所示。请求提交到 Action 处理后返回的视图页面为 showText.jsp，代码如例 2-62 所示。另外还需要配置 web.xml,代码和 1.3.1 节中例 1-3 相同;配置 struts.xml 文件,代码如例 2-64 所示。该项目文件结构如图 2-48 所示。

2)在 web.xml 中配置核心控制器 FilterDispatcher

参考 1.3.1 节中的例 1-3。

3)编写视图组件(JSP 页面)

【例 2-61】 aFormId.jsp。

```
<%@page contentType="text/html" pageEncoding="UTF-8"%>
<%@taglib prefix="s" uri="/struts-tags"%>
<%@taglib prefix="sx" uri="/struts-dojo-tags"%>
<html>
    <head>
        <meta http-equiv="Content-Type" content=
        "text/html; charset=UTF-8">
        <title>a 标签属性的应用</title>
        <sx:head/>
    </head>
    <body>
        <s:url id="getText" value="getText.action"/>
        <hr/>
        将 form 表单中的数据提交到服务器:
        <sx:div id="t1"></sx:div>
        <hr/>
        <s:form id="form1">
            <s:textfield label="请输入数据" name="text"/>
        </s:form>
        <hr/>
        <sx:a id="a1" href="%{getText}" targets="t1" formId="form1">
            单击此处将更新 id 为 t1 的 div 元素内容</sx:a>
    </body>
</html>
```

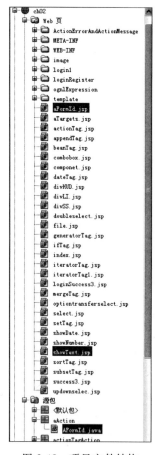

图 2-48 项目文件结构

【例 2-62】 showText.jsp。

```
<%@page contentType="text/html" pageEncoding="UTF-8"%>
<%@taglib prefix="s" uri="/struts-tags"%>
<html>
    <head>
        <meta http-equiv="Content-Type" content="text/html; charset=UTF-8">
        <title>JSP Page</title>
    </head>
    <body>
        输入的数据是:<s:property value="text"/>
    </body>
</html>
```

4)编写业务控制器 Action

【例 2-63】 业务控制器类(AFormId.java)。

```java
package aAction;
import com.opensymphony.xwork2.ActionSupport;

public class AFormId extends ActionSupport{
    private String text;
    public String getText() {
        return text;
    }
    public void setText(String text) {
        this.text =text;
    }
    public String execute(){
        return SUCCESS;
    }
}
```

5)修改 struts.xml 配置 Action

【例 2-64】 在 struts.xml 中配置 Action(struts.xml)。

```xml
<!DOCTYPE struts PUBLIC
"-//Apache Software Foundation//DTD Struts Configuration 2.0//EN"
"http://struts.apache.org/dtds/struts-2.0.dtd">
<struts>
    ⋮
    <package name="aTaglib" extends="struts-default" >
        <action name="getText" class="aAction.AFormId">
            <result name="success">/showText.jsp</result>
        </action>
    </package>
    ⋮
</struts>
```

6)项目部署和运行

项目运行后效果如图 2-49 所示。输入数据单击超链接后的页面效果如图 2-50 所示。

图 2-49 运行后的效果

图 2-50　输入数据后单击超链接的效果

4.＜s：submit＞标签

submit 标签用于向服务器发送异步请求，和 a 标签类似。

常用属性如下所示。

（1）afterNotifyTopics：指定在请求之后（请求成功）发表的话题清单，话题之间使用英文状态下的逗号分隔。

（2）ajaxAfterValidation：指定如果验证成功是否发送一个异步请求，该属性只在 validation 属性值为 true 时有效。

（3）beforNotifyTopics：指定请求之前发表的话题清单，话题之间使用英文状态下的逗号分隔。

（4）errorNotifyTopics：指定在请求之后（请求失败）发表的话题清单，话题之间使用英文状态下的逗号分隔。

（5）errorText：指定获取数据发生错误时的提示信息。

（6）executeScripts：指定是否在本页面执行服务器响应的 JavaScript 脚本代码，默认值为 false。

（7）formFilter：指定过滤表单字段的函数。

（8）fromId：请求参数的表单。

（9）handler：指定本页面的脚本函数作为处理函数。如果指定了该属性，将不会向服务器发送 Ajax 请求。

（10）highlightColor：指定突出显示颜色，对 targets 属性所指定的元素进行突出显示。

（11）highlightDuration：指定 targets 所指定元素突出显示的持续时间，单位为 ms。如果 highlightColor 属性无值，该属性无效。

（12）javascriptTooltip：指定是否使用 JavaScript 生成浮动提示框。

（13）loadingText：指定内容正在装载过程中的提示信息，主要用来提示用户正在装载的内容。

（14）method：对应 HTML 中的 method 属性。

（15）parseContent：指定是否分析返回的动态 Web 内容以查询组件。

（16）separateScripts：指定是否需要为每个标签单独创建一个应用范围来运行脚本代码。

（17）showLoadingText：指定是否在封装内容时显示提示信息。

（18）src：指定当按钮类型为 Image 时，按钮的图片来源。

（19）targets：指定 HTML 元素的 ID，设置服务器响应并加载到该属性指定的几个 HTML 元素上。

（20）transport：指定传送请求参数的传输对象。

（21）type：指定提交按钮的类型，如 input、button 和 image。

（22）validation：指定是否进行 Ajax 验证。

5. ＜s：tabbedPanel＞标签

tabbedPanel 标签用来在 HTML 页面中生成类似于 Windows 程序的 Tab 页，从而可以在有限的空间中放置更多的内容。tabbedPanel 标签生成的 Tab 页的内容可以是静态的，也可以是动态的。如果是静态的，则直接指定 Tab 页面的内容；如果是动态的，则可以使用 Ajax 方式来动态加载 Tab 页的内容。

tabbedPanel 标签生成整个 Tab 框架，而 tabbedPanel 标签类的 div 子标签则生成单独的 Tab 页，每个 div 标签生成一个 Tab 页。因为 div 标签本身是一个 Ajax 标签，允许内容动态改变，因此每个 Tab 页的内容可以动态改变。

常用属性如下所示。

（1）afterNotifyTopics：指定在请求之后（请求成功）发表的话题清单，话题之间使用英文状态下的逗号分隔。

（2）ajaxAfterValidation：指定如果验证成功是否发送一个异步请求，该属性只在 validation 属性值为 true 时有效。

（3）beforNotifyTopics：指定请求之前发表的话题清单，话题之间使用英文状态下的逗号分隔。

（4）closeButton：指定 Tab 页面上关闭按钮的位置，可选值有 tab 和 pane。

（5）doLayout：指定 abbedPanel 是否为固定高度，如果该属性值为 false，其高度将随着 Tab 页面的大小而改变。

（6）errorNotifyTopics：指定在请求之后（请求失败）发表的话题清单，话题之间使用英文状态下的逗号分隔。

（7）errorText：指定获取数据发生异常时的提示信息。

（8）executeScripts：指定是否在本页面执行服务器响应的 JavaScript 脚本代码，默认值为 false。

（9）formFilter：指定过滤表单字段的函数。

（10）fromId：请求参数的表单。

（11）handler：指定本页面的脚本函数作为处理函数。如果指定了该属性，将不会向服务器发送 Ajax 请求。

（12）highlightColor：指定突出显示颜色，对 targets 属性所指定的元素进行突出显示。

（13）highlightDuration：指定 targets 所指定元素突出显示的持续时间，单位为 ms。如果 highlightColor 属性无值，该属性无效。

（14）javascriptTooltip：指定是否使用 JavaScript 生成浮动提示框。

（15）labelposition：指定 Tab 页面中标签的位置，可选值有 top（默认值）、right、bottom 和 left。

（16）loadingText：指定内容正在装载过程中的提示信息，主要用来提示用户正在装载的内容。

(17) method：对应 HTML 中的 method 属性。

(18) parseContent：指定是否分析返回的动态 Web 内容以查询组件。

(19) selectedTab：指定加载页面时，初始状态下显示哪个 Tab 页面，默认显示第一个。

(20) separateScripts：指定是否需要为每个标签单独创建一个应用范围来运行脚本代码。

(21) showLoadingText：指定是否在封装内容时显示提示信息。

(22) targets：指定 HTML 元素的 ID，设置服务器响应并加载到该属性指定的几个 HTML 元素上。

(23) transport：指定传送请求参数的传输对象。

(24) validation：指定是否进行 Ajax 验证。

下面通过项目来介绍一下 tabbedPanel 标签属性的综合使用。该实例与 div 标签的 href 属性、updateFreq 属性和 delay 属性实例类似，只需将例 2-51 中的 divHUD.jsp 页面修改为 tabbedPanel.jsp，代码如例 2-65 所示，其他文件保持不变。

【例 2-65】 tabbedPanel.jsp。

```
<%@page contentType="text/html" pageEncoding="UTF-8"%>
<%@taglib prefix="s" uri="/struts-tags"%>
<%@taglib prefix="sx" uri="/struts-dojo-tags"%>
<html>
    <head>
        <meta http-equiv="Content-Type" content="text/html; charset=UTF-8">
        <title>tabbedPanel 标签的应用</title>
        <sx:head/>
    </head>
    <body>
        <sx:tabbedpanel id="t1" closeButton="pane" doLayout="true"
            selectedTab="second" labelposition="left"
            cssStyle="width:600px;height:100px;">
        <hr/>
        <s:url id="getDate" value="getDate"/>
        <sx:div id="first" theme="ajax" label="第一个 Tab 页面"></sx:div>
        这是第一个 Tab 页面
        <sx:div id="second" theme="ajax" href="%{getDate}"
            label="第二个 Tab 页面"></sx:div>
        </sx:tabbedpanel>
    </body>
</html>
```

tabbedPanel.jsp 运行效果如图 2-51 所示。

6. <s：autocompleter>标签

autocompleter 标签会生成一个带下拉按钮的单行文本输入框，当用户单击下拉按钮时，将看到一系列的选项，单击某个选项时可以将该选项填入单行文本框。

下拉列表框的选项会在页面加载时自动加载，而且随着用户在单行文本框中的输入而

图 2-51　tabbedPanel 标签属性运行效果

改变,当用户输入字符串时,下拉列表框中的选项总是和单行文本框中的内容以某种方式匹配。此时,用户也可以通过上、下箭头来选择合适的选项,并将指定选项填入单行文本框。

如果设置 autocompleter 标签的 autoComplete=true(默认是 false),该标签将会在单行文本框中生成输入提示。如果希望强制用户只能输入下拉列表中的列表项,则可以设置 forceValidOption=true(默认是 false)。

常用属性如下所示。

(1) afterNotifyTopics:指定在请求之后(请求成功)发表的话题清单,话题之间使用英文状态下的逗号分隔。

(2) autoComplete:指定是否在单行文本框中显示输入提示。

(3) beforNotifyTopics:指定请求之前发表的话题清单,话题之间使用英文状态下的逗号分隔。

(4) dataFieldName:指定被返回的 JSON 对象里包含着字段的名字。

(5) delay:指定延迟时间。

(6) dropdownHeight:指定下拉列表的高度。

(7) dropdownWidth:指定下拉列表的宽度,默认与单行文本框一致。

(8) emptyOption:指定是否插入一个空选项。

(9) errorNotifyTopics:指定在请求之后(请求失败)发表的话题清单,话题之间使用英文状态下的逗号分隔。

(10) forceValidOption:指定单行文本框是否接受下拉列表中的选择。

(11) formFilter:指定过滤表单字段的函数。

(12) fromId:请求参数的表单。

(13) handler:指定本页面的脚本函数作为处理函数。如果指定了该属性,将不会向服务器发送 Ajax 请求。

(14) headerValue:指定选项清单中第一项的键。

(15) headerKey:指定选项清单中第一项的值。

(16) iconPath:指定下拉列表的图标文件路径。

(17) javascriptTooltip:指定是否使用 JavaScript 生成浮动提示框。

(18) keyName:指定将被选中的键赋给哪一个属性。

(19) list:指定下拉列表选项的集合。

(20) listKey:指定列表中用来提供选项标号的对象的属性。

(21) listValue:指定列表中用来提供选项值的对象的属性。

(22) loadingText:指定内容正在装载过程中的提示信息,主要用来提示用户正在装载的内容。

(23) loadMinimumCount：当 loadOnTextChange 属性值为 true 时，用来指定输入多少个字符后，才能重新加载下拉列表的选项。

(24) maxLength：指定最大长度。

(25) preload：指定是否在加载页面的同时重新加载清单。

(26) resultsLimit：指定选项最多可以有多少个，如果该属性值为－1，表示可以有无限多。

(27) searchType：指定下拉列表与单行文本框的字符串匹配模式，startstring 值是默认值，显示以文本框中字符串开头的选项，startword 值表示显示以文本框中单词开头的选项，substring 表示显示包含文本框中字符串的选项。

(28) showDownArrow：指定是否显示下拉箭头，默认值为 true。

autocompleter 标签使用的选项既可以赋值给它的 list 属性，也可以通过一个 JSON (JavaScript Object Notation)对象动态地发送给它。JSON 是一种轻量级的数据交换格式。它是基于 JavaScript 的一个子集。JSON 采用完全独立于语言的文本格式，但是也沿袭了类似于 C 语言家族的习惯(包括 C、C++、C♯、Java、JavaScript、Perl、Python 等)。这些特性使 JSON 成为理想的数据交换语言。易于阅读和编写，同时也易于机器解析和生成。如果想了解更多关于 JSON 对象的信息，请访问 http：//json.org。

下面通过项目来了解一下 autocompleter 标签属性的使用。

1) 项目介绍

项目有一个页面 autocompleter.jsp，代码如例 2-66 所示，该页面对应的 Action 为 AutocompleterAction，代码如例 2-67 所示。请求提交到 Action 处理后返回一个 JSON 文件 GetList.js，代码如例 2-68 所示。另外还需要配置 web.xml，代码和 1.3.1 节中例 1-3 相同；配置 struts.xml 文件，代码如例 2-69 所示。该项目 Web 文件结构如图 2-52 所示。该项目 Web 源代码以及 struts.xml 结构如图 2-53 所示。

图 2-52　项目 Web 文件结构

图 2-53　项目源文件以及配置文件

2)在 web.xml 中配置核心控制器 FilterDispatcher

参考 1.3.1 节中的例 1-3。

3)编写视图组件(JSP 页面)

【例 2-66】 autocompleter.jsp。

```jsp
<%@page contentType="text/html" pageEncoding="UTF-8"%>
<%@taglib prefix="s" uri="/struts-tags"%>
<%@taglib prefix="sx" uri="/struts-dojo-tags"%>
<html>
    <head>
        <meta http-equiv="Content-Type" content="text/html; charset=UTF-8">
        <title>autocompleter 标签属性的应用</title>
        <sx:head/>
    </head>
    <body>
        <s:url id="getList" value="getList"/>
        1. 获取服务器端的 List
        <hr/>
        选择你喜欢的旅游城市:
        <br/>
        <sx:autocompleter name="a1" href="%{getList}"></sx:autocompleter>
        <hr/>
        2. 设置 autoComplete="false"
        <br/>
        选择你喜欢的旅游城市:
        <br/>
        <sx:autocompleter name="a1" href="%{getList}"
            autoComplete="false"></sx:autocompleter>
        <hr/>
        3. 设置 showDownArrow="false"
        <br/>
        选择你喜欢的旅游城市:
        <br/>
        <sx:autocompleter name="a1" href="%{getList}"
            showDownArrow="false"></sx:autocompleter>
        <hr/>
        4. 设置 searchType="substring"
        <br/>
        选择你喜欢的旅游城市:
        <br/>
        <sx:autocompleter name="a1" href="%{getList}"
            searchType="substring"></sx:autocompleter>
        <hr/>
    </body>
</html>
```

4）编写业务控制器 Action

【例 2-67】 业务控制器类（AutocompleterAction.java）。

```java
package autocompleterAction;
import com.opensymphony.xwork2.ActionSupport;

public class AutocompleterAction extends ActionSupport{
    private String a1;
    public String getA1() {
        return a1;
    }
    public void setA1(String a1) {
        this.a1 = a1;
    }
    public String execute(){
        return SUCCESS;
    }
}
```

5）JSON 文件

【例 2-68】 JSON 文件（GetList.js）。

```
[
    ["北京"],
    ["上海"],
    ["青岛"],
]
```

6）修改 struts.xml 配置 Action

下面的 struts.xml 文件可用于配置本章所有的 Action。

【例 2-69】 在 struts.xml 中配置 Action（struts.xml）。

```xml
<!DOCTYPE struts PUBLIC
"-//Apache Software Foundation//DTD Struts Configuration 2.0//EN"
"http://struts.apache.org/dtds/struts-2.0.dtd">
<struts>
    <package name="login" extends="struts-default" >
        <action name="login1" class="loginAction1.LoginAction1">
            <!--定义两个逻辑视图和物理资源之间的映射,name 值是 Action 中返回的结果-->
            <result name="input">/login1/login1.jsp</result>
            <result name="success">/login1/success1.jsp</result>
        </action>
         <!--不指定 method 属性的 Action 配置-->
        <action name="loginReg" class="loginRegisterAction.LoginRegisterAction">

            <result name="input">/loginRegister/loginRegister.jsp</result>
            <result name="success">/loginRegister/success2.jsp</result>
```

```xml
            </action>
        </package>
        <package name="ognl" extends="struts-default" >
            <action name="ognl" class="ognl.OGNLAction">
                <result name="success">/ognlExpression/registerSuccess.jsp</result>
            </action>
        </package>
        <package name="notform" extends="struts-default" >
            <action name="em"
                class="actionerrorAndactionmessage.ActionErrorActionMessage">
                <result name="success">
                    /ActionErrorAndActionMessage/showActionErrorMessage.jsp
                </result>
            </action>
        </package>
        <package name="actionTag" extends="struts-default" >
            <action name="tag1" class="actionTagAction.ActionTagAction">
                <result name="success">success3.jsp</result>
            </action>
            <action name="tag2" class="actionTagAction.ActionTagAction"
                method="login">
                <result name="success">loginSuccess3.jsp</result>
            </action>
        </package>
        <package name="divTaglib" extends="struts-default" >
            <action name="getDate" class="divAction.DivHUD">
                <result name="success">/showDate.jsp</result>
            </action>
            <action name="getNumber" class="divAction.DivLI">
                <result name="success">/showNumber.jsp</result>
            </action>
        </package>
        <package name="aTaglib" extends="struts-default" >
            <action name="getText" class="aAction.AFormId">
                <result name="success">/showText.jsp</result>
            </action>
        </package>
        <package name="autocompleterTaglib" extends="struts-default" >
            <action name="getList" class="autocompleterAction.AutocompleterAction" >
                <result name="success">/js/GetList.js</result>
            </action>
        </package>
    </struts>
```

7) 项目部署和运行

项目运行后效果如图 2-54 所示。选择数据后效果如图 2-55 所示。

图 2-54　运行后的效果

图 2-55　选择数据

2.6　本章小结

　　本章详细介绍了 Struts2 的核心组件，通过本章的学习应对 Struts2 框架有深入的了解，应掌握以下内容。

　　（1）struts.xml 文件的配置。

　　（2）核心控制器。

　　（3）业务控制器。

（4）OGNL。

（5）Struts2 的常用标签。

2.7 习　　题

2.7.1 选择题

1. Struts2 扩展组件是通过配置文件和（　　）来管理的。
   A. 核心控制器　　　B. IoC　　　　　　C. AOP　　　　　　D. Action
2. 在 struts.xml 配置文件中，能够把其他配置文件包含进来的元素是（　　）。
   A. <package>　　　B. <action>　　　C. <include>　　　D. <result>
3. 在 struts.xml 配置文件中，对业务控制器进行配置的元素是（　　）。
   A. <package>　　　B. <action>　　　C. <include>　　　D. <result>
4. 在 struts.xml 配置文件中，配置逻辑视图和物理视图映射关系的元素是（　　）。
   A. <package>　　　B. <action>　　　C. <include>　　　D. <result>
5. 在 Struts2 中 Action 接口提供的一个实现类是（　　）。
   A. ActionContext　　　　　　　　　　B. ActionSupport
   C. ActionMessage　　　　　　　　　　D. ServletActionContext
6. 在 Struts2 中常用的表达式语言是（　　）。
   A. HTML　　　　　B. JavaScript　　　C. JSP　　　　　　D. OGNL

2.7.2 填空题

1. Struts2 框架有两种文件配置格式：_____ 和 _____。
2. Struts2 加载常量的顺序是 _____、_____ 和 _____。
3. 在 Struts2 框架中，通过包配置来管理 _____ 和 _____。
4. Struts2 中 Action 与 Servlet 是 _____。
5. Struts2 中直接访问 Servlet 有 _____ 和 _____ 两种方式。
6. Struts2 提供两种动态方法的调用方式：_____ 和 _____。
7. OGNL 有三个参数，分别是 _____、_____ 和 _____。
8. 按标签库提供的功能可将 Struts2 标签库分为三大类：_____、_____ 和 Ajax 标签。
9. 用户界面标签可分为 _____ 和 _____。
10. 非用户界面标签可分为 _____ 和 _____。

2.7.3 简答题

1. 简述 struts.xml 配置文件的作用。
2. 简述 Struts2 核心控制器 FilterDispatcher 的作用。
3. 简述 Struts2 业务控制器 Action 的作用。

2.7.4 实训题

1. 使用 Action 访问 ActionContext 方式,编写一个网站计数器。
2. 将 2.3.3 节中的登录系统改为 IoC 方式或者非 IoC 方式。
3. 将 2.3.5 节中的程序改为指定 method 属性方式实现。
4. 使用 OGNL 编写一个对集合操作的 Java Web 应用程序。
5. 使用 Struts2 的标签库开发一个注册页面并实现注册功能。

第 3 章　Struts2 的高级组件

本章将介绍 Struts2 中比较常用的高级组件。通过这些高级组件的学习,将有助于进一步了解和使用 Struts2 框架。

本章主要内容如下所示。

(1) Struts2 对国际化的支持。
(2) Struts2 常用拦截器的使用。
(3) Struts2 的数据验证功能。
(4) Struts2 对文件上传和下载的支持。

3.1　Struts2 的国际化

"国际化"是指一个应用程序在运行时能够根据客户端请求所来自国家或地区语言的不同而显示不同的用户界面。例如,请求来自于一台中文操作系统的客户端计算机,则应用程序响应界面中的各种标签、错误提示和帮助信息均使用中文文字;如果客户端计算机采用英文操作系统,则应用程序也应能识别并自动以英文界面做出响应。

引入国际化机制的目的在于提供自适应的、更友好的用户界面,而不必改变程序的其他功能或者业务逻辑。人们常用 I18N 这个词作为国际化的简称,其来源是英文单词 Internationalization 的首末字母 I 和 N 及它们之间的字符数 18。

3.1.1　Struts2 实现国际化的流程

Struts2 国际化是建立在 Java 国际化基础上的,Java 对国际化进行了优化和封装,从而简化了国际化的实现过程。Struts2 国际化流程如图 3-1 所示。

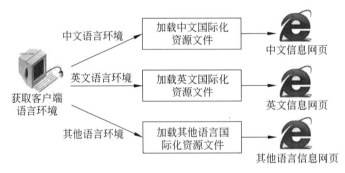

图 3-1　Struts2 国际化流程

具体流程如下所示。

(1) 不同地区使用的操作系统环境不同,如中文操作系统、英文操作系统等。获得客户端地区的语言环境后,在 struts.xml 文件中会找到相应的国际化资源文件,如果操作系统环境是中文语言环境,就加载中文国际化资源文件。所以国际化需要编写支持多个语言的国际化资源文件,并且在 struts.xml 文件中配置。

(2) 根据选择的语言加载相应的国际化资源文件,视图通过 Struts2 标签读取国际化资源文件并把数据输出到页面上,完成页面的显示。

下面介绍在国际化流程中用到的文件。

1. 国际化资源文件或者资源文件

国际化资源文件又称资源文件,是以 properties 为扩展名的文件,新建一个文本文件把扩展名改为 properties 即可,该文件以"键=值"对的形式存储资源数据。

例如:

```
key=value
loginName=用户名称
loginPassword=用户密码
```

当需要多个资源文件为不同语言版本提供国际化服务时,可以为资源文件命名,命名的格式有以下两种形式:

资源文件名.properties
资源文件名_语言种类.properties

文件名后缀必须是 properties,语言种类必须是有效的 ISO(国际标准化组织)语言代码,ISO-639 标准定义的这些代码格式为英文小写、双字符,具体如表 3-1 所示。

表 3-1 常用标准语言代码

语言	编码	语言	编码	语言	编码
汉语(Chinese)	zh	法语(French)	fr	日语(Japanese)	ja
英语(English)	en	德语(German)	de	意大利语(Italian)	it

资源文件如果使用第一种命名方式,即默认语言代码,当系统找不到与客户端请求的语言环境匹配的资源文件时,就使用默认的属性文件。

例如,如果要对前面介绍的登录系统进行国际化处理,要求根据不同的语言环境显示中文和英文用户界面,那么就需要创建中文和英文的资源文件,分别取名为 globalMessages_GBK.properties 和 globalMessages_en_US.properties,内容分别如例 3-1 和例 3-2 所示。

说明:例 3-2 中有 en,例 3-1 中是原始文件在 116 页中转化为了 zh 格式,这样写没有问题。

【例 3-1】 中文资源文件(globalMessages_GBK.properties)。

```
loginTitle=用户登录
loginName=用户名称
loginPassword=用户密码
loginSubmit=登录
```

【例 3-2】 英文资源文件(globalMessages_en_US.properties)。

```
loginTitle=UserLogin
```

```
loginName=UserName
loginPassword=UserPassword
loginSubmit=Login
```

从例3-1和例3-2中可以看出，国际化资源文件的内容都是以 key＝value 形式存在的。资源文件中 key 部分是相同的，即等号左边部分相同，value 部分不同，即等号右边部分不同。

在国际化时，所有的字符都要使用标准的编码方式，需要把中文字符转换为 Unicode 代码，否则在国际化处理时页面将会出现乱码。如例 3-1 中的中文资源文件是不能直接使用的，必须转换为指定的编码方式。可以使用 JDK 自带的 native2ascii 工具进行中文资源文件编码方式的转换。具体操作如下：选择"开始"→"运行"菜单，输入 cmd，出现如图 3-2 所示的命令行窗口。

图 3-2　命令行窗口

如果开发平台使用的是 NetBeans 7.2，将资源文件放在"D：\Web 框架技术（Struts2＋Hibernate＋Spring3）教程\ch03\src\java"，即和 struts.xml 文件一起放在默认的资源包中，其中"D：\Web 框架技术（Struts2＋Hibernate＋Spring3）教程"为工作区，"\ch03"为项目名，"\src\java"是资源文件存放的目录；如果使用的是 MyEclipse 10.6 或 Eclipse 4.2，将资源文件放在 struts.xml 文件所在的文件夹，位置是"D：\Web 框架技术（Struts2＋Hibernate＋Spring3）教程\ch03\src"。

在图 3-2 中通过 cd 命令进入到资源文件所在的文件夹下，输入 native2ascii -encoding UTF-8 globalMessages_GBK.properties globalMessages_zh_CN.properties，回车，如图 3-3 所示。命令执行后在该文件夹下会生成一个 globalMessages_zh_CN.properties 文件，该文件的内容如例 3-3 所示。

图 3-3　编码转换命令

【例 3-3】　编译后对应的 Unicode 代码（globalMessages_zh_CN.properties）。

```
loginTitle=\u7528\u6237\u767b\u5f55
loginName=\u7528\u6237\u540d\u79f0
loginPassword=\u7528\u6237\u5bc6\u7801
```

```
loginSubmit=\u767b\u5f55
```

2. 在 struts.xml 文件中配置

编写完国际化资源文件后,需要在 struts.xml 配置文件中配置国际化资源文件的名称,从而使 Struts2 的 I18N 拦截器在需要加载国际化资源文件的时候能找到这些国际化资源文件,在 struts.xml 中的配置很简单,代码如例 3-4 所示。

【例 3-4】 在 struts.xml 中配置资源文件。

```xml
<!DOCTYPE struts PUBLIC
    "-//Apache Software Foundation//DTD Struts Configuration 2.0//EN"
    "http://struts.apache.org/dtds/struts-2.0.dtd">
<struts>
    <!--使用 Struts2 中的 I18N 拦截器,并通过 constant 元素配置常量,指定国际资源文件名
    字,value 的值就是常量值,即国际化资源文件的名字-->
    <constant name="struts.custom.i18n.resources" value="globalMessages" />
    <constant name="struts.i18n.encoding" value="UTF-8" />
    <package name="I18N" extends="struts-default">
        <action name="checkLogin" class="loginAction.LoginAction">
            <result name="success">/I18N/loginSuccess.jsp</result>
            <result name="error">/I18N/login.jsp</result>
        </action>
    </package>
</struts>
```

3. 输出国际化信息

国际化资源文件中的 value 值要根据语言环境通过 Struts2 标签输出到页面上。可以在页面上输出国际化信息也可以在 Struts2 表单标签的 label 标签上输出国际化信息。

3.1.2 Struts2 国际化应用实例

本项目开发一个中英文登录页面,通过该项目的开发可以帮助了解 Struts2 国际化应用的开发过程。

1. 项目介绍

该项目为中英文登录系统,项目有一个登录页面(login.jsp),代码如例 3-5 所示;对应的资源文件如例 3-1 和例 3-2 所示;登录页面对应的业务控制器是 LoginAction 类,代码如例 3-7 所示;如果登录成功(用户名、密码正确)转到 loginSuccess.jsp 页面,代码如例 3-6 所示;如果登录失败(用户名、密码不正确)则重新回到登录页面(login.jsp)。此外,还需要配置 web.xml,代码如例 1-3 所示;配置 struts.xml,代码如例 3-4 所示。项目的文件结构如图 3-4 所示。

2. 在 web.xml 中配置核心控制器 FilterDispatcher

参考 1.3.1 节中的例 1-3。

图 3-4 项目文件结构

3. 编写国际化资源文件并进行编码转换

编写中、英文国际化资源文件，参考例 3-1 和例 3-2；编码转换参考图 3-3。

4. 编写视图组件（JSP 页面）输出国际化消息

编写一个如图 3-5 或者图 3-6 所示的登录页面。

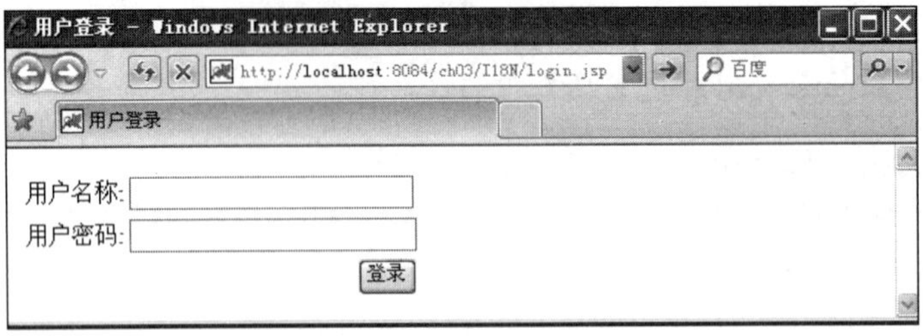

图 3-5 中文登录页面

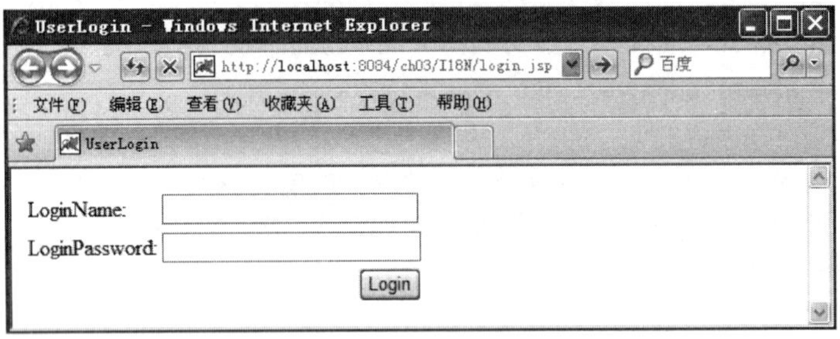

图 3-6 英文登录页面

登录页面是一个 JSP 页面，代码如例 3-5 所示。

【例 3-5】 中英文登录页面（login.jsp）。

```
<%@page contentType="text/html; charset=UTF-8" %>
<%@taglib prefix="s" uri="/struts-tags" %>
<html>
    <head>
        <!--使用 text 标签输出国际化消息 -->
        <title><s:text name="loginTitle"/></title>
    </head>
    <body>
        <s:form action="checkLogin" method="post">
            <!--表单元素的 key 值与资源文件的 key 对应-->
            <s:textfield name="name" key="loginName" size="20"/>
            <s:password name="password" key="loginPassword" size="22"/>
            <s:submit key="loginSubmit"/>
        </s:form>
    </body>
</html>
```

登录成功页面，代码如例 3-6 所示。

【例3-6】 登录成功页面(loginSuccess.jsp)。

```
<%@page contentType="text/html; charset=UTF-8" %>
<%@taglib prefix="s" uri="/struts-tags" %>
<html>
    <head>
        <!--使用text标签输出国际化消息 -->
        <title><s:text name="successPage"/></title>
    </head>
    <body>
        <hr>
        <s:text name="loginName"/>:<s:property value="name"/><br>
        <s:text name="loginPassword"/>:<s:property value="password"/>
    </body>
</html>
```

5. 编写业务控制器Action

登录页面(login.jsp)对应的业务控制器是如例3-7所示的LoginAction类。

【例3-7】 中英文登录页面对应的业务控制器(LoginAction.java)。

```
package loginAction;
import com.opensymphony.xwork2.ActionContext;
import com.opensymphony.xwork2.ActionSupport;

public class LoginAction extends ActionSupport{
    private String name;
    private String password;
    //用于定义标题信息
    private String tip;
    public String getName() {
        return name;
    }
    public void setName(String name) {
        this.name =name;
    }
    public String getPassword() {
        return password;
    }
    public void setPassword(String password) {
        this.password =password;
    }
    public String getTip() {
        return tip;
    }
    public void setTip(String tip) {
        this.tip =tip;
    }
    public String execute() throws Exception
```

```
    {
        if (getName().equals("QQ")&&getPassword().equals("123") )
        {
            ActionContext.getContext().getSession().put("name",getName());

            return SUCCESS;
        }
        else
        {
            return ERROR;
        }
    }
}
```

6. 在 struts.xml 中配置 Action 与国际资源文件

修改配置文件 struts.xml，在配置文件中配置 Action 和国际化资源文件，参考例 3-4。

7. 项目部署和运行

项目部署后运行，如果操作系统是中文系统，运行 login.jsp 后，出现如图 3-5 所示的页面，在"用户名称"中输入 QQ、"用户密码"中输入 123 后，单击"登录"按钮，将跳转到如图 3-7所示的登录成功页面。

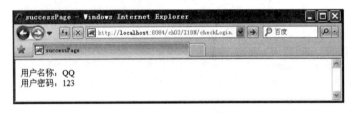

图 3-7　中文操作系统下的登录成功页面

如果使用的是英文操作系统或者设置浏览器语言为英文，运行后将出现如图 3-6 所示的英文登录页面。把浏览器设置为英文语言的步骤如下。

（1）选择"IE 浏览器"→"工具"→"Internet 选项"，如图 3-8 所示。

图 3-8　选择"Internet 选项"

（2）单击图 3-8 中的"Internet 选项"，出现如图 3-9 所示的对话框。

（3）单击图 3-9 中的"语言"按钮，出现如图 3-10 的对话框，其中"语言"区域默认为"中文（中国）[zh-cn]"，即在中文操作系统下，IE 浏览器默认的首选语言是中文，也可以添加其他语言。

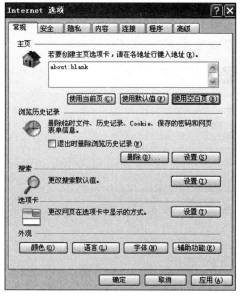

图 3-9　"Internet 选项"对话框

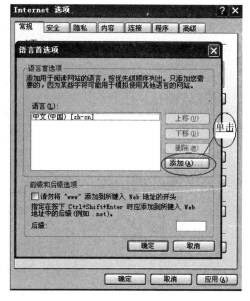

图 3-10　"语言首选项"对话框

（4）单击图 3-10 所示对话框中的"添加"按钮，弹出"添加语言"对话框，找到"英语（美国）[en-US]"项，如图 3-11 所示。

（5）在图 3-11 所示对话框中选定所需的语言后单击"确定"按钮，返回到"语言首选项"对话框，如图 3-12 所示，此时语言区域中已添加了"英语（美国）[en-US]"，选定"英语（美国）

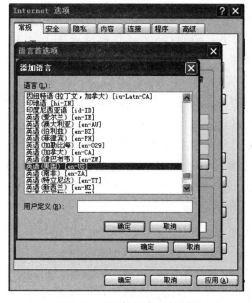

图 3-11　"添加语言"对话框

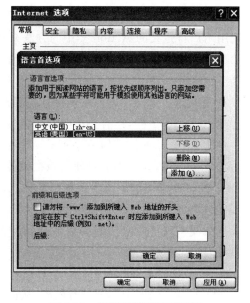

图 3-12　"语言首选项"对话框

[en-US]"项,单击"上移"按钮,把英语作为浏览器的首选语言,最后单击"确定"按钮完成设置。

经过以上步骤配置后,浏览器的首选语言已是"英语",浏览器在运行 JSP 页面(login.jsp)时,将优先使用"英语",如果有英文国际化资源文件,将显示英文的登录页面,如图 3-6 所示。在图 3-6 所示页面中的 UserName 中输入 QQ 并在 UserPassword 中输入 123 后,单击 Login 按钮,页面将跳转到如图 3-13 所示的登录成功页面。

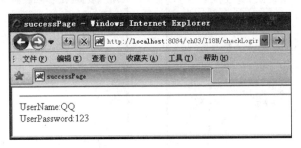

图 3-13 英文版的登录成功页面

3.2 Struts2 的拦截器

拦截器是 Struts2 的核心组件,Struts2 的绝大部分功能都是通过拦截器来完成的,所以 Struts2 的很多功能都构建在拦截器的基础上。

3.2.1 Struts2 拦截器的基础知识

拦截器(Interceptor)体系是 Struts2 的一个重要组成部分,正是大量的内置拦截器实现了 Struts2 的大部分操作。

当 FilterDispatcher 拦截到用户请求后,大量的拦截器将会对用户请求进行处理,然后才调用用户自定义的 Action 类中的方法来处理请求。例如,params 拦截器将 HTTP 请求中的参数解析出来,并将这些解析出来的参数设置为 Action 的属性;servlet-config 拦截器直接将 HTTP 请求中的 HttpServletRequest 实例和 HttpServletResponse 实例传给 Action;国际化拦截器 I18N 对国际化资源进行操作;文件上传拦截器 fileUpload 将文件信息传给 Action。另外,还有数据校验拦截器对数据校验信息进行拦截。

对于 Struts2 的拦截器体系而言,当需要使用某个拦截器时,只需在配置文件 struts.xml中配置即可;如果不需要使用该拦截器,只需在 struts.xml 配置文件中取消配置即可。Struts2 的拦截器可以理解为一种可插拔式的设计思想,所以 Struts2 框架具有非常好的可扩展性。

拦截器实现了面向切面编程(Aspect-Oriented Programming,AOP)的设计思想,拦截是 AOP 的一种实现策略。

AOP 是目前软件开发中的一个热点,也是 Spring 框架中的一个重要内容。利用 AOP 可以对业务逻辑的各个部分进行隔离,从而使得业务逻辑各部分之间的耦合度降低,提高程序的可重用性,同时提高开发的效率。

3.2.2 Struts2 拦截器实现类

在项目开发中,Struts2 内置的拦截器可以完成项目的大部分功能,但有些与项目业务逻辑相关的通用功能则需要通过自定义拦截器来实现,如权限控制、用户输入内容的控制等。

自定义拦截器需要实现 Struts2 提供的 Interceptor 接口。通过实现该接口可以开发拦截器类。该接口的代码如例 3-8 所示。

【例 3-8】 Struts2 的拦截器接口(Interceptor.java)。

```java
import com.opensymphony.xwork2.ActionInvocation;
import java.io.Serializable;

public interface Interceptor extends Serializable {
    void destroy();
    void init();
    String intercept(ActionInvocation invocation) throws Exception;
}
```

该接口提供了 3 个方法。

(1) destroy()方法:与 init()方法对应,用于在拦截器执行完之后释放 init()方法里打开的资源。

(2) init()方法:由拦截器在执行之前调用,主要用于初始化系统资源,如打开数据库资源等。

(3) intercept(ActionInvocation Invocation)方法:该方法是拦截器的核心方法,实现具体的拦截操作,返回一个字符串作为逻辑视图。与 Action 一样,如果拦截器能够成功调用 Action,则 Action 中的 execute()方法返回一个字符串类型值,将其作为逻辑视图返回,否则,返回开发者自定义的逻辑视图。

在 Java 语言中,有时候通过一个抽象类来实现一个接口,在抽象类中提供该接口的空实现。这样在编写类的时候,可以直接继承该抽象类,不用实现那些不需要的方法。Struts2 框架也提供了一个抽象拦截器类(AbstractInterceptor),该类对 init()和 destroy()方法进行空实现,因为很多时候实现拦截器都不需要申请资源。在开发自定义拦截器时,通过继承这个类开发起来简单方便。AbstractInterceptor 类的代码如例 3-9 所示。

【例 3-9】 Struts2 的抽象拦截器(AbstractInterceptor.java)。

```java
import com.opensymphony.xwork2.ActionInvocation;

public abstract class AbstractInterceptor implements Interceptor {
    public void init()
    {
    }
    public void destroy()
    {
    }
```

```
public abstract String intercept(ActionInvocation invocation) throws Exception;
}
```

3.2.3 Struts2 拦截器应用实例

对内置的拦截器,如国际化、数据校验、文件上传和下载,将在别的章节介绍。本节项目是对自定义拦截器使用的练习。下面通过一个文字过滤项目来熟悉自定义拦截器的使用。

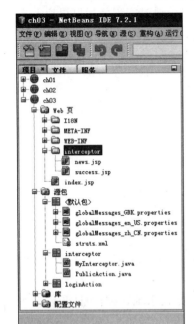

图 3-14 项目文件结构

1. 项目介绍

该项目开发一个网上论坛过滤系统,如果网友发表不文明或者不和谐的语言,将通过拦截器对这些文字进行自动替代。该项目编写一个自定义拦截器(MyInterceptor.java),代码如例 3-10 所示;项目有一个发表新闻评论的页面(news.jsp),代码如例 3-11 所示,其对应的业务控制器是 PublicAction 类,代码如例 3-13 所示;评论成功跳转到 success.jsp 页面,代码如例 3-12 所示。此外还需要配置 web.xml,代码如例 1-3 所示;配置 struts.xml,代码如例 3-14 所示。项目的文件结构如图 3-14 所示。

2. 在 web.xml 中配置核心控制器 FilterDispatcher

参考 1.3.1 节中的例 1-3。

3. 编写自定义拦截器

编写一个自定义拦截器用于对发表的评论内容进行过滤,代码如例 3-10 所示。

【例 3-10】 自定义拦截器的编写(MyInterceptor.java)。

```
package interceptor;
import com.opensymphony.xwork2.Action;
import com.opensymphony.xwork2.ActionInvocation;
import com.opensymphony.xwork2.interceptor.AbstractInterceptor;

public class MyInterceptor extends AbstractInterceptor {
    public String intercept(ActionInvocation ai) throws Exception {
        // 获取 Action 的实例
        Object object =ai.getAction();
        if (object!=null)
        {
            if(object instanceof PublicAction)
            {
                PublicAction ac= (PublicAction)object;
                //获取用户提交的评论内容
                String content =ac.getContent();
                //判断用户提交的评论内容是否有要过滤的内容
                if(content.contains("讨厌"))
```

```
            {
                //以"喜欢"代替要过滤的"讨厌"
                content =content.replaceAll("讨厌","喜欢");
                //把替代后的评论内容设置为 Action 的评论内容
                ac.setContent(content);
            }
            //对象不空,继续执行
            return ai.invoke();
        }else{
            //返回 Action 中的 LOGIN 逻辑视图字符串,参考 2.3.1 节
            return Action.LOGIN;
        }
    }
    return Action.LOGIN;
    }
}
```

4．编写视图组件(JSP 页面)发表评论

编写一个如图 3-15 所示的评论页面,代码如例 3-11 所示。

图 3-15 评论页面

【例 3-11】 评论页面(news.jsp)。

```
<%@page language="java" import="java.util.*" pageEncoding="utf-8"%>
<%@taglib prefix="s" uri="/struts-tags"%>
<html>
    <head>
        <title>评论</title>
    </head>
    <body>
        请发表你的评论!
        <hr>
        <s:form action="public" method="post">
            <s:textfield name="title" label="评论标题" maxLength="36"/>
            <s:textarea name="content" cols="36" rows="6" label="评论内容"/>
```

```
            <s:submit value="提交"/>
        </s:form>
    </body>
</html>
```

评论成功页面代码如例 3-12 所示。

【例 3-12】 评论成功页面(success.jsp)。

```
<%@page contentType="text/html; charset=UTF-8" %>
<%@taglib prefix="s" uri="/struts-tags" %>
<html>
    <head>
        <title>评论成功</title>
    </head>
    <body>
        评论如下：
        <hr>
        评论标题：<s:property value="title"/>
        <br>
        评论内容：<s:property value="content"/>
    </body>
</html>
```

5. 编写业务控制器 Action

news.jsp 对应的业务控制器是如例 3-13 所示的 PublicAction 类。

【例 3-13】 评论页面对应的业务控制器(PublicAction.java)。

```java
package interceptor;
import com.opensymphony.xwork2.ActionSupport;

public class PublicAction extends ActionSupport{
    private String title;
    private String content;
    public void setTitle(String title) {
        this.title =title;
    }
    public String getTitle() {
        return title;
    }
    public void setContent(String content) {
        this.content =content;
    }
    public String getContent() {
        return content;
    }
    public String execute()
```

```
    {
        return SUCCESS;
    }
}
```

6. 在 struts.xml 中配置自定义拦截器和 Action

修改配置文件 struts.xml,在配置文件中配置拦截器和 Action,代码如例 3-14 所示。

【例 3-14】 配置拦截器和 Action(struts.xml)。

```xml
<!DOCTYPE struts PUBLIC
    "-//Apache Software Foundation//DTD Struts Configuration 2.0//EN"
    "http://struts.apache.org/dtds/struts-2.0.dtd">
<struts>
    <!--Configuration for the default package. -->
    <constant name="struts.custom.i18n.resources" value="globalMessages" />
    <constant name="struts.i18n.encoding" value="utf-8" />
    <package name="I18N" extends="struts-default">
        <!--Struts2 规定,对拦截器的配置应放到 Action 的配置之前.例如在 3.1 节中对
        Action 已经配置,在 3.2 节中对拦截器配置时应把对拦截器的配置放到所有 Action 配置
        的前面,即在 package 中,拦截器的配置总是放到前面,Action 的配置放到后面-->
        <interceptors>
            <!--文字过滤拦截器配置,replace 是拦截器的名字-->
            <interceptor name="replace" class="interceptor.MyInterceptor" />
        </interceptors>
        <action name="checkLogin" class="loginAction.LoginAction">
            <result name="success">/I18N/loginSuccess.jsp</result>
            <result name="error">/I18N/login.jsp</result>
        </action>
        <!--文字过滤 Action 的配置-->
        <action name="public" class="interceptor.PublicAction">
            <result name="success">/interceptor/success.jsp</result>
            <result name="login">/interceptor/success.jsp</result>
            <!--Struts2 系统默认拦截器-->
            <interceptor-ref name="defaultStack" />
            <!--使用自定义拦截器-->
            <interceptor-ref name="replace"/>
        </action>
    </package>
</struts>
```

7. 项目部署和运行

项目部署后运行,运行效果如图 3-15 所示。输入完评论后单击"提交"按钮,如图 3-16 所示,跳转到评论成功页面,如图 3-17 所示。可以看出,经过拦截器拦截后文字被过滤。

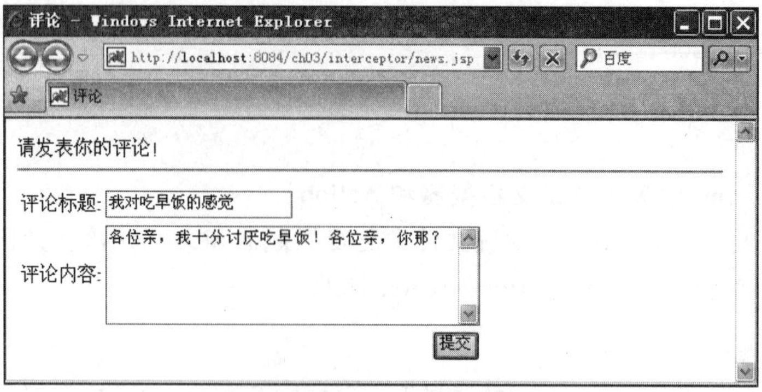

图 3-16　输入评论

图 3-17　经过拦截器拦截以后的数据

3.3　Struts2 的输入校验

在网络上,由于用户对要输入数据格式理解的多样性,导致用户输入的数据与开发者的意图不一致。为了保证数据在存储过程中的正确性和一致性,必须对用户输入的信息进行校验。只有通过了严格校验的数据才能提高系统的安全性、健壮性,保证系统的正常运行。Struts2 框架内置了许多功能强大的输入校验器,内置的输入校验器基本上可以实现主要的输入验证功能,程序员不需要编写任何校验代码即可完成项目开发中遇到的常用输入校验。另外,为了扩展 Struts2 框架的输入校验功能,Struts2 框架允许使用 validate()方法、validateXxx()方法和自定义校验器,实现对 Struts2 框架中输入验证的扩展。

3.3.1　Struts2 输入验证的基础知识

网络上,为了保证网站的稳定运行,良好的输入验证机制是前提,也是一个成熟系统的必备条件。对数据进行验证通常可以分为两部分:首先,验证用户输入数据的有效性;其次,在用户输入无效的数据后为用户提供友好的提示信息。

1. 需要输入校验的原因

在互联网上,Web 站点是对外提供服务的,由于站点的开放性,Web 站点保存的数据主要是从客户端接收到的。输入数据的用户来自不同的行业,他们有着不同的生活习惯、教育背景,从而不能绝对保证输入内容的正确性。例如,用户操作计算机不熟练、输入出错、网络问题或者恶意输入等,这些都可能导致数据的异常。如果对数据不加校验,有可能导致系统阻塞甚至系统崩溃。如图 3-18 所示页面展示的就是用户随意输入数据的一种表现。

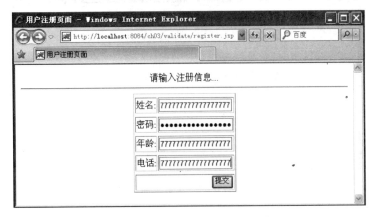

图 3-18　无输入验证的用户注册页面

在图 3-18 所示页面中，存在用户名和密码的位数有可能过长、年龄和电话数据的输入不符合事实等数据异常情况。如果在数据库表的设计中没有定义那么长的字段，就可能会导致数据库异常。所以说，必须对客户端输入的信息进行校验。

Java Web 项目中系统的输入校验方式有客户端校验和服务器端校验两种。Struts2 框架对用户输入数据的校验也可以分为两种：客户端校验和服务器端校验。为了实现完全的数据输入校验，需要将这两种验证方式紧密结合、互相协作。

2. 客户端校验

客户端校验可以在客户端通过 JavaScript 脚本或者 Ajax 对用户输入的数据进行基本校验。下面通过使用 JavaScript 脚本语言对客户端输入数据进行验证。

1）项目介绍

本项目使用 JavaScript 脚本对注册页面进行验证。JavaScript 的脚本写在＜head＞中，注册页面（register.jsp）的代码如例 3-15 所示，对应的业务控制器为 RegistAction，代码如例 3-17 所示，如果输入的数据验证成功，进入验证成功页面（success.jsp），代码如例 3-16 所示。此外还需要配置 web.xml，代码如例 1-3 所示；配置 struts.xml，代码如例 3-18 所示。项目的文件结构如图 3-19 所示。

2）在 web.xml 中配置核心控制器 FilterDispatcher
参考 1.3.1 节中的例 1-3。

3）编写视图组件（JSP 页面）

图 3-19　注册系统文件结构图

注册页面如图 3-20 所示。代码如例 3-15 所示。输入的注册数据如果通过验证则跳转到验证成功页面，代码如例 3-16 所示。

【**例 3-15**】　具有 JavaScript 脚本验证功能的注册页面（register.jsp）。

图 3-20 注册页面

```jsp
<%@page contentType="text/html; charset=UTF-8" %>
<%@taglib prefix="s" uri="/struts-tags" %>
<html>
    <head>
        <title>用户注册页面</title>
        <!--检验输入表单数据的函数-->
        <script language="JavaScript">
            function trim(str)
            {
                //使用正则式去掉字符的前后空格
                return str.replace(/^\s*/,"").replace(/\s*$/,"");
            }
            function check(form)
            {
                //定义错误标志字符串
                var errorStr="";
                //获取表单的 4 个数据
                var userName=trim(form.userName.value);
                var userPassword=trim(form.userPassword.value);
                var userAge=trim(form.userAge.value);
                var userTelephone=trim(form.userTelephone.value);
                var pattern =/^\d{8}$/;
                //判断用户名是否为空
                if(userName==null||userName=="")
                {
                    errorStr="用户名不能为空!";
                }
                else if(userPassword.length>16||userPassword.length<6)
                {
                    errorStr="密码长度必须在 6~16 之间";
                }
                else if(userAge>130||userAge<0)
```

```
            {
                errorStr="年龄必须在 0~130 之间";
            }
            else if(!pattern.test(userTelephone)){
                errorStr="电话号码为 8 位阿拉伯数字组成!";
            }
            if(errorStr=="")
            {
                return true;
             }else
            {
                alert(errorStr);
                return false;
            }
        }
    </script>
</head>
<body>
    <center>
        请输入注册信息…
        <hr>
        <s:form action="register.action" method=post onSubmit=
            "return check(this);">
            <table border="1">
                <tr>
                    <td>
                        <s:textfield name="userName" label="姓名"
                        size="16"/>
                    </td>
                </tr>
                <tr>
                    <td>
                        <s:password name="userPassword" label="密码"
                        size="18"/>
                    </td>
                </tr>
                <tr>
                    <td>
                        <s:textfield name="userAge" label="年龄"
                        size="16"/>
                    </td>
                </tr>
                <tr>
                    <td>
                        <s:textfield name="userTelephone" label="电话"
```

```
                    size="16"/>
                </td>
            </tr>
            <tr>
                <td><s:submit value="提交"/></td>
            </tr>
        </table>
    </s:form>
</center>
</body>
</html>
```

【例3-16】 验证成功后的页面(success.jsp)。

```
<%@page contentType="text/html; charset=UTF-8" %>
<%@taglib prefix="s" uri="/struts-tags" %>
<html>
    <head>
        <title>校验成功</title>
    </head>
    <body>
        校验通过,用户信息如下:
        <hr>
        姓名:<s:property value="userName"/>
        <br>
        密码:<s:property value="userPassword"/>
        <br>
        年龄:<s:property value="userAge"/>
        <br>
        电话:<s:property value="userTelephone"/>
    </body>
</html>
```

4)编写业务控制器Action

注册页面对应的业务控制器是RegistAction类,代码如例3-17所示。

【例3-17】 注册页面对应的业务控制器(RegistAction.java)。

```
package validate;
import com.opensymphony.xwork2.ActionSupport;

public class RegistAction extends ActionSupport{
    private String userName;
    private String userPassword;
    private int userAge;
    private String userTelephone;
    public String getUserName() {
        return userName;
```

```java
        }
        public void setUserName(String userName) {
            this.userName =userName;
        }
        public String getUserPassword() {
            return userPassword;
        }
        public void setUserPassword(String userPassword) {
            this.userPassword =userPassword;
        }
        public int getUserAge() {
            return userAge;
        }
        public void setUserAge(int userAge) {
            this.userAge =userAge;
        }
        public String getUserTelephone() {
            return userTelephone;
        }
        public void setUserTelephone(String userTelephone) {
            this.userTelephone = userTelephone;
        }
        public String execute(){
            return SUCCESS;
        }
}
```

5) 修改 struts.xml 配置 Action

修改配置文件 struts.xml,代码如例 3-18 所示。

【例 3-18】 在 struts.xml 配置 Action(struts.xml)。

```xml
<!DOCTYPE struts PUBLIC
    "-//Apache Software Foundation//DTD Struts Configuration 2.0//EN"
    "http://struts.apache.org/dtds/struts-2.0.dtd">
<struts>
    <constant name="struts.custom.i18n.resources" value="globalMessages" />
    <constant name="struts.i18n.encoding" value="utf-8" />
    <package name="I18N" extends="struts-default">
      <interceptors>
            <!--文字过滤拦截器配置,replace 是拦截器的名字-->
            <interceptor name="replace" class="interceptor.MyInterceptor" />
      </interceptors>
      <action name="checkLogin" class="loginAction.LoginAction">
          <result name="success">/I18N/loginSuccess.jsp</result>
          <result name="error">/I18N/login.jsp</result>
      </action>
```

```xml
<!--文字过滤Action配置-->
<action name="public" class="interceptor.PublicAction">
    <result name="success">/interceptor/success.jsp</result>
    <result name="login">/interceptor/success.jsp</result>
    <!--Struts2系统默认拦截器-->
    <interceptor-ref name="defaultStack" />
    <!--使用自定义拦截器-->
    <interceptor-ref name="replace"/>
</action>
<action name="register" class="validate.RegistAction">
    <result name="input">/validate/register.jsp</result>
    <result name="success">/validate/success.jsp</result>
</action>
</package>
</struts>
```

6）项目部署和运行

注册页面运行效果如图3-20所示,如果输入的年龄不符合要求将出现如图3-21所示的页面。如果数据符合验证要求,将出现如图3-22所示的页面。

图3-21 输入的数据不符合验证要求

图3-22 校验成功页面

从上面的实例可以看出,当客户端访问注册页面时,如果输入的数据不符合验证要求就无法实现注册功能。但是服务器发送到客户端的页面是静态页面,可以很方便地在浏览器的"查看"->"源文件"中查看这些代码并修改代码,这样就能轻易绕过客户端校验。由此可见,要绕过客户端校验很容易,所以还需要服务器端验证。

3. 服务器端校验

服务器端校验就是将数据校验放在服务器端进行。例如,数据库中设置限制条件、用 Java 代码进行校验等。

Struts2 框架中,可以在 Action 的 execute()中对输入数据实现服务器端的校验。

例如:

```
public String execute(){
    if(userName==null||userName==""){
        return INPUT;
    }
    else if(userPassword.length()>16||userPassword.length()<6){
        return INPUT;
    }
    else if(userAge>130||userAge<0){
        return INPUT;
    }
    else if(userTelephone.length()!=8){
        return INPUT;
    }
    else{
        return SUCCESS;
    }
}
```

上述代码可以完成注册页面的数据验证。但是一般并不会在 execute()方法中进行数据验证,因为 execute()方法主要的功能是调用业务组件和返回逻辑视图。另外,在软件设计中,一般要求一个方法尽量完成单一的任务,而不推荐实现两个以上的功能,否则就违背了软件设计中"高内聚,低耦合"的思想。

3.3.2 Struts2 的手工验证

常见的基于 MVC 模式的 Web 框架中都会提供规范的数据校验部分,专门完成数据校验工作。Struts2 也提供了校验器,本节主要介绍手工验证,下一节将介绍自动验证(即内置验证器)。

ActionSupport 类实现了 Action、Validateable、ValidationAware、TextProvider、LocaleProvider 和 Serializable 接口。其中 Validateable 接口就是验证器接口,该接口有一个 validate()方法,所以只要用户在编写一个 Action 类时重写了该方法就可以实现验证功能。

1. 重写 validate()方法

在 Struts2 框架中,validate()方法是专门用来验证数据的,实现的时候需要继承

ActionSupport 类,并重写 validate()方法来完成输入验证。

下面对客户端验证的项目进行修改,即改为服务器端的手工验证,使用 validate()方法。

1) 项目介绍

本项目是对 3.3.1 节客户端验证项目的改进,不过本节是使用 validate()方法实现服务器端验证。项目介绍请参考 3.3.1 节。

2) 在 web.xml 中配置核心控制器 FilterDispatcher

参考 1.3.1 节中的例 1-3。

3) 编写视图组件(JSP 页面)

注册页面(register1.jsp)如 3.3.1 中图 3-20 所示,只是对代码进行了简单的修改,修改后代码如例 3-19 所示,验证通过页面跳转到成功页面,成功页面(success.jsp)与例 3-16 类似。

【例 3-19】 修改后的注册页面(register1.jsp)。

```jsp
<%@page contentType="text/html; charset=UTF-8" %>
<%@taglib prefix="s" uri="/struts-tags" %><html>
    <head>
        <title>用户注册页面</title>
    </head>
    <body>
        <center>
            请输入注册信息…
            <hr>
            <s:form action="register1.action" method="post" >
                <table border="1">
                    <tr>
                        <td>
                            <s:textfield name="userName" label="姓名"
                            size="16"/>
                        </td>
                    </tr>
                    <tr>
                        <td>
                            <s:password name="userPassword" label="密码"
                            size="18"/>
                        </td>
                    </tr>
                    <tr>
                        <td>
                            <s:textfield name="userAge" label="年龄"
                             size="16"/>
                        </td>
                    </tr>
                    <tr>
                        <td>
                            <s:textfield name="userTelephone" label="电话"
```

```
                        size="16"/>
                </td>
            </tr>
            <tr>
                <td><s:submit value="提交"/></td>
            </tr>
        </table>
    </s:form>
    </center>
</body>
</html>
```

4) 编写业务控制器 Action

注册页面(register1.jsp)对应的业务控制器是 RegistAction1,要定义该类只需在例 3-17 的 RegistAction 类中添加一个方法,即覆盖该 ActionSupport 类中的 validate()方法,其他部分不变,代码如例 3-20 所示。

【例 3-20】 注册页面对应的业务控制器(RegistAction1.java)。

```java
package validate;
import com.opensymphony.xwork2.ActionSupport;

public class RegistAction1 extends ActionSupport{
    private String userName;
    private String userPassword;
    private int userAge;
    private String userTelephone;
    public String getUserName() {
        return userName;
    }
    public void setUserName(String userName) {
        this.userName =userName;
    }
    public String getUserPassword() {
        return userPassword;
    }
    public void setUserPassword(String userPassword) {
        this.userPassword =userPassword;
    }
    public int getUserAge() {
        return userAge;
    }
    public void setUserAge(int userAge) {
        this.userAge =userAge;
    }
    public String getUserTelephone() {
```

```java
        return userTelephone;
    }
    public void setUserTelephone(String userTelephone) {
        this.userTelephone =userTelephone;
    }
    public void validate(){
        if(userName==null ||userName.length()<6 || userName.length()>16){
            addFieldError("userName","用户姓名的长度不符合要求,6~16位!");
        }
        if(userPassword.length()>16||userPassword.length()<6){
            addFieldError("userPassword","密码长度不符合要求,6~16位!");
        }
        if(userAge>130||userAge<1){
            addFieldError("userAge","年龄不符合要求,1~130 岁");
        }
        if(userTelephone.length()!=8){
            addFieldError("userTelephone","电话号码不符合要求,8 位");
        }
    }
    public String execute(){
        return SUCCESS;
    }
}
```

5）修改 struts.xml 配置 Action

修改配置文件 struts.xml,代码如例 3-21 所示。

【例 3-21】 在 struts.xml 配置 Action(struts.xml)。

⋮
```xml
<action name="register1" class="validate.RegistAction1">
    <result name="input">/validate/register1.jsp</result>
    <result name="success">/validate/success.jsp</result>
</action>
```
⋮

6）项目部署和运行

运行效果如图 3-20 所示,若不输入数据直接单击"提交"按钮,效果如图 3-23 所示,提示输入的数据不符合要求。如果输入的数据符合要求则进入验证成功页面,效果如图 3-22 所示。

2. 重写 validateXxx()方法

在 Struts2 框架中,一个 Action 可以包含多个业务处理逻辑(可参考 2.3.5 节),也就是类似于多个 execute()方法,只是方法名字不同,使用的时候只需在 struts.xml 文件中配置 Action 就行,而且也可以指定 method 属性,Struts2 框架将根据属性值来执行相应的逻辑处理。Struts2 中允许提供 validateXxx()方法,专门校验 Action 中对应的 xxx()方法。例如,在 Action 中有一个 login()方法,在 Action 中就可以使用 validateLogin()方法来进行验

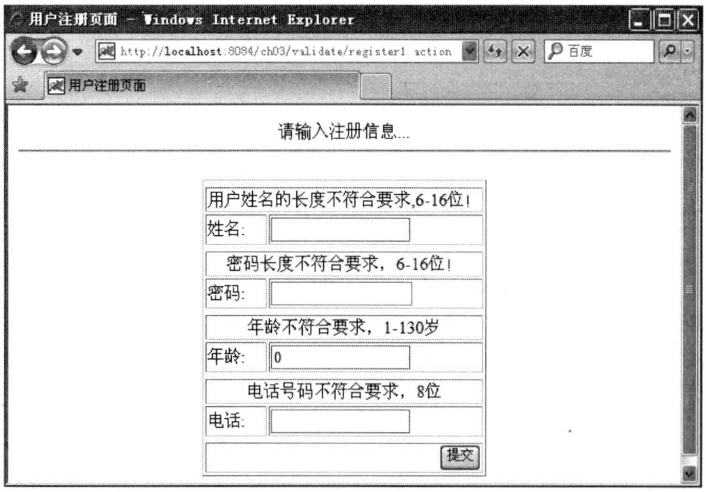

图 3-23　输入的数据不符合验证要求

证处理。

如果 Action 中有 validate()方法和 validateXxx()方法,数据校验时将先执行 validateXxx()方法后执行 validate()方法,对数据进行两次校验。

3.3.3　Struts2 内置校验器的使用

Struts2 框架中提供了大量的内置校验器。在 Java Web 项目开发中,大部分校验功能都可以通过内置校验器来完成,使用时只需要简单配置即可。常用的内置校验器有必填校验器、必填字符串校验器、字符串长度校验器、整数校验器、日期校验器、邮件地址校验器、网址校验器、表达式校验器、字段表达式校验器等。下面将分别介绍常用的内置校验器。在介绍内置校验器的使用之前,首先介绍校验器的配置风格。

1. 校验器的配置风格

Struts2 框架提供两种配置校验器的方式:字段校验器配置风格和非字段校验器配置风格。这两种配置风格没有本质的区别,只是在组织方式和关注点方面不同。

1) 字段校验器配置风格

使用字段校验器配置风格时,校验文件以<field>元素为基本元素,由于这个基本元素的 name 属性值为被校验的字段,所以是字段优先,因此称为字段校验器配置风格。该风格的配置格式如例 3-22 所示。

【例 3-22】　字段校验器配置风格。

```
<!--内置校验器配置使用的元素,内置校验器配置在其中-->
<validators>
    <!--内置校验器配置一个字段的元素-->
    <field name="被校验的字段">
        <!--用来指定校验器的类型-->
        <field-validator type="校验器的类型">
            <!--用来向校验器传递参数,可以包含多个 param-->
            <param name="参数名">参数值</param>
```

```
            <!--用来指定校验失败的提示信息-->
            <message>校验失败提示的信息</message>
        </field-validator>
    </field>
    <!--下一个要验证的字段-->
    ⋮
</validators>
```

2) 非字段校验器配置风格

非字段校验器配置风格是以校验器优先的配置方式。以<validator>为基本元素,在元素<validators>中可以配置多个<validator>。该风格的配置格式如例 3-23 所示。

【例 3-23】 非字段校验器配置风格。

```
<validators>
    <!--用来指定校验器的类型-->
    <validator type="校验器的类型" >
        <!--用来指定要校验的字段-->
        <param name="fildName">需要被校验的字段</param>
        <!--用来向校验器传递参数,可以包含多个 param-->
        <param name="参数名">参数值</param>
        <!--用来指定校验失败的提示信息-->
        <message>校验失败提示的信息</message>
    </validator>
    <!--下一个要验证的字段-->
    ⋮
</validators>
```

在 Struts2 中使用内置校验器时需要在验证文件中配置校验器。验证文件的命名规则是：Action 类名-别名-validation.xml 或者 Action 类名-validation.xml。如果校验器对应的 Action 类名为 Register2Action,那么该验证文件可命名为 Register2Action-validation.xml。验证文件一般都与 Action 类保存在相同的目录下,这样对于不同的 Action 处理请求将会自动加载不同的校验文件。

2. 必填校验器

该校验器的名称为 required,校验字段是否为空,用于要求字段必须有值。在实际项目开发中对字段进行校验时,一般使用字符串长度校验器。常用参数如下所示。

fieldName：指定校验字段的名称,如果是字段校验风格的配置,则不用指定该参数。

3. 必填字符串校验器

该校验器的名称为 requiredstring,要求字段为一个非空字符串,并且长度需要大于 0。在实际项目开发中对字段进行校验时,一般使用字符串长度校验器。常用参数如下所示。

(1) fieldName：指定校验字段的名称,如果是字段校验风格的配置,则不用指定该参数。

(2) trim：指定是否在校验之前对字符串进行整理,截取字符串前后空格,默认值为 true。

4．字符串长度校验器

该校验器的名称为 stinglength,用于校验字段中字符串长度是否在指定的范围内。常用参数如下所示。

(1) fieldname：指定校验字段的名称,如果是字段校验风格的配置,则不用指定该参数。

(2) maxLength：指定字符串的最大长度,可选项,不选则最大长度不限制。

(3) minLength：指定字符串的最小长度,可选项,不选则最小长度不限制。

(4) trim：指定是否在校验之前对字符串进行整理,截取字符串前后空格,默认值为 true。

该校验器既可以使用字段校验,也可以使用非字段校验,其格式如下：

```
<validators>
    <validator type="stringlength">
        <param name="fieldName">userName</param>
        <param name="maxLength">16</param>
        <param name="minLength">6</param>
        <message>姓名长度为${minLength}到${maxLength}个字符！</message>
    </validator>
    ⋮
    <field name="userName">
        <field-validator type="stringlength" >
            <param name="maxLength">16</param>
            <param name="minLength">6</param>
            <message>姓名长度为${minLength}到${maxLength}个字符！
            </message>
        </field-validator>
    </field>
    ⋮
</validators>
```

5．整数校验器

该校验器的名称为 int,用于要求被校验的整数值在指定范围内,否则校验失败。常用参数如下所示。

(1) fieldname：指定校验字段的名称,如果是字段校验风格的配置,则不用指定该参数。

(2) max：指定整数的最大值,可选项,不选则最大值不限制。

(3) min：指定整数的最小值,可选项,不选则最小值不限制。

该校验器既可以使用字段校验,也可以使用非字段校验,其格式如下：

```
<validators>
    <validator type="int">
        <param name="fieldName">userAge</param>
        <param name="min">1</param>
```

```xml
            <param name="max">130</param>
            <message>年龄必须在${min}到${max}之间</message>
        </validator>
         ⋮
        <field name="userAge">
            <field-validator type="int" >
                <param name="min">1</param>
                <param name="max">130</param>
                <message>年龄必须在${min}到${max}之间</message>
            </field-validator>
        </field>
         ⋮
</validators>
```

6．日期校验器

该校验器的名称为 date，该校验器要求字段的日期值在指定的范围内。常用参数如下所示。

（1）fieldname：指定校验字段的名称，如果是字段校验风格的配置，则不用指定该参数。

（2）max：指定日期的最大值，可选项，不选则最大值不限制。

（3）min：指定日期的最小值，可选项，不选则最小值不限制。

该校验器既可以使用字段校验，也可以使用非字段校验，其格式如下：

```xml
<validators>
    <validator type="date">
        <param name="fieldName" >birthday</param>
        <param name="min">1990-12-30</param>
        <param name="max">2020-01-01</param>
        <message>出生日期必须在${min}到${max}之间</message>
    </validator>
     ⋮
    <field name="birthday">
        <field-validator type="date" >
            <param name="min">1990-12-30</param>
            <param name="max">2020-01-01</param>
            <message>出生日期必须在${min}到${max}之间</message>
        </field-validator>
    </field>
     ⋮
</validators>
```

使用日期校验器时，在 JSP 页面中要遵循以下格式：

```jsp
<s:form action="date.action" method="post">
     ⋮
```

```
        <s:datetimepicker displayFormat="yyyy-MM-dd" label="生日" name="birthday" >
        </s:datetimepicker>
        ⋮
        <s:submit value="提交"/>
</s:form>
```

7. 邮件地址校验器

该校验器的名称为 email，该校验器要求指定字段必须满足邮件地址规则，系统的邮件地址正则表达式如下：

\\b(^[_A-Za-z0-9-](\\.[_A-Za-z0-9-]))*@([_A-Za-z0-9-]+(\\.com)|(\\.net)|(\\.org)|(\\.info)|(\\.end)|(\\.mil)|(\\.gov)|(\\.ws)|(\\.biz)|(\\.us)|(\\.tv)|(\\.cc)|(\\.aero)|(\\.arpa)|(\\.coop)|(\\.int)|(\\.museum)|(\\.name)|(\\.pro)|(\\.trave)|(\\.nato)|(\\...{2,3})|(\\...{2,3})$|)\\b

随着网络技术的发展，邮件地址格式越来越丰富，上面的正则表达式有可能并没有覆盖实际的所有电子邮件地址，建议开发者自己定义邮件地址的正则表达式。

该校验器既可以使用字段校验，也可以使用非字段校验，其格式如下：

```
<validators>
    <validator type="email">
        <param name="fieldName">userEmail</param>
        <message>请使用正确的邮件格式！</message>
    </validator>
    ⋮
    <field name="userEmail">
        <field-validator type="email" >
            <message>请使用正确的邮件格式！</message>
        </field-validator>
    </field>
    ⋮
</validators>
```

8. 网址校验器

该校验器的名称为 url，该校验器要求被校验字段必须为合法的 URL 地址。

该校验器既可以使用字段校验，也可以使用非字段校验，其格式如下：

```
<validators>
    <validator type="url">
        <param name="fieldName">netURL</param>
        <message>无效的网络地址！</message>
    </validator>
    ⋮
    <field name="netURL">
        <field-validator type="url" >
            <message>无效的网络地址！</message>
```

```
            </field-validator>
        </field>
        ⋮
</validators>
```

9. 表达式校验器

该校验器的名称为expression,该校验器是一个非字段校验器,所以不能以字段校验器的配置风格来配置。表达式校验器要求OGNL表达式返回的值为true,否则校验失败。常用参数如下所示。

expression:该参数为一个逻辑表达式,使用OGNL表达式。

该校验器使用非字段校验风格,其格式如下:

```
<validators>
    <validator type="expression">
        <param name="expression" >OGNL 表达式 </param>
        <message>无效的 OGNL 表达式</message>
    </validator>
    ⋮
</validators>
```

10. 字段表达式校验器

该校验器的名称为fieldexpression,要求字段必须满足一个逻辑表达式。常用参数如下所示。

(1) fieldname:指定校验字段的名称,如果是字段校验风格的配置,则不用指定该参数。

(2) expression:该参数为一个逻辑表达式,使用OGNL表达式。

该校验器既可以使用字段校验,也可以使用非字段校验,其格式如下:

```
<validators>
    <validator type="fieldexpression">
        <param name="fieldName">userPassword </param>
        <param name="expression">
            <!--验证两次输入的密码是否相同-->
            <![CDATA[userPassword==ruserPassword]]>
        </param>
        <message>校验失败!</message>
    </validator>
    ⋮
    <field name="userPassword">
        <field-validator type="fieldexpression">
            <param name="expression">
                <!--验证两次输入的密码是否相同-->
                <![CDATA[userPassword==ruserPassword]]>
            </param>
            <message>校验失败!</message>
```

```
        </field-validator>
    </field>
    ⋮
</validators>
```

3.3.4 Struts2 内置校验器应用实例

下面以注册项目为例,介绍 Struts2 中常用内置校验器在项目开发中的应用。

1. 项目介绍

本注册项目中使用了 4 个 Struts2 内置校验器,分别为 stringlength、int、email、fieldexpression。项目有一个注册页面(register2.jsp),代码如例 3-24 所示,对应的业务控制器为 Register2Action 类,代码如例 3-26 所示。如果输入的数据经内置验证器验证成功,进入验证成功页面(success1.jsp),代码如例 3-25 所示。此外,还需要配置 web.xml,代码如例 1-3 所示;配置 struts.xml,代码如例 3-27 所示;编写验证规则文件(Register2Action-validation.xml),代码如例 3-28 所示。项目的文件结构如图 3-24 所示。

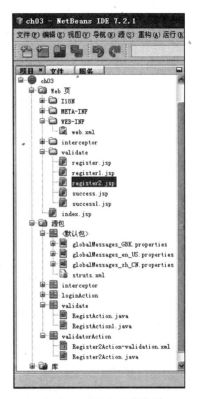

图 3-24 项目文件结构图

2. 在 web.xml 中配置核心控制器 FilterDispatcher

参考 1.3.1 节中的例 1-3。

3. 编写视图组件(JSP 页面)

编写注册页面,代码如例 3-24 所示。注册页面如图 3-25 所示。在注册页面中输入数据,如果数据能够通过验证则跳转到验证成功页面,代码如例 3-25 所示。

图 3-25 注册页面

【例 3-24】 注册页面(register2.jsp)。

```jsp
<%@page contentType="text/html; charset=UTF-8" %>
<%@taglib prefix="s" uri="/struts-tags" %>
<html>
    <head>
        <title>用户注册页面</title>
    </head>
    <body>
        <center>
            请输入注册信息…
            <hr>
            <s:form action="register2.action" method="post" >
                <table border="1">
                    <tr>
                        <td><s:textfield name="userName" label="姓名"
                            size="16"/></td>
                    </tr>
                    <tr>
                        <td><s:password name="userPassword" label="密码"
                            size="18"/></td>
                    </tr>
                    <tr>
                        <td><s:password name="ruserPassword"
                            label="再次输入密码" size="18"/></td>
                    </tr>
                    <tr>
                        <td><s:textfield name="userAge" label="年龄"
                            size="16"/></td>
                    </tr>
                    <tr>
                        <td><s:textfield name="userTelephone" label="电话"
                            size="16"/></td>
                    </tr>
                    <tr>
                        <td><s:textfield name="userEmail" label="邮箱"
                            size="16"/></td>
                    </tr>
                    <tr>
                        <td><s:submit value="提交"/></td>
                    </tr>
                </table>
            </s:form>
        </center>
    </body>
</html>
```

【例 3-25】 验证成功页面（success1.jsp）。

```jsp
<%@page contentType="text/html; charset=UTF-8" %>
```

```jsp
<%@ taglib prefix="s" uri="/struts-tags" %>
<html>
    <head>
        <title>校验成功</title>
    </head>
    <body>
        校验通过,用户信息如下:
        <hr>
        姓名:<s:property value="userName"/>
        <br>
        密码:<s:property value="userPassword"/>
        <br>
        年龄:<s:property value="userAge"/>
        <br>
        电话:<s:property value="userTelephone"/>
        <br>
        邮箱:<s:property value="userEmail"/>
    </body>
</html>
```

4. 编写业务控制器 Action

注册页面对应的业务控制器是 Register2Action,代码如例 3-26 所示。

【例 3-26】 注册页面对应的业务控制器(Register2Action.java)。

```java
package validatorAction;
import com.opensymphony.xwork2.ActionSupport;

public class Register2Action  extends ActionSupport{
    private String userName;
    private String userPassword;
    private String ruserPassword;
    private int userAge;
    private int userTelephone;
    private String userEmail;
    public String getUserName() {
        return userName;
    }
    public void setUserName(String userName) {
        this.userName = userName;
    }
    public String getUserPassword() {
        return userPassword;
    }
    public void setUserPassword(String userPassword) {
        this.userPassword = userPassword;
    }
    public String getRuserPassword() {
```

```java
        return ruserPassword;
    }
    public void setRuserPassword(String ruserPassword) {
        this.ruserPassword = ruserPassword;
    }
    public int getUserAge() {
        return userAge;
    }
    public void setUserAge(int userAge) {
        this.userAge = userAge;
    }
    public int getUserTelephone() {
        return userTelephone;
    }
    public void setUserTelephone(int userTelephone) {
        this.userTelephone = userTelephone;
    }
    public String getUserEmail() {
        return userEmail;
    }
    public void setUserEmail(String userEmail) {
        this.userEmail = userEmail;
    }
    public String execute(){
        return SUCCESS;
    }
}
```

5. 修改 struts.xml 配置 Action

配置 struts.xml，代码如例 3-27 所示。

【例 3-27】 在 struts.xml 配置 Action(struts.xml)。

```xml
<!DOCTYPE struts PUBLIC
    "-//Apache Software Foundation//DTD Struts Configuration 2.0//EN"
    "http://struts.apache.org/dtds/struts-2.0.dtd">
<struts>
    <constant name="struts.custom.i18n.resources" value="globalMessages" />
    <constant name="struts.i18n.encoding" value="utf-8" />
    <package name="I18N" extends="struts-default">
      <interceptors>
            <!--文字过滤拦截器配置，replace 是拦截器的名字-->
            <interceptor name="replace" class="interceptor.MyInterceptor" />
      </interceptors>
      <action name="checkLogin" class="loginAction.LoginAction">
          <result name="success">/I18N/loginSuccess.jsp</result>
          <result name="error">/I18N/login.jsp</result>
      </action>
```

```xml
        <!--文字过滤Action配置-->
        <action name="public" class="interceptor.PublicAction">
            <result name="success">/interceptor/success.jsp</result>
            <result name="login">/interceptor/success.jsp</result>
            <!--Struts2系统默认拦截器-->
            <interceptor-ref name="defaultStack" />
            <!--使用自定义拦截器-->
            <interceptor-ref name="replace"/>
        </action>
        <action name="register" class="validate.RegistAction">
            <result name="input">/validate/register.jsp</result>
            <result name="success">/validate/success.jsp</result>
        </action>
        <action name="register1" class="validate.RegistAction1">
            <result name="input">/validate/register1.jsp</result>
            <result name="success">/validate/success.jsp</result>
        </action>
        <action name="register2" class="validatorAction.Register2Action">
            <result name="input">/validate/register2.jsp</result>
            <result name="success">/validate/success1.jsp</result>
        </action>
    </package>
</struts>
```

6. 内置验证器的验证文件

在 Struts2 中使用内置校验器需要在验证文件中进行配置,并把该验证文件与 Action 类放在同一个文件夹下,如图 3-24 所示。验证文件 Register2Action-validation.xml 的代码如例 3-28 所示。

【例 3-28】 验证文件编写(Register2Action-validation.xml)。

```xml
<!DOCTYPE validators PUBLIC
    "-//OpenSymphony Group//XWork Validator 1.0.2//EN"
    "http://www.opensymphony.com/xwork/xwork-validator-1.0.2.dtd">
<validators>
    <validator type="stringlength">
        <param name="fieldName">userName</param>
        <param name="maxLength">16</param>
        <param name="minLength">6</param>
        <message>姓名长度为${minLength}到${maxLength}个字符!</message>
    </validator>
    <validator type="stringlength">
        <param name="fieldName">userPassword</param>
        <param name="maxLength">16</param>
        <param name="minLength">6</param>
        <message>密码长度为${minLength}到${maxLength}个字符!</message>
    </validator>
    <validator type="fieldexpression">
```

```xml
        <param name="fieldName">userPassword</param>
        <param name="expression">
            <!--验证两次输入的密码是否相同-->
            <![CDATA[userPassword==ruserPassword]]>
        </param>
        <message>两次密码不一致!</message>
    </validator>
    <validator type="int">
        <param name="fieldName">userAge</param>
        <param name="min">1</param>
        <param name="max">130</param>
        <message>年龄必须在${min}到${max}之间!</message>
    </validator>
    <validator type="int">
        <param name="fieldName">userTelephone</param>
        <param name="min">22222222</param>
        <param name="max">99999999</param>
        <message>电话必须是${min}到${max}之间的八位号码!</message>
    </validator>
    <validator type="email">
        <param name="fieldName">userEmail</param>
        <message>请使用正确的邮件格式!</message>
    </validator>
</validators>
```

7. 项目部署和运行

项目部署运行后,页面运行效果如图3-25所示,单击"提交"按钮后,在执行Action前将读取验证文件RequiredAction-validation.xml对数据进行验证。在图3-25所示页面中输入数据,如图3-26所示;输入的数据如果不符合验证要求,将出现如图3-27所示页面。如果数据验证成功,将跳转到验证成功页面,如图3-28所示。

图3-26 在注册页面中输入数据

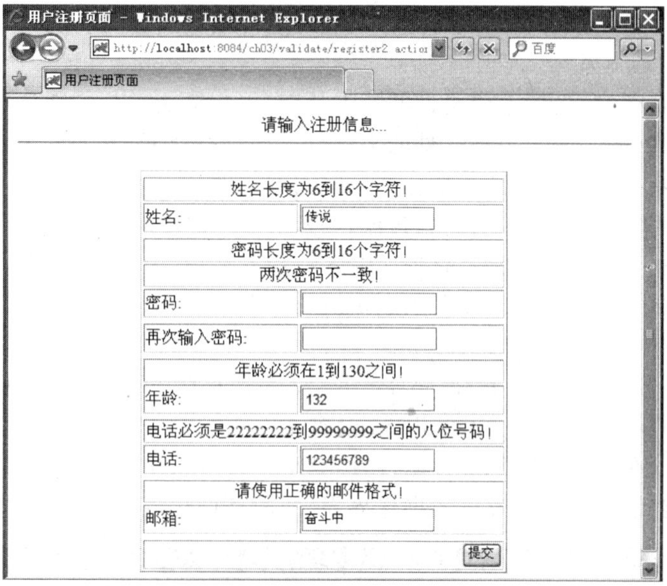

图 3-27 数据不符合要求

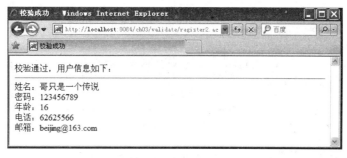

图 3-28 数据校验成功页面

3.4 Struts2 的文件上传和下载

在项目开发中经常会遇到文件上传和下载问题,下面将分别介绍 Struts2 中的文件上传和文件下载功能的实现。

3.4.1 文件上传

文件上传是很多 Web 项目中提供的功能。在 Struts2 框架中,提供了文件上传功能。Struts2 框架的类库中实现文件上传和下载所需的 JAR 文件是 commons-fileupload-1.2.2.jar 和 commons-io-2.0.1.jar,把这两个 JAR 加载到类库中即可,有关 JAR 文件的下载和加载请参考 1.1.2 节。

使用 Struts2 的上传文件功能时,只需要使用普通的 Action 类即可,但是为了获取一些上传文件的信息,如上传文件名、文件类型,需要按照一定的规则在 Action 中增加一些 getter 和 setter 方法。

在上传文件时,可以对上传文件的大小、文件类型等进行控制。在 Struts2 中提供了文件上传拦截器(fileUpload),该拦截器能够实现对上传文件的过滤功能。

fileUpload拦截器常用属性如下所示。

(1) maximumSize:设置上传文件的最大长度(以字节为单位),默认值为2MB。

(2) allowedTypes:设置上传文件的类型,以","为分隔符,可以上传多种类型的文件,如text/html,如果不设置该属性就是允许任何类型文件。

如果使用了拦截器对上传文件进行过滤,一旦上传的文件格式不符合要求,将在页面中提示异常信息,提示的异常信息是Struts2框架中格式为"键=值"的信息,常用的"键"如下所示。

① struts.messages.error.content.type.not.allowed:设置上传的文件类型不匹配的提示信息。

② struts.messages.error.file.too.large:设置上传的文件太大时的提示信息。

③ struts.messages.error.uploading:设置文件不能上传时的通用提示信息。

这些信息可以在国际化资源文件中设置,如果不设置将出现如图3-29所示异常提示信息。所以为了能够给用户一个友好的操作界面,一般都使用国际化资源文件。

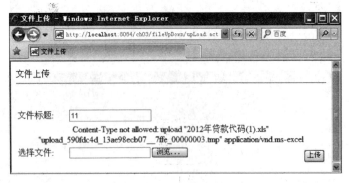

图3-29 上传异常

图3-30 项目文件结构图

下面通过一个项目来理解在Struts2框架中怎样实现文件上传功能。

1. 项目介绍

本项目实现文件上传功能。有一个文件上传的页面(fileUp.jsp),代码如例3-29所示,上传文件页面对应的业务控制器为UploadAction,该Action要封装上传的文件名、文件类型等,代码如例3-31所示,如果上传的文件能够经过过滤,进入文件上传成功页面(fileUpSuccess.jsp),代码如例3-30所示。还需要配置web.xml,代码如例1-3所示;配置struts.xml文件,代码如例3-32所示,该配置文件中使用了Struts2框架提供的fileUpload拦截器,而且使用了国际化资源文件;修改3.1.2节中用到的国际化资源文件globalMessages_GBK.properties,代码如例3-33所示,修改过国际化资源文件后,使用native2ascii工具进行编码,将globalMessages_GBK.properties文件编码成为globalMessages_zh_CN.properties,可参考3.1.1节。项目的文件结构如图3-30所示。

2. 在 web.xml 中配置核心控制器 FilterDispatcher

参考 1.3.1 节中的例 1-3。

3. 编写视图组件（JSP 页面）

上传文件页面如图 3-31 所示,代码如例 3-29 所示。上传成功的页面代码如例 3-30 所示。

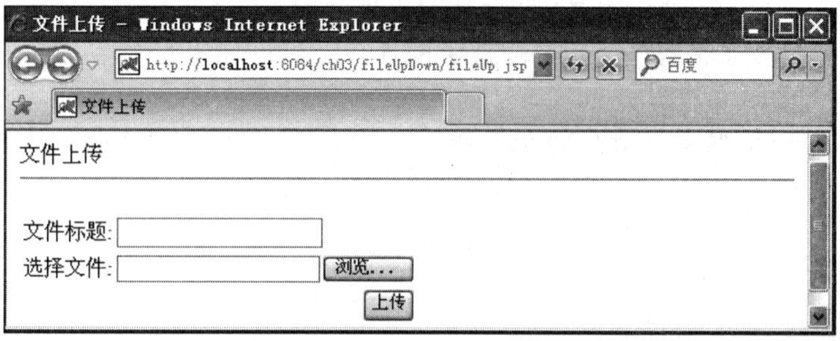

图 3-31 文件上传页面

【例 3-29】 文件上传页面（fileUp.jsp）。

```
<%@page contentType="text/html; charset=UTF-8" %>
<%@taglib prefix="s" uri="/struts-tags" %>
<html>
    <head>
        <title>文件上传</title>
    </head>
    <body>
        文件上传
        <hr/>
        <!--enctype 设置为 multipart/form-data,该属性用来设置浏览器采用二进制的方式
        来处理表单数据,上传文件时需要使用该属性-->
        <s:form action="upLoad"  enctype="multipart/form-data">
            <s:textfield name="title" label="文件标题"/><br/>
            <!--Struts2 使用拦截器 fileUpload 显示国际化信息.这里不必使用 key 值,但是
            在 3.1.2 节中需要使用-->
            <s:file name="upload" label="选择文件" /><br/>
            <s:submit value="上传"/>
        </s:form>
    </body>
</html>
```

【例 3-30】 文件上传成功页面（fileUpSuccess.jsp）。

```
<%@page contentType="text/html; charset=UTF-8" %>
<%@taglib prefix="s" uri="/struts-tags" %>
<html>
    <head>
        <title>文件上传成功</title>
```

```
    </head>
    <body>
        <h3>文件上传成功</h3>
        <hr/>
        文件标题:<s:property value="+title"/><br/>
        <s:property value="uploadFileName"/><br>
        <!--save 是在项目目录下创建的文件夹,用来保存上传的文件.上传后文件将被保存在
        Tomcat/webapps/ch03/save 目录下;在开发工具中使用时需在 ch03/Web/中新建一个文
        件夹 save,参考图 3-30-->
        <img src="<s:property value="'../save/'+uploadFileName"/>"/>
        <br/>
    </body>
</html>
```

4. 编写业务控制器 Action

文件上传页面对应的业务控制器类是 RequiredAction,代码如例 3-31 所示。

【例 3-31】 文件上传页面对应的业务控制器(UploadAction.java)。

```java
package fileUpDown;
import com.opensymphony.xwork2.ActionSupport;
import java.io.File;
import java.io.FileInputStream;
import java.io.FileOutputStream;
import org.apache.struts2.ServletActionContext;

public class UploadAction extends ActionSupport{
    //文件标题
    private String title;
    //上传文件对象
    private File upload;
    //上传文件名
    private String uploadFileName;
    //获取在 struts.xml 文件中配置的文件保存路径
    private String savePath;
    public void setTitle(String title) {
        this.title =title;
    }
    public String getTitle(){
        return this.title;
    }
    public void setUpload(File upload){
        this.upload =upload;
    }
    public File getUpload() {
        return this.upload;
    }
```

```java
    public void setUploadFileName(String uploadFileName) {
        this.uploadFileName = uploadFileName;
    }
    public String getUploadFileName(){
        return this.uploadFileName;
    }
    public void setSavePath(String value){
        this.savePath = value;
    }
    private String getSavePath() throws Exception {
        return ServletActionContext.getServletContext().getRealPath(savePath);
    }
    public String execute() throws Exception{
        //以服务器的文件保存地址和原文件名建立上传文件输出流
        FileOutputStream fos = new FileOutputStream(getSavePath()
            +"\\" +getUploadFileName());
        //定义输出流对象
        FileInputStream fis = new FileInputStream(getUpload());
        byte[] buffer = new byte[1024];
        int len = 0;
        while ((len = fis.read(buffer))>0){
            fos.write(buffer , 0 , len);
        }
        fos.close();
        return SUCCESS;
    }
}
```

5. 修改 struts.xml 配置 Action

修改配置文件 struts.xml,代码如例 3-32 所示。

【例 3-32】 在 struts.xml 配置 Action(struts.xml)。

```xml
<!DOCTYPE struts PUBLIC
    "-//Apache Software Foundation//DTD Struts Configuration 2.0//EN"
    "http://struts.apache.org/dtds/struts-2.0.dtd">
<struts>
    <constant name="struts.custom.i18n.resources" value="globalMessages" />
    <constant name="struts.i18n.encoding" value="utf-8" />
    <package name="I18N" extends="struts-default">
        <interceptors>
            <!--文字过滤拦截器配置,replace 是拦截器的名字-->
            <interceptor name="replace" class="interceptor.MyInterceptor" />
        </interceptors>
        <action name="checkLogin" class="loginAction.LoginAction">
            <result name="success">/I18N/loginSuccess.jsp</result>
            <result name="error">/I18N/login.jsp</result>
```

```xml
</action>
<!--文字过滤 Action 配置-->
<action name="public" class="interceptor.PublicAction">
    <result name="success">/interceptor/success.jsp</result>
    <result name="login">/interceptor/success.jsp</result>
    <!--Struts2 系统默认拦截器-->
    <interceptor-ref name="defaultStack" />
    <!--使用自定义拦截器-->
    <interceptor-ref name="replace"/>
</action>
<action name="register" class="validate.RegistAction">
    <result name="input">/validate/register.jsp</result>
    <result name="success">/validate/success.jsp</result>
</action>
<action name="register1" class="validate.RegistAction1">
    <result name="input">/validate/register1.jsp</result>
    <result name="success">/validate/success.jsp</result>
</action>
 <action name="register2" class="validatorAction.Register2Action">
    <result name="input">/validate/register2.jsp</result>
    <result name="success">/validate/success1.jsp</result>
</action>
<action name="upLoad" class="fileUpDown.UploadAction">
    <!--fileUpload拦截器配置-->
    <interceptor-ref name="fileUpload">
        <!--设置上传文件的最大字节数-->
        <param name="maximumSize">10000000</param>
        <!--设置上传文件的类型-->
        <param name="allowedTypes">
            image/gif,image/png,image/jpeg,image/jpg,image/pjpeg
        </param>
    </interceptor-ref>
    <interceptor-ref name="defaultStack" />
    <!--设置上传文件保存的文件夹-->
    <param name="savePath">./save</param>
    <result name="input">/fileUpDown/fileUp.jsp</result>
    <result name="success">/fileUpDown/fileUpSuccess.jsp</result>
</action>
</package>
</struts>
```

6. 编写国际化资源文件

修改 3.1.2 节中用到的国际化资源文件 globalMessages_GBK.properties,代码如例 3-33 所示。修改后需要重新编译为 globalMessages_zh_CN.properties,原有同名文件应先删除。

【例 3-33】 国际化资源文件（globalMessages_GBK.properties）。

```
loginTitle=用户登录
loginName=用户名称
loginPassword=用户密码
loginSubmit=登录
struts.messages.error.content.type.not.allowed=你只能上传图片资料,请重新上传!
struts.messages.error.file.too.large=你上传的文件太大,请重新上传!
```

7. 项目部署和运行

项目部署后,页面运行效果如图 3-31 所示。在图 3-31 所示页面中单击"浏览…"按钮将弹出"选择文件"对话框,可在该对话框中查找要上传的文件,如图 3-32 所示。

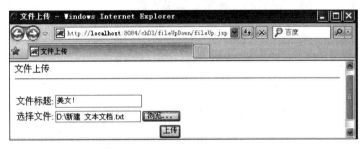

图 3-32　选择上传的文件

如果上传的文件不符合拦截器中配置的要求,将使用国际化资源在页面上提示,提示信息如图 3-33 所示。

图 3-33　国际化信息提示

如果上传的文件符合要求,页面跳转到文件上传成功页面,并把标题、文件名和图片显示出来,如图 3-34 所示。

3.4.2　文件下载

要在 Struts2 中实现文件下载,需要在 struts.xml 配置文件中先配置用于下载的拦截器 download,然后配置＜result name="success" type="stream"＞中 stream 的参数值。stream 的常用参数如下所示。

(1) contentType：用于指定下载文件的类型,该文件类型应与互联网 MIME 标准中的规定一致,如 text/xml 表示 XML 类型的文件,text/gif 表示 gif 图片,text/plain 表示纯文

图 3-34　文件上传成功页面

本类型。

（2）inputName：用于指定下载文件的输入流入口，在 Action 中需要指定该输入流入口。如果在 Action 中声明的是 getInputStream()方法，应在配置文件 struts.xml 中配置为＜param name＝"inputName"＞inputStream＜/param＞；如果在 Action 中声明的是 getTargetFile()方法，应在配置文件 struts.xml 中配置为＜param name＝"inputName"＞targetFile＜/param＞。

（3）contentDisposition：用于指定文件下载的处理方式，有内联（Inline）和附件（Attachment）两种方式。内联方式表示浏览器会尝试直接显示文件，附件方式会弹出文件保存对话框，默认值为内联。

（4）bufferSize：用于设置下载文件时的缓存大小。

文件在下载时也可以进行权限控制，例如，如果用户没有登录就不能下载，需要先登录后下载。

下面通过一个项目来理解在 Struts2 框架中是怎样实现文件下载功能的。

1. 项目介绍

本项目实现文件下载功能。有一个下载文件页面（fileDown.jsp），代码如例 3-34 所示，下载页面对应的业务控制器类为 FileDownload，该 Action 中覆盖了 getInputStream()方法，代码如例 3-35 所示。还需要配置 web.xml，代码如例 1-3 所示；配置 struts.xml，代码如例 3-36 所示，该配置文件中使用了 Struts2 框架提供的 download 拦截器。项目的文件结构如图 3-35 所示。

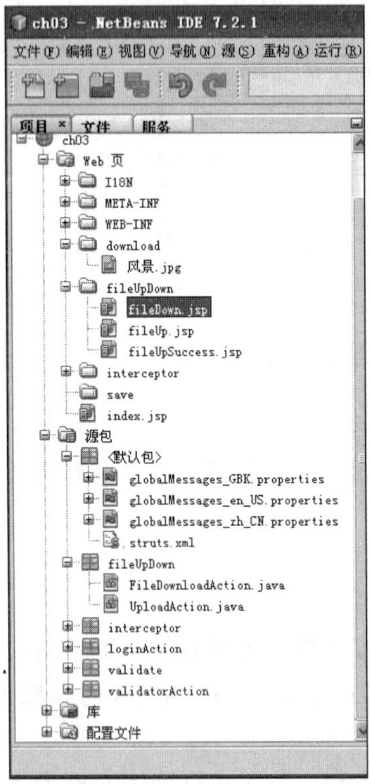

图 3-35　项目文件结构图

2. 在 web.xml 中配置核心控制器 FilterDispatcher

参考 1.3.1 节中的例 1-3。

3. 编写视图组件(JSP 页面)

下载文件页面效果如图 3-36 所示,代码如例 3-34 所示。

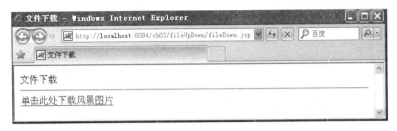

图 3-36 文件下载页面

【例 3-34】 文件下载页面(fileDown.jsp)。

```jsp
<%@ page contentType="text/html; charset=UTF-8" %>
<%@ taglib prefix="s" uri="/struts-tags" %>
<html>
    <head>
        <title>文件下载</title>
    </head>
    <body>
        文件下载
        <hr/>
        <a href="fileDownload.action">单击此处下载风景图片</a>
    </body>
</html>
```

4. 编写业务控制器 Action

文件下载页面对应的业务控制器是 FileDownload,代码如例 3-35 所示。

【例 3-35】 文件下载页面对应的业务控制器(FileDownload.java)。

```java
package fileUpDown;
import java.io.InputStream;
import org.apache.struts2.ServletActionContext;
import com.opensymphony.xwork2.ActionSupport;

public class FileDownload extends ActionSupport {
    //指定文件的下载路径
    private String path;
    public String getPath() {
        return path;
    }
    public void setPath(String path) {
        this.path = path;
    }
    //该方法返回一个 InputStream 类型的输入流,是下载目标文件的入口
```

```java
    public InputStream getInputStream() throws Exception {
        return ServletActionContext.getServletContext().getResourceAsStream(path);
    }
    public String execute() throws Exception {
        return SUCCESS;
    }
}
```

5. 修改 struts.xml 配置 Action

修改配置文件 struts.xml,代码如例 3-36 所示。

【例 3-36】 在 struts.xml 配置 Action(struts.xml)。

```xml
<!DOCTYPE struts PUBLIC
    "-//Apache Software Foundation//DTD Struts Configuration 2.0//EN"
    "http://struts.apache.org/dtds/struts-2.0.dtd">
<struts>
    <constant name="struts.custom.i18n.resources" value="globalMessages" />
    <constant name="struts.i18n.encoding" value="utf-8" />
    <package name="I18N" extends="struts-default">
        <interceptors>
            <!--文字过滤拦截器配置,replace 是拦截器的名字-->
            <interceptor name="replace" class="interceptor.MyInterceptor" />
        </interceptors>
        <action name="checkLogin" class="loginAction.LoginAction">
            <result name="success">/I18N/loginSuccess.jsp</result>
            <result name="error">/I18N/login.jsp</result>
        </action>
        <!--文字过滤 Action 配置-->
        <action name="public" class="interceptor.PublicAction">
            <result name="success">/interceptor/success.jsp</result>
            <result name="login">/interceptor/success.jsp</result>
            <!--Struts2 系统默认拦截器-->
            <interceptor-ref name="defaultStack" />
            <!--使用自定义拦截器-->
            <interceptor-ref name="replace"/>
        </action>
        <action name="register" class="validate.RegistAction">
            <result name="input">/validate/register.jsp</result>
            <result name="success">/validate/success.jsp</result>
        </action>
        <action name="register1" class="validate.RegistAction1">
            <result name="input">/validate/register1.jsp</result>
            <result name="success">/validate/success.jsp</result>
```

```xml
        </action>
         <action name="register2" class="validatorAction.Register2Action">
            <result name="input">/validate/register2.jsp</result>
            <result name="success">/validate/success1.jsp</result>
        </action>
        <action name="upLoad" class="fileUpDown.UploadAction">
            <!--fileUpload拦截器配置-->
            <interceptor-ref name="fileUpload">
                <!--设置上传文件的最大字节数-->
                <param name="maximumSize">10000000</param>
                <!--设置上传文件的类型-->
                <param name="allowedTypes">
                    image/gif,image/png,image/jpeg,image/jpg,image/pjpeg
                </param>
            </interceptor-ref>
            <interceptor-ref name="defaultStack" />
            <!--设置上传文件保存的文件夹-->
            <param name="savePath">./save</param>
            <result name="input">/fileUpDown/fileUp.jsp</result>
            <result name="success">/fileUpDown/fileUpSuccess.jsp</result>
        </action>
        <action name="fileDownload" class="fileUpDown.FileDownloadAction">
            <!--设置文件所在的位置,需要在项目中添加一个名为download的文件夹,位置参
            考图3-35,该文件夹中有一个风景.jpg文件-->
            <param name="path">/download/风景.jpg</param>
            <!--设置stream属性-->
            <result name="success" type="stream">
                <!--设置stream对应的参数-->
                <param name="contentType">image/jpg</param>
                <param name="inputName">inputStream</param>
                <param name="contentDisposition">attachment;
                    filename="hlm.jpg"</param>
                <param name="bufferSize">40960</param>
            </result>
        </action>
    </package>
</struts>
```

6. 项目部署和运行

项目部署后,运行效果如图3-36所示,单击"单击此处下载风景图片"将弹出如图3-37所示的文件下载页面,单击"保存"按钮,将弹出如图3-38所示对话框,选择要保存的路径。单击图3-37中的"打开"按钮将打开文件,如图3-39所示。

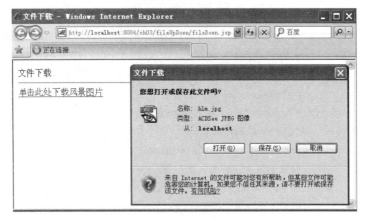

图 3-37 文件下载

图 3-38 "另存为"对话框

图 3-39 打开文件

3.5 本章小结

本章主要介绍了 Struts2 的高级组件,通过对高级组件的学习,能够为今后的基于 Struts2 项目开发提供更多的帮助,也为第 4 章的基于 Struts2 项目开发奠定基础。

通过本章的学习应了解和掌握以下内容。

(1) Struts2 的国际化及其应用。
(2) Struts2 的拦截器及其应用。
(3) Struts2 的输入校验及其应用。
(4) Struts2 的文件上传和下载及其应用。

3.6 习　　题

3.6.1 选择题

1. Struts2 中根据用户语言环境在页面显示不同语言的是(　　)。
   A. 国际化　　　　　B. 输入验证　　　C. 文件上传　　　D. 文件下载
2. 加载国际化资源文件时使用的拦截器是(　　)。
   A. I18N　　　　　　B. fileUpload　　　C. download　　　D. params
3. 加载文件上传时使用的拦截器是(　　)。
   A. I18N　　　　　　B. fileUpload　　　C. download　　　D. params
4. 加载文件下载时使用的拦截器是(　　)。
   A. I18N　　　　　　B. fileUpload　　　C. download　　　D. params
5. Struts2 框架中的抽象拦截器类是(　　)。
   A. Interceptor　　　　　　　　　　　B. FileUploadInterceptor
   C. AbstractInterceptor　　　　　　　D. DownloadInterceptor

3.6.2 填空题

1. Struts2 国际化资源文件的后缀是_____。
2. 编译 Struts2 的资源文件使用的工具是_____。
3. 在 Struts2 框架中,拦截器的设计思路来源于_____。
4. Struts2 框架中,对用户输入数据的校验分为两种:客户端校验和_____。

3.6.3 简答题

1. 什么是国际化,为什么使用国际化?
2. 简述 Struts2 中实现国际化的过程。
3. 什么是拦截器,拦截器的作用是什么?
4. 简述 Struts2 中输入校验的作用。

3.6.4 实训题

1. 使用自定义拦截器实现登录系统的权限控制,即如果输入的名字为空,提示"没有登录名,请输入登录名"。
2. 使用重写 validateXxx() 的方法来实现 3.3.2 节中的实例功能。
3. 编写一个拦截器实现对文件下载权限的控制。

第4章 基于Struts2的个人信息管理系统项目实训

本章实训是对前3章所学Struts2框架知识的综合应用,通过本项目的学习应能够熟练运用Struts2框架的基础知识开发Java Web项目。

4.1 项目需求说明

在日常办公中有许多常用的个人数据,如朋友电话、邮件地址、日程安排、日常记事、个人文件等,这些都可以用一个信息管理系统进行管理。个人信息管理系统也可以内置于掌上数字助理器中,以提供电子名片、便条、行程管理等功能。本实训项目基于B/S设计,也可以发布到网上,用户可以随时存取个人信息。

用户可以在系统中任意添加、修改、删除个人数据,包括个人信息管理、通讯录管理、日程安排管理、个人文件管理。

要实现的功能包括5个方面。

(1) 登录与注册。系统的登录和注册功能。

(2) 个人基本信息管理模块。系统中对个人基本信息的管理包括用户的姓名、性别、出生日期、民族、学历、职称、登录名、密码、电话、家庭住址等。

(3) 用户通讯录模块。系统的用户通讯录用于保存用户的通讯录信息,包括联系人的姓名、电话、邮箱、工作单位、地址、QQ等。可以自由添加联系人的信息,查询或删除联系人。

(4) 日程安排模块。日程模块记录用户的活动安排或者其他有关事项,如添加从某一时间到另一时间要做什么事、日程标题、内容、开始时间、结束时间。可以自由查询、修改、删除。

(5) 个人文件管理模块。该模块实现用户在网上存储临时文件的功能。用户可以新建文件夹、修改、删除、移动文件夹;上传文件、修改文件名、下载文件、删除文件、移动文件等。

4.2 项目系统分析

系统功能描述如下。

1. 用户登录与注册

通过用户名和密码登录系统;注册信息包含个人的信息。

2. 查看个人信息

主页面显示个人基本信息:用户登录名、用户密码、用户姓名、用户性别、出生日期、用户民族、用户学历、用户职称、用户电话、用户住址、用户邮箱等。

3. 修改个人信息

用户可以修改个人的基本信息。如果修改了登录名,下次登录时应使用新的登录名。

4. 修改登录密码

用户可以修改登录密码。

5. 查看通讯录

用户可以浏览通讯录列表,按照姓名检索等。

6. 维护通讯录

用户可以增加、修改、删除联系人。

7. 查看日程安排

用户可以查看日程安排列表,可以查看某一日程的内容、时间等。

8. 维护日程

一个新的日程安排包括日程标题、内容。用户可以对日程执行添加、修改、删除等操作。

9. 浏览下载文件

用户可以任意浏览文件、文件夹,并可以下载文件到本地。

10. 维护文件

用户可以新建文件夹,修改、删除、移动文件夹,移动文件到文件夹,修改文件名、上传文件、下载文件、删除文件等。

系统模块结构如图 4-1 所示。

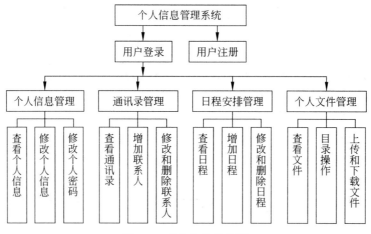

图 4-1 系统模块结构图

4.3 项目数据库设计

如果已经学过相应的 DBMS,请按照数据库优化的思想自行设计相应的表结构。本系统提供的表设计仅供参考,读者可根据自己所学知识选择相应 DBMS 对表进行设计和优化。本项目在数据库中可建立如下表,用于存放相关信息。

用户表(user)用于管理 index.jsp 页面中用户登录的信息以及用户注册(register.jsp)的信息。具体表设计如表 4-1 所示。

联系人表(friends)用于管理通讯录,即管理联系人(好友)。具体表设计如表 4-2 所示。

表 4-1 用户表(user)

字段名称	字段类型	字段长度	字段说明
userName	varchar	30	用户登录名
password	varchar	30	用户登录密码
name	varchar	30	用户真实姓名
sex	varchar	2	用户性别
birth	varchar	10	出生日期
nation	varchar	10	用户民族
edu	varchar	10	用户学历
work	varchar	30	用户职称
phone	varchar	10	用户电话
place	varchar	30	用户住址
email	varchar	30	用户邮箱

表 4-2 联系人表(friends)

字段名称	字段类型	字段长度	字段说明
userName	varchar	30	用户登录名
name	varchar	30	好友名称
phone	varchar	10	好友电话
email	varchar	30	好友邮箱
workplace	varchar	30	好友工作单位
place	varchar	30	好友住址
QQ	varchar	10	好友 QQ 号

备注：表 friends 中的用户登录名字段 userName 用于关联用户的好友信息列表。
日程安排管理表(date)用于管理用户的日程安排。具体表设计如表 4-3 所示。

表 4-3 日程安排管理表(date)

字段名称	字段类型	字段长度	字段说明
userName	varchar	30	用户登录名
date	varchar	30	日程时间
thing	varchar	255	日程内容

备注：表 date 中的用户登录名字段 userName 用于关联用户的日程信息。
个人文件管理表(file)用于管理个人文件管理。具体表设计如表 4-4 所示。

表 4-4 个人文件管理表（file）

字段名称	字段类型	字段长度	字段说明
userName	varchar	30	用户登录名
title	varchar	30	文件标题
name	varchar	30	文件名字
contentType	varchar	30	文件类型
size	varchar	30	文件大小
filePath	varchar	30	用户操作

备注：表 file 中的用户登录名字段 userName 用于关联用户的文件管理信息。

该项目使用 MySQL 数据库系统，如果需要使用该数据库可在 www.oracle.com 下载。读者也可以选择自己熟悉的数据库。安装完 MySQL 以后，可以再安装一个 MySQL 的工具，该工具能够在使用 MySQL 时提供可视化、友好的图形用户界面。该项目数据库名为 personMessage，该数据库中的表包括 date、file、friends 和 user，如图 4-2 所示。在项目"库"中添加 MySQL 所需驱动，本书使用的是 MySQL 5.0，在"库"中添加 5.0 的驱动 mysql-connector-java-5.0.6.jar（如果使用 MySQL 5.1，或者其他版本的 MySQL，就需要添加和该版本对应的驱动）。

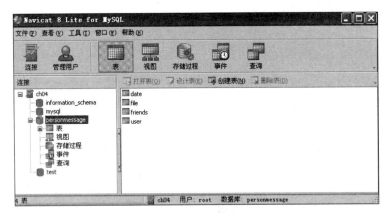

图 4-2 项目中数据库和表

4.4 项目实现

4.4.1 项目文件结构

项目的页面文件结构如图 4-3 所示。项目的源包文件结构如图 4-4 所示。

登录页面（index.jsp）和注册页面（register.jsp）在文件夹 login 中，该文件夹中页面对应的业务控制器 Action 在源包 edu.login.Action 中，对应的 Action 分别为 loginAction 类和 registerAction 类；登录和注册时需要连接数据库，对数据库的操作封装到 DBJavaBean

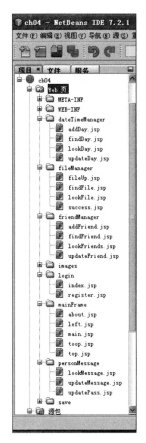

图 4-3 项目的页面文件结构图　　　　图 4-4 项目的源包文件结构图

包中的 DB 类中,该类提供了项目中数据库操作需要的所有方法。

图 4-3 所示文件结构中,dateTimeManager 文件夹中的页面是日程安排管理功能相关页面,其对应的 Action 在图 4-4 中 edu.dateTimeManager.Action 包里。fileManager 文件夹中的页面是个人文件管理功能相关页面,其对应的 Action 在 edu.fileManager.Action 包里。friendManager 文件夹中的页面是通讯录管理功能用到的页面,其对应的 Action 在 edu.friendManager.Action 包里。images 文件夹中保存项目中用到的图片。mainFrame 文件夹中的页面是主页面相关页面。personMessage 文件夹中的页面是个人信息管理功能相关页面,其对应的 Action 在 edu.personManager.Action 包里。另外,源包 JavaBean 包中的 5 个类封装了修改个人信息、修改个人密码、修删联系人、修改日程、文件下载用到的数据,如果不使用这几个类,可以在 JSP 页面中使用<s:action>直接调用 Action 类中保存的数据,一般不建议在 JSP 页面中调用 Action,所以本项目使用 JavaBean 在 JSP 页面上调用数据。

4.4.2 用户登录和注册功能的实现

本系统提供登录界面,如果用户没有注册,需要先注册后登录。登录页面如图 4-5 所示。

图 4-5 系统登录页面

登录页面(index.jsp)代码如下：

```
<%@page contentType="text/html" pageEncoding="UTF-8"%>
<%@taglib prefix="s" uri="/struts-tags" %>
<html>
    <head>
        <meta http-equiv="Content-Type" content="text/html; charset=UTF-8">
        <title><s:text name="个人信息管理系统"/></title>
    </head>
    <body bgcolor="#CCCCFF">
        <s:form action="loginAction" method="post">
            <table align="center" width="100%">
                <tr>
                    <td align="right" width="50%">
                      <img src="../images/cc.gif" alt="为之则易,不为则难！"
                        height="80"/>
                    </td>
                    <td align="left" width="50%">
                        <h1>个人信息管理系统</h1>
                    </td>
                </tr>
                <tr>
                    <td colspan="2">
                        <hr align="center" width="100%" size="20"
                          color="green"/>
                    </td>
                </tr>
                <tr>
```

```
            <td width="30%" align="center">
                <image src="../images/a.jpg" alt="长城" height="280"/>
            </td>
            <td width="70%">
                <table border="5" align="center" bgcolor="#99aadd">
                    <tr>
                        <td>
                            <s:textfield name="userName"
                                label="登录名" size="16"/>
                        </td>
                    </tr>
                    <tr>
                        <td>
                            <s:password name="password"
                                label="登录密码" size="18"/>
                        </td>
                    </tr>
                    <tr>
                        <td colspan="2" align="center">
                            <input type="submit" value="确定"/>
                                 
                            <input type="reset" value="清空"/>
                        </td>
                    </tr>
                    <tr>
                        <td colspan="2" align="center">
                            <s:a href="http://localhost:8084/ch04
                                /login/register.jsp">注册</s:a>
                        </td>
                    </tr>
                </table>
            </td>
        </tr>
    </table>
    </s:form>
    </body>
</html>
```

单击图4-5所示页面中的"注册"按钮,出现如图4-6所示的注册页面。
用户需先注册后登录,注册页面(register.jsp)代码如下:

```
<%@page contentType="text/html" pageEncoding="UTF-8"%>
<%@taglib prefix="s" uri="/struts-tags" %>
<html>
    <head>
        <meta http-equiv="Content-Type" content="text/html; charset=UTF-8">
```

图 4-6 注册页面

```
<title><s:text name="个人信息管理系统->注册"/></title>
</head>
<body bgcolor="#CCCCFF">
    <s:form action="registerAction" method="post">
        <table align="center">
            <tr>


                <td width="40%">
                    <table border="2" bgcolor="#AABBCCDD"
                        width="100%" align="center">
                        <tr>
                            <td colspan="2" align="center">
                                <font color="yellow">
                                    <s:text name="请填写以下注册信息"/>
                                </font>
                            </td>
                        </tr>
                        <tr>
                            <td>
                                <s:textfield name="userName"
                                    label="登录名"/>
                            </td>
                        </tr>
                        <tr>
                            <td>
```

```
                <s:password name="password1"
                    label="密码" size="21"/>
            </td>
        </tr>
        <tr>
            <td>
                <s:password name="password2"
                    label="再次输入密码" size="21"/>
            </td>
        </tr>
        <tr>
            <td>
                <s:textfield name="name"
                    label="用户真实姓名"/>
            </td>
        </tr>
        <tr>
            <td>
                <s:text name="用户性别:"></s:text>
            </td>
            <td>
                <input type="radio" name="sex"
                    value="男" checked/>男
                <input type="radio" name="sex"
                    value="女"/>女
            </td>
        </tr>
        <tr>
            <td>
                <s:textfield name="birth" label="出生日期"/>
            </td>
        </tr>
        <tr>
            <td>
                <s:textfield name="nation"
                    label="用户民族"/>
            </td>
        </tr>
        <tr>
            <td>
                <s:select name="edu" label="用户学历"
                    headerValue="-----请选择-------"
                    headerKey="1" list="{'博士',' 硕士','本科',
                    '专科','高中','初中','小学','其他'}">
                </s:select>
```

```
                            </td>
                        </tr>
                        <tr>
                            <td>
                                <s:select name="work" label="用户职称"
                                    headerValue="--------请选择--------"
                                    headerKey="1" list="{'软件测试工程师',
                                    '软件开发工程师','教师','学生','职员','经理',
                                    '老板','公务员','其他'}">
                                </s:select>
                            </td>
                        </tr>
                        <tr>
                            <td>
                                <s:textfield name="phone"
                                    label="用户电话"/>
                            </td>
                        </tr>
                        <tr>
                            <td>
                                <s:textfield name="place"
                                    label="用户住址"/>
                            </td>
                        </tr>
                        <tr>
                            <td>
                                <s:textfield name="email"
                                    label="用户邮箱"/>
                            </td>
                        </tr>
                        <tr>
                            <td colspan="2" align="center">
                                <input type="submit" value="确定"/>
                                     
                                <input type="reset" value="清空"/>
                                     
                                <s:a href="http://localhost:8084/ch04/
                                    login/index.jsp">返回</s:a>
                            </td>
                        </tr>
                    </table>
                </td>
            </tr>
        </table>
    </s:form>
```

```
        </body>
</html>
```

登录页面对应的业务控制器类是 LoginAction，注册页面对应的业务控制器类是 RegisterAction。

LoginAction.java 的代码如下：

```
package edu.login.Action;
import DBJavaBean.DB;
import com.opensymphony.xwork2.ActionSupport;
import java.sql.*;
import javax.servlet.http.HttpServletRequest;
import org.apache.struts2.interceptor.ServletRequestAware;

public class LoginAction extends ActionSupport implements ServletRequestAware{
    private String userName;
    private String password;
    private ResultSet rs=null;
    private String message=ERROR;
    private HttpServletRequest request;
    public String getUserName() {
        return userName;
    }
    public void setUserName(String userName) {
        this.userName=userName;
    }
    public String getPassword() {
        return password;
    }
    public void setPassword(String password) {
        this.password=password;
    }
    public void setServletRequest(HttpServletRequest hsr) {
        request=hsr;
    }
    public void validate(){
        if(this.getUserName()==null||this.getUserName().length()==0){
            addFieldError("username","请输入登录名字!");
        }else{
            try{
                DB mysql=new DB();
                rs=mysql.selectMess(request, this.getUserName());
                if(!rs.next()){
                    addFieldError("username","此用户尚未注册!");
                }
            }catch(Exception e){
```

```java
                e.printStackTrace();
            }
        }
        if(this.getPassword()==null||this.getPassword().length()==0){
            addFieldError("password","请输入登录密码!");
        }else{
            try{
                DB mysql=new DB();
                rs=mysql.selectMess(request, this.getUserName());
                if(rs.next()){
                    rs=mysql.selectLogin(request, this.getUserName(),
                                    this.getPassword());
                    if(!rs.next()){
                        addFieldError("password","登录密码错误!");
                    }
                }
            }catch(Exception e){
                e.printStackTrace();
            }
        }
    }
    public String execute() throws Exception {
        //实例化对数据库操作的封装类
        DB mysql=new DB();
        //调用 DB 类中的方法,实现登录有关操作
        String add=mysql.addList(request, this.getUserName());
        if(add.equals("ok")){
            message=SUCCESS;
        }
        return message;
    }
}
```

RegisterAction.java 的代码如下:

```java
package edu.login.Action;
import DBJavaBean.DB;
import com.opensymphony.xwork2.ActionSupport;
import java.sql.ResultSet;
import java.sql.SQLException;
import javax.servlet.http.HttpServletRequest;
import org.apache.struts2.interceptor.ServletRequestAware;

public class RegisterAction extends ActionSupport implements ServletRequestAware{
    private String userName;
    private String password1;
```

```java
    private String password2;
    private String name;
    private String sex;
    private String birth;
    private String nation;
    private String edu;
    private String work;
    private String phone;
    private String place;
    private String email;
    private ResultSet rs=null;
    private String message="error";
    private HttpServletRequest request;
public String getUserName() {
        return userName;
    }
    public void setUserName(String userName) {
        this.userName=userName;
    }
    public String getPassword1() {
        return password1;
    }
    public void setPassword1(String password1) {
        this.password1=password1;
    }
    public String getPassword2() {
        return password2;
    }
    public void setPassword2(String password2) {
        this.password2=password2;
    }
    public String getName() {
        return name;
    }
    public void setName(String name) {
        this.name=name;
    }
    public String getSex() {
        return sex;
    }
    public void setSex(String sex) {
        this.sex=sex;
    }
    public String getBirth() {
        return birth;
```

```java
    }
    public void setBirth(String birth) {
        this.birth=birth;
    }
    public String getNation() {
        return nation;
    }
    public void setNation(String nation) {
        this.nation=nation;
    }
    public String getEdu() {
        return edu;
    }
    public void setEdu(String edu) {
        this.edu=edu;
    }
    public String getWork() {
        return work;
    }
    public void setWork(String work) {
        this.work=work;
    }
    public String getPhone() {
        return phone;
    }
    public void setPhone(String phone) {
        this.phone=phone;
    }
    public String getPlace() {
        return place;
    }
    public void setPlace(String place) {
        this.place=place;
    }
    public String getEmail() {
        return email;
    }
    public void setEmail(String email) {
        this.email=email;
    }
    public void setServletRequest(HttpServletRequest hsr) {
        request=hsr;
    }
    public void validate(){
        if(getUserName()==null||getUserName().length()==0){
```

```java
        addFieldError("userName","登录名字不允许为空!");
    }else{
        try {
            DB mysql=new DB();
            rs=mysql.selectMess(request, this.getUserName());
            if(rs.next()){
                addFieldError("userName","此登录名字已存在!");
            }
        } catch (SQLException ex) {
            ex.printStackTrace();
        }
    }
    if(getPassword1()==null||getPassword1().length()==0){
        addFieldError("password1","登录密码不允许为空!");
    }
    if(getPassword2()==null||getPassword2().length()==0){
        addFieldError("password2","重复密码不允许为空!");
    }
    if(!(getPassword1().equals(getPassword2()))){
        addFieldError("password2","两次密码不一致!");
    }
    if(getName()==null||getName().length()==0){
        addFieldError("name","用户姓名不允许为空!");
    }
    if(getBirth()==null||getBirth().length()==0||
        getBirth().equals("yyyy-mm-dd")){
        addFieldError("birth","用户生日不允许为空!");
    }else{
        if(getBirth().length()!=10){
            addFieldError("birth","用户生日格式为'yyyy-mm-dd'!");
        }else{
            String an=this.getBirth().substring(4, 5);
            String bn=this.getBirth().substring(7, 8);
            if(!(an.equals("-"))||!(bn.equals("-"))){
                addFieldError("birth","用户生日格式为'yyyy-mm-dd'!");
            }
        }
    }
    if(getNation()==null||getNation().length()==0){
        addFieldError("nation","用户民族不允许为空!");
    }
    if(getEdu().equals("1")){
        addFieldError("edu","请选择用户学历!");
    }
    if(getWork().equals("1")){
```

```
            addFieldError("work","请选择用户工作!");
        }
        if(getPhone()==null||getPhone().length()==0){
            addFieldError("phone","用户电话不允许为空!");
        }
        if(getPlace()==null||getPlace().length()==0){
            addFieldError("place","用户地址不允许为空!");
        }
        if(getEmail()==null||getEmail().length()==0){
            addFieldError("email","用户 email 不允许为空!");
        }
    }
    public String execute() throws Exception{
        DB mysql=new DB();
        String mess=mysql.insertMess(request, this.getUserName(),
                                    this.getPassword1(), this.getName(),
                                    this.getSex(), this.getBirth(),
                                    this.getNation(), this.getEdu(),
                                    this.getWork(), this.getPhone(),
                                    this.getPlace(), this.getEmail());
        if(mess.equals("ok")){
            message=SUCCESS;
        }else if(mess.equals("one")){
            message=INPUT;
        }
        return message;
    }
}
```

Action 需要在 struts.xml 中进行配置,项目中用到的配置文件 struts.xml 代码如下(该配置文件包含对项目中所有 Action 的配置):

```
<!DOCTYPE struts PUBLIC
"-//Apache Software Foundation//DTD Struts Configuration 2.0//EN"
"http://struts.apache.org/dtds/struts-2.0.dtd">
<struts>
    <include file="example.xml"/>
    <!--Configuration for the default package.-->
    <package name="default" extends="struts-default">
        <action name="loginAction" class="edu.login.Action.LoginAction">
            <result name="success">/mainFrame/main.jsp</result>
            <result name="input">/login/index.jsp</result>
            <result name="error">/login/index.jsp</result>
        </action>
        <action name="registerAction" class="edu.login.Action.RegisterAction">
            <result name=" success">/login/index.jsp</result>
```

```xml
        <result name="input">/login/register.jsp</result>
        <result name="error">/login/register.jsp</result>
</action>
<action name="upMessAction"
        class="edu.personManager.Action.UpdateMessAction">
        <result name=" success">/personMessage/lookMessage.jsp</result>
        <result name="input">/personMessage/updateMessage.jsp</result>
        <result name="error">/personMessage/updateMessage.jsp</result>
</action>
<action name="upPassAction"
        class="edu.personManager.Action.UpdatePassAction">
        <result name="success">/personMessage/lookMessage.jsp</result>
        <result name="input">/personMessage/updatePass.jsp</result>
</action>
<action name="addFriAction"
        class="edu.friendManager.Action.AddFriAction">
        <result name="success">/friendManager/lookFriends.jsp</result>
        <result name="input">/friendManager/addFriend.jsp</result>
</action>
<action name="findFriAction"
        class="edu.friendManager.Action.FindFriAction">
        <result name="success">/friendManager/findFriend.jsp</result>
        <result name="error">/friendManager/lookFriends.jsp</result>
        <result name="input">/friendManager/lookFriends.jsp</result>
</action>
<action name="upFriAction"
        class="edu.friendManager.Action.UpdateFriAction">
        <result name="success">/friendManager/lookFriends.jsp</result>
        <result name="input">/friendManager/updateFriend.jsp</result>
</action>
<action name="deleteFriAction"
        class="edu.friendManager.Action.DeleteFriAction">
        <result name="success">/friendManager/lookFriends.jsp</result>
</action>
<action name="addDayAction"
        class="edu.dateTimeManager.Action.AddDayAction">
        <result name="success">/dateTimeManager/lookDay.jsp</result>
        <result name="input">/dateTimeManager/addDay.jsp</result>
        <result name="error">/dateTimeManager/addDay.jsp</result>
</action>
<action name="findDayAction"
        class="edu.dateTimeManager.Action.FindDayAction">
        <result name="success">/dateTimeManager/findDay.jsp</result>
        <result name="input">/dateTimeManager/lookDay.jsp</result>
        <result name="error">/dateTimeManager/lookDay.jsp</result>
```

```xml
        </action>
        <action name="upDayAction"
            class="edu.dateTimeManager.Action.UpdateDayAction">
            <result name="success">/dateTimeManager/lookDay.jsp</result>
            <result name="input">/dateTimeManager/updateDay.jsp</result>
            <result name="error">/dateTimeManager/updateDay.jsp</result>
        </action>
        <action name="deleteDayAction"
            class="edu.dateTimeManager.Action.DeleteDayAction">
            <result name="success">/dateTimeManager/lookDay.jsp</result>
        </action>
        <action name="addFileAction"
            class="edu.fileManager.Action.AddFileAction">
            <interceptor-ref name="fileUpload">
                <param name="maximumSize">1024000000</param>
            </interceptor-ref>
            <interceptor-ref name="defaultStack"/>
                <param name="savePath">/save</param>
            <result name="success">/fileManager/success.jsp</result>
            <result name="input">/fileManager/fileUp.jsp</result>
            <result name="error">/fileManager/fileUp.jsp</result>
        </action>
        <action name="findFileAction"
            class="edu.fileManager.Action.FindFileAction">
            <result name="success">/fileManager/findFile.jsp</result>
            <result name="input">/fileManager/lookFile.jsp</result>
            <result name="error">/fileManager/lookFile.jsp</result>
        </action>
        <action name="deleteFileAction"
            class="edu.fileManager.Action.DeleteFileAction">
            <result name="success">/fileManager/lookFile.jsp</result>
            <result name="error">/fileManager/findFile.jsp</result>
        </action>
        <action name="downFileAction"
            class="edu.fileManager.Action.DownFileAction">
            <param name="path">/save/${downloadFileName}</param>
            <result name="success" type="stream">
                <param name="contentType">
                    application/octet-stream;charset=ISO8859-1
                </param>
                <param name="inputName">InputStream</param>
                <param name="contentDisposition">
                    attachment;filename="${downloadFileName}"
                </param>
                <param name="bufferSize">40960</param>
```

```
            </result>
        </action>
    </package>
</struts>
```

为了实现登录和注册,在登录和注册对应的 Action 中,都要用到 DB 类中的方法进行数据库连接并通过 DB 类中的方法对数据库中的数据进行操作,即 DB 类封装了项目中所有与数据库操作有关的方法。

DB.java 类的代码如下:

```
package DBJavaBean;
import JavaBean.UserNameBean;
import JavaBean.MyDayBean;
import JavaBean.MyFileBean;
import JavaBean.MyFriBean;
import JavaBean.MyMessBean;
import java.sql.*;
import java.util.ArrayList;
import javax.servlet.http.HttpServletRequest;
import javax.servlet.http.HttpSession;
import javax.swing.JOptionPane;
import org.apache.struts2.interceptor.ServletRequestAware;

//以 IoC 方式直接访问 Servlet,通过 request 获取 session 对象
public class DB implements ServletRequestAware{
    private String driverName="com.mysql.jdbc.Driver";
    /*url 后面的"?useUnicode=true&characterEncoding=gbk",可以处理向数据库中添加
       数据时出现的乱码问题*/
    private String url="jdbc:mysql://localhost:3306/personmessage?
    useUnicode=true&characterEncoding=gbk";
    private String user="root";
    private String password="root";
    private Connection con=null;
    private Statement st=null;
    private ResultSet rs=null;
    private HttpServletRequest request;
    public DB(){
    }
    public String getDriverName() {
        return driverName;
    }
    public void setDriverName(String driverName) {
        this.driverName=driverName;
    }
    public String getUrl() {
        return url;
```

```java
    }
    public void setUrl(String url) {
        this.url=url;
    }
    public String getUser() {
        return user;
    }
    public void setUser(String user) {
        this.user=user;
    }
    public String getPassword() {
        return password;
    }
    public void setPassword(String password) {
        this.password=password;
    }
    public void setServletRequest(HttpServletRequest hsr) {
        request=hsr;
    }
    //完成连接数据库操作,生成容器并返回
    public Statement getStatement(){
        try{
            Class.forName(getDriverName());
            con=DriverManager.getConnection(getUrl(), getUser(),
                                    getPassword());
            return con.createStatement();
        }catch(Exception e){
            e.printStackTrace();
            return null;
        }
    }
    //完成注册,把用户的注册信息录入到数据库中
    public String insertMess(HttpServletRequest request,String userName,
                    String password,String name,String sex,String birth,
                    String nation,String edu,String work,String phone,
                    String place,String email){
        try{
            String sure=null;
            rs=selectMess(request,userName);
            //判断用户名是否已存在,如果存在返回 one
            if(rs.next()){
                sure="one";
            }else{
                String sql="insert into user"
                    +"(userName,password,name,sex,birth,nation,edu,work,
```

```java
                    phone,place,email)"+"values("+"'"+userName+"'"+",
                    "+"'"+password+"'"+","+"'"+name+"'"+","+"'"+sex+
                    "'"+","+"'"+birth+"'"+","+"'"+nation+"'"+","+
                    "'"+edu+"'"+","+"'"+work+"'"+","+"'"+phone+"'"+",
                    "+"'"+place+"'"+","+"'"+email+"'"+")";
            st=getStatement();
            int row=st.executeUpdate(sql);
            if(row==1){
                //调用myMessage()方法,更新session中保存的用户信息
                String mess=myMessage(request,userName);
                if(mess.equals("ok")){
                    sure="ok";
                }else{
                    sure=null;
                }
            }else{
                sure=null;
            }
        }
        return sure;
    }catch(Exception e){
        e.printStackTrace();
        return null;
    }
}
//更新注册的个人信息
public String updateMess(HttpServletRequest request,String userName,
                    String name,String sex,String birth,String nation,
                    String edu,String work,String phone,String place,
                    String email){
    try{
        String sure=null;
        String sql="update user set name='"
                +name+"',sex='"+sex+"',birth='"+birth+"',nation=
                '"+nation+"',edu='"+edu+"',work='"+work+"',phone=
                '"+phone+"',place='"+place+"',email='"+email+
                "' where userName='"+userName+"'";
        st=getStatement();
        int row=st.executeUpdate(sql);
        if(row==1){
            //调用myMessage()方法,更新session中保存的用户信息
            String mess=myMessage(request,userName);
            if(mess.equals("ok")){
                sure="ok";
            }else{
```

```
                    sure=null;
                }
            }else{
                sure=null;
            }
            return sure;
        }catch(Exception e){
            e.printStackTrace();
            return null;
        }
    }
    //查询个人信息,并返回结果集 rs
    public ResultSet selectMess(HttpServletRequest request,String userName){
        try{
            String sql="select * from user where userName='"+userName+"'";
            st=getStatement();
            return st.executeQuery(sql);
        }catch(Exception e){
            e.printStackTrace();
            return null;
        }
    }
    //把个人信息通过 myMessBean 保存到 session 对象中
    public String myMessage(HttpServletRequest request,String userName){
        try{
            ArrayList listName=null;
            HttpSession session=request.getSession();
            listName=new ArrayList();
            rs=selectMess(request,userName);
            while(rs.next()){
                MyMessBean mess=new MyMessBean();
                mess.setName(rs.getString("name"));
                mess.setSex(rs.getString("sex"));
                mess.setBirth(rs.getString("birth"));
                mess.setNation(rs.getString("nation"));
                mess.setEdu(rs.getString("edu"));
                mess.setWork(rs.getString("work"));
                mess.setPhone(rs.getString("phone"));
                mess.setPlace(rs.getString("place"));
                mess.setEmail(rs.getString("email"));
                listName.add(mess);
                session.setAttribute("MyMess", listName);
            }
            return "ok";
        }catch(Exception e){
```

```java
            e.printStackTrace();
            return null;
        }
    }
    //添加联系人
    public String insertFri(HttpServletRequest request,String userName,
        String name,String phone,String email,String workplace,String place,
        String QQ){
        try{
            String sure=null;
            rs=selectFri(request,userName,name);
            //判断联系人姓名是否已存在
            if(rs.next()){
                sure="one";
            }else{
                String sql="insert into friends"+
                        "(userName,name,phone,email,workplace,place,QQ)"+
                        "values("+"'"+userName+"'"+","+"'"+name+"'"+","
                        +"'"+phone+"'"+","+"'"+email+"'"+","+"'"+
                        workplace+"'"+","+"'"+place+"'"+","+"'"+QQ+"'"+")";
                st=getStatement();
                int row=st.executeUpdate(sql);
                if(row==1){
                    //调用 myFridnds()方法,更新 session 中通讯录中的信息
                    String fri=myFriends(request,userName);
                    if(fri.equals("ok")){
                        sure="ok";
                    }else{
                        sure=null;
                    }
                }else{
                    sure=null;
                }
            }
            return sure;
        }catch(Exception e){
            e.printStackTrace();
            return null;
        }
    }
    //删除联系人
    public String deleteFri(HttpServletRequest request,String userName,
        String name){
        try{
            String sure=null;
```

```java
            String sql="delete from friends where userName='"+userName+"' and
                    name='"+name+"'";
            st=getStatement();
            int row=st.executeUpdate(sql);
            if(row==1){
                //调用myFridnds()方法,更新session中保存的通讯录中的信息
                String fri=myFriends(request,userName);
                if(fri.equals("ok")){
                    sure="ok";
                }else{
                    sure=null;
                }
            }else{
                sure=null;
            }
            return sure;
        }catch(Exception e){
            e.printStackTrace();
            return null;
        }
    }
    //修改联系人
    public String updateFri(HttpServletRequest request,String userName,
                        String friendName,String name,String phone,
                        String email,String workplace,String place,
                        String QQ){
        try{
            String sure=null;
            //先删除该联系人的信息
            String del=deleteFri(request,userName,friendName );
            if(del.equals("ok")){
                //重新录入修改后的信息
                String in=insertFri(request,userName,
                                name,phone,email,workplace,place,QQ);
                if(in.equals("ok")){
                    //调用myFridnds()方法,更新session中的通讯录信息
                    String fri=myFriends(request,userName);
                    if(fri.equals("ok")){
                        sure="ok";
                    }else{
                        sure=null;
                    }
                }else{
                    sure=null;
                }
```

```java
            }else{
                sure=null;
            }
            return sure;
        }catch(Exception e){
            e.printStackTrace();
            return null;
        }
    }
    //查询联系人
    public ResultSet selectFri(HttpServletRequest request,String userName,
            String name){
        try{
            String sql="select * from friends where userName='"+userName+"' and
                    name='"+name+"'";
            st=getStatement();
            return st.executeQuery(sql);
        }catch(Exception e){
            e.printStackTrace();
            return null;
        }
    }
    //获取通讯录中所有联系人的信息
    public ResultSet selectFriAll(HttpServletRequest request,String userName){
        try{
            String sql="select * from friends where userName='"+userName+"'";
            st=getStatement();
            return st.executeQuery(sql);
        }catch(Exception e){
            e.printStackTrace();
            return null;
        }
    }
    //获取通讯录中所有联系人的信息,并把它们保存到session对象中
    public String myFriends(HttpServletRequest request,String userName){
        try{
            ArrayList listName=null;
            HttpSession session=request.getSession();
            listName=new ArrayList();
            rs=selectFriAll(request,userName);
            if(rs.next()){
                rs=selectFriAll(request,userName);
                while(rs.next()){
                    MyFriBean mess=new MyFriBean();
                    mess.setName(rs.getString("name"));
```

```java
                    mess.setPhone(rs.getString("phone"));
                    mess.setEmail(rs.getString("email"));
                    mess.setWorkplace(rs.getString("workplace"));
                    mess.setPlace(rs.getString("place"));
                    mess.setQQ(rs.getString("QQ"));
                    listName.add(mess);
                    session.setAttribute("friends", listName);
                }
            }else{
                session.setAttribute("friends", listName);
            }
            return "ok";
        }catch(Exception e){
            e.printStackTrace();
            return null;
        }
    }
    //添加日程
    public String insertDay(HttpServletRequest request,String userName,String
        date,String thing){
        try{
            String sure=null;
            rs=selectDay(request,userName,date);
            //判断是否日程已有安排
            if(rs.next()){
                sure="one";
            }else{
                String sql="insert into date"+
                        "(userName,date,thing)"+"values ("+"'"+userName+
                        "'"+","+"'"+date+"'"+","+"'"+thing+"'"+")";
                st=getStatement();
                int row=st.executeUpdate(sql);
                if(row==1){
                    //调用myDayTime()方法,更新session对象中的日程信息
                    String day=myDayTime(request,userName);
                    if(day.equals("ok")){
                        sure="ok";
                    }else{
                        sure=null;
                    }
                }else{
                    sure=null;
                }
            }
            return sure;
```

```java
        }catch(Exception e){
            e.printStackTrace();
            return null;
        }
    }
    //删除日程
    public String deleteDay(HttpServletRequest request,String userName,String
            date){
        try{
            String sure=null;
            String sql="delete from date where userName='"+userName+"' and
                    date='"+date+"'";
            st=getStatement();
            int row=st.executeUpdate(sql);
            if(row==1){
                //调用myDayTime()方法,更新session对象中保存的日程信息
                String day=myDayTime(request,userName);
                if(day.equals("ok")){
                    sure="ok";
                }else{
                    sure=null;
                }
            }else{
                sure=null;
            }
            return sure;
        }catch(Exception e){
            e.printStackTrace();
            return null;
        }
    }
    //修改日程
    public String updateDay(HttpServletRequest request,String userName,String
            Day,String date,String thing){
        try{
            String sure=null;
            //先删除该日程
            String del=deleteDay(request,userName,Day);
            if(del.equals("ok")){
                //重新录入修改后的信息
                String in=insertDay(request,userName,date,thing);
                if(in.equals("ok")){
                    //调用myDayTime()方法,更新session对象中的日程信息
                    String day=myDayTime(request,userName);
                    if(day.equals("ok")){
```

```java
                    sure="ok";
                }else{
                    sure=null;
                }
            }else{
                sure=null;
            }
        }else{
            sure=null;
        }
        return sure;
    }catch(Exception e){
        e.printStackTrace();
        return null;
    }
}
//查询日程
public ResultSet selectDay(HttpServletRequest request,String userName,String
        date){
    try{
        String sql="select * from date where userName='"+userName+"' and
                date='"+date+"'";
        st=getStatement();
        return st.executeQuery(sql);
    }catch(Exception e){
        e.printStackTrace();
        return null;
    }
}
//查询所有的日程信息
public ResultSet selectDayAll(HttpServletRequest request,String userName){
    try{
        String sql="select * from date where userName='"+userName+"'";
        st=getStatement();
        return st.executeQuery(sql);
    }catch(Exception e){
        e.printStackTrace();
        return null;
    }
}
//查询所有的日程信息,并把它们保存到session对象中
public String myDayTime(HttpServletRequest request,String userName){
    try{
        ArrayList listName=null;
        HttpSession session=request.getSession();
```

```java
            listName=new ArrayList();
            rs=selectDayAll(request,userName);
            if(rs.next()){
                rs=selectDayAll(request,userName);
                while(rs.next()){
                    MyDayBean mess=new MyDayBean();
                    mess.setDay(rs.getString("date"));
                    mess.setThing(rs.getString("thing"));
                    listName.add(mess);
                    session.setAttribute("day", listName);
                }
            }else{
                session.setAttribute("day", listName);
            }
            return "ok";
        }catch(Exception e){
            e.printStackTrace();
            return null;
        }
    }
    //保存上传文件的信息
    public String insertFile(HttpServletRequest request,String userName,String
            title,String name,String contentType,String size,String filePath){
        try{
            String sure=null;
            //查询文件标题是否已存在
            rs=selectFile(request,userName,"title",title);
            if(rs.next()){
                sure="title";
            }else{
                //查询文件名是否已存在
                rs=selectFile(request,userName,"name",name);
                if(rs.next()){
                    sure="name";
                }else{
                    String sql="insert into file"+
                            "(userName,title,name,contentType,size,filePath)"
                            +"values("+"'"+userName+"'"+","+"'"+title+"'"
                            +","+"'"+name+"'"+","+"'"+contentType+"'"+",
                            "+"'"+size+"'"+","+"'"+filePath+"'"+")";
                    st=getStatement();
                    int row=st.executeUpdate(sql);
                    if(row==1){
                        //调用myFile()方法,更新session中保存的文件信息
                        String file=myFile(request,userName);
```

```java
                    if(file.equals("ok")){
                        sure="ok";
                    }else{
                        sure=null;
                    }
                }else{
                    sure=null;
                }
            }
            return sure;
        }catch(Exception e){
            e.printStackTrace();
            return null;
        }
    }
    //删除文件
    public String deleteFile(HttpServletRequest request,String userName,String
            title){
        try{
            String sure=null;
            String sql="delete from file where userName='"+userName+"' and
                title='"+title+"'";
            st=getStatement();
            int row=st.executeUpdate(sql);
            if(row==1){
                //调用 myFile()方法,更新 session 中保存的文件信息
                String file=myFile(request,userName);
                if(file.equals("ok")){
                    sure="ok";
                }else{
                    sure=null;
                }
            }else{
                sure=null;
            }
            return sure;
        }catch(Exception e){
            e.printStackTrace();
            return null;
        }
    }
    //修改文件
    public String updateFile(HttpServletRequest request,String userName,String
        Title,String title,String name,String contentType,String size,
```

```java
        String filePath){
    try{
        String sure=null;
        //先删除该文件
        String del=deleteFile(request,userName,Title);
        if(del.equals("ok")){
            //重新录入修改后的信息
            String in=insertFile(request,userName,title,
                name,contentType,size,filePath);
            if(in.equals("ok")){
                //调用myFile()方法,更新session中保存的文件信息
                String file=myFile(request,userName);
                if(file.equals("ok")){
                    sure="ok";
                }else{
                    sure=null;
                }
            }else{
                sure=null;
            }
        }else{
            sure=null;
        }
        return sure;
    }catch(Exception e){
        e.printStackTrace();
        return null;
    }
}
//查询文件
public ResultSet selectFile(HttpServletRequest request,
        String userName,String type,String name){
    try{
        String sql="select * from file where userName='"+userName+"' and
            "+type+"='"+name+"'";
        st=getStatement();
        return st.executeQuery(sql);
    }catch(Exception e){
        e.printStackTrace();
        return null;
    }
}
//查询所有的文件信息
public ResultSet selectFileAll(HttpServletRequest request,String userName){
    try{
```

```java
        String sql="select * from file where userName='"+userName+"'";
        st=getStatement();
        return st.executeQuery(sql);
    }catch(Exception e){
        e.printStackTrace();
        return null;
    }
}
//查询所有的文件信息,并把它们保存到 session 对象中
public String myFile(HttpServletRequest request,String userName){
    try{
        ArrayList listName=null;
        HttpSession session=request.getSession();
        listName=new ArrayList();
        rs=selectFileAll(request,userName);
        if(rs.next()){
            rs=selectFileAll(request,userName);
            while(rs.next()){
                MyFileBean mess=new MyFileBean();
                mess.setTitle(rs.getString("title"));
                mess.setName(rs.getString("name"));
                mess.setContentType(rs.getString("contentType"));
                mess.setSize(rs.getString("size"));
                listName.add(mess);
                session.setAttribute("file", listName);
            }
        }else{
            session.setAttribute("file", listName);
        }
        return "ok";
    }catch(Exception e){
        e.printStackTrace();
        return null;
    }
}
//查询登录名和密码是否存在
public ResultSet selectLogin(HttpServletRequest request,
        String userName,String password){
    try{
        String sql="select * from user where userName='"+userName+"' and
                password='"+password+"'";
        st=getStatement();
        return st.executeQuery(sql);
    }catch(Exception e){
        e.printStackTrace();
```

```
        return null;
    }
}
//把登录用户的信息保存到 session 对象中
public String myLogin(HttpServletRequest request,String userName){
    try{
        ArrayList listName=null;
        HttpSession session=request.getSession();
        listName=new ArrayList();
        rs=selectMess(request,userName);
        if(rs.next()){
            rs=selectMess(request,userName);
            while(rs.next()){
                UserNameBean mess=new UserNameBean();
                mess.setUserName(rs.getString("userName"));
                mess.setPassword(rs.getString("password"));
                listName.add(mess);
                session.setAttribute("userName", listName);
            }
        }else{
            session.setAttribute("userName", listName);
        }
        return "ok";
    }catch(Exception e){
        e.printStackTrace();
        return null;
    }
}
//返回登录用户的用户名
public String returnLogin(HttpServletRequest request){
    String LoginName=null;
    HttpSession session=request.getSession();
    ArrayList login= (ArrayList)session.getAttribute("userName");
        if(login==null||login.size()==0){
            LoginName=null;
        }else{
            for(int i=login.size()-1;i>=0;i--){
                UserNameBean nm=(UserNameBean)login.get(i);
                LoginName=nm.getUserName();
            }
        }
        return LoginName;
}
/*调用 myLogin()、myMessage()、myFriends()、myDayTime()、myFile()方法,把所有和用
    户有关的信息全部保存到 session 对象中。该方法在登录成功后调用*/
```

```java
    public String addList(HttpServletRequest request,String userName){
        String sure=null;
        String login=myLogin(request,userName);
        String mess=myMessage(request,userName);
        String fri=myFriends(request,userName);
        String day=myDayTime(request,userName);
        String file=myFile(request,userName);
        if(login.equals("ok")&&mess.equals("ok")&&fri.equals("ok")
            &&day.equals("ok")&&file.equals("ok")){
            sure="ok";
        }else{
            sure=null;
        }
        return sure;
    }
    //修改用户密码
    public String updatePass(HttpServletRequest request,String userName,String
            password){
        try{
            String sure=null;
            String sql="update user set password='"+password+"' where
                    userName='"+userName+"'";
            st=getStatement();
            int row=st.executeUpdate(sql);
            if(row==1){
                String mess=myLogin(request,userName);
                if(mess.equals("ok")){
                    sure="ok";
                }else{
                    sure=null;
                }
            }else{
                sure=null;
            }
            return sure;
        }catch(Exception e){
            e.printStackTrace();
            return null;
        }
    }
    //查找联系人,并将其信息保存到session对象中
    public String findFri(HttpServletRequest request,String userName,String name){
        try{
            ArrayList listName=null;
            HttpSession session=request.getSession();
```

```java
            listName=new ArrayList();
            rs=selectFri(request,userName,name);
            if(rs.next()){
                rs=selectFri(request,userName,name);
                while(rs.next()){
                    MyFriBean mess=new MyFriBean();
                    mess.setName(rs.getString("name"));
                    mess.setPhone(rs.getString("phone"));
                    mess.setEmail(rs.getString("email"));
                    mess.setWorkplace(rs.getString("workplace"));
                    mess.setPlace(rs.getString("place"));
                    mess.setQQ(rs.getString("QQ"));
                    listName.add(mess);
                    session.setAttribute("findfriend", listName);
                }
            }else{
                session.setAttribute("findfriend", listName);
            }

            return "ok";
        }catch(Exception e){
            e.printStackTrace();
            return null;
        }
    }
    //从查找到的联系人session对象中获取联系人姓名,并返回
    public String returnFri(HttpServletRequest request){
        String FriendName=null;
        HttpSession session=request.getSession();
        ArrayList login= (ArrayList)session.getAttribute("findfriend");
            if(login==null||login.size()==0){
                FriendName=null;
            }else{
                for(int i=login.size()-1;i>=0;i--){
                    MyFriBean nm= (MyFriBean)login.get(i);
                    FriendName=nm.getName();
                }
            }
            return FriendName;
    }
    //查找日程,并把日程信息保存到session对象中
    public String findDay(HttpServletRequest request,String userName,String date){
        try{
            ArrayList listName=null;
            HttpSession session=request.getSession();
```

```java
            listName=new ArrayList();
            rs=selectDay(request,userName,date);
            if(rs.next()){
                rs=selectDay(request,userName,date);
                while(rs.next()){
                    MyDayBean mess=new MyDayBean();
                    mess.setDay(rs.getString("date"));
                    mess.setThing(rs.getString("thing"));
                    listName.add(mess);
                    session.setAttribute("findday",listName);
                }
            }else{
                session.setAttribute("findday",listName);
            }

            return "ok";
        }catch(Exception e){
            e.printStackTrace();
            return null;
        }
    }
    //从查找到的日程 session 中获取日程信息,并返回
    public String returnDay(HttpServletRequest request){
        String date=null;
        HttpSession session=request.getSession();
        ArrayList login= (ArrayList)session.getAttribute("findday");
            if(login==null||login.size()==0){
                date=null;
            }else{
                for(int i=login.size()-1;i>=0;i--){
                    MyDayBean nm= (MyDayBean)login.get(i);
                    date=nm.getDay();
                }
            }
            return date;
    }
    //查找文件信息,并把文件的信息保存到 session 对象中
    public String findFile(HttpServletRequest request,String userName,String
            title){
        try{
            ArrayList listName=null;
            HttpSession session=request.getSession();
            listName=new ArrayList();
            rs=selectFile(request,userName,"title",title);
            if(rs.next()){
```

```java
                rs=selectFile(request,userName,"title",title);
                while(rs.next()){
                    MyFileBean mess=new MyFileBean();
                    mess.setTitle(rs.getString("title"));
                    mess.setName(rs.getString("name"));
                    mess.setContentType(rs.getString("contentType"));
                    mess.setSize(rs.getString("size"));
                    mess.setFilePath(rs.getString("filePath"));
                    listName.add(mess);
                    session.setAttribute("findfile", listName);
                }
            }else{
                session.setAttribute("findfile", listName);
            }

            return "ok";
        }catch(Exception e){
            e.printStackTrace();
            return null;
        }
    }
//根据不同的条件,从查找到的文件session对象中获取相应的文件信息
    public String returnFile(HttpServletRequest request,String face){
        String file=null;
        HttpSession session=request.getSession();
        ArrayList login= (ArrayList)session.getAttribute("findfile");
            if(login==null||login.size()==0){
                file=null;
            }else{
                for(int i=login.size()-1;i>=0;i--){
                    MyFileBean nm= (MyFileBean)login.get(i);
                    if(face.equals("title")){
                        file=nm.getTitle();
                    }else if(face.equals("filePath")){
                        file=nm.getFilePath();
                    }if(face.equals("fileName")){
                        file=nm.getName();
                    }
                }
            }
            return file;
    }
//一个带参数的信息提示框,供调试使用
    public void message(String msg){
        int type=JOptionPane.YES_NO_OPTION;
```

```
        String title="信息提示";
        JOptionPane.showMessageDialog(null,msg,title,type);
    }
}
```

4.4.3 系统主页面功能的实现

如果注册成功,将返回登录页面。在图 4-5 所示页面中输入登录名和登录密码,单击"登录"按钮后进入"个人信息管理系统"的主页面(main.jsp),如图 4-7 所示。

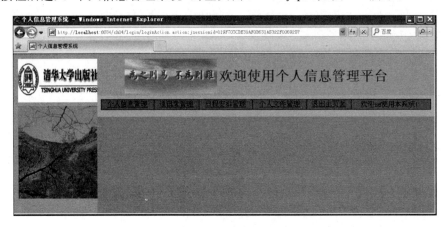

图 4-7 系统主页面

主页面(main.jsp)代码如下:

```
<%@page contentType="text/html" pageEncoding="UTF-8"%>
<%@taglib prefix="s" uri="/struts-tags" %>
<html>
    <head>
        <meta http-equiv="Content-Type" content="text/html; charset=UTF-8">
        <title><s:text name="个人信息管理系统"/></title>
    </head>
    <frameset cols="20%,*" framespacing="0" border="no" frameborder="0">
        <frame src="../mainFrame/left.jsp" name="left" scrolling="no">
        <frameset rows="20%,10%,*">
            <frame src="../mainFrame/top.jsp" name="top" scrolling="no">
            <frame src="../mainFrame/toop.jsp" name="toop" scrolling="no">
            <frame src="../mainFrame/about.jsp" name="main">
        </frameset>
    </frameset>
</html>
```

图 4-7 中的页面是使用框架进行分割的,子窗口分别链接 left.jsp、top.jsp、toop.jsp 和 about.jsp 页面。

left.jsp 页面代码如下:

```jsp
<%@page contentType="text/html" pageEncoding="UTF-8"%>
<%@taglib prefix="s" uri="/struts-tags" %>
<html>
    <head>
        <meta http-equiv="Content-Type" content="text/html; charset=UTF-8">
        <title><s:text name="个人管理系统"/></title>
    </head>
    <body bgcolor="#CCCCFF">
        <table>
            <tr align="center">
                <td>
                    <img src="../images/top.jpg" alt="清华大学出版社"
                        height="100" width="200">
                </td>
            </tr>
            <tr>
                <td>
                    <img src="../images/a.jpg" alt="长城" width="200">
                </td>
            </tr>
        </table>
    </body>
</html>
```

top.jsp 页面代码如下：

```jsp
<%@page contentType="text/html" pageEncoding="UTF-8"%>
<%@taglib prefix="s" uri="/struts-tags" %>
<html>
    <head>
        <meta http-equiv="Content-Type" content="text/html; charset=UTF-8">
        <title><s:text name="个人管理系统"/></title>
    </head>
    <body bgcolor="#CCCCFF">
        <table width="100%" align="center">
          <tr align="center">
            <td align="right">
                <img src="../images/cc.gif" alt="为之则易,不为则难!"
                    height="80">
            </td>
            <td align="left">
                <h1>
                    <font color="blue">欢迎使用个人信息管理平台</font>
                </h1>
            </td>
          </tr>
```

```
        </table>
    </body>
</html>
```

toop.jsp 页面代码如下：

```jsp
<%@page import="JavaBean.UserNameBean"%>
<%@page import="java.util.ArrayList"%>
<%@page contentType="text/html" pageEncoding="UTF-8"%>
<%@taglib prefix="s" uri="/struts-tags" %>
<html>
    <head>
        <meta http-equiv="Content-Type" content="text/html; charset=UTF-8">
        <title><s:text name="个人管理系统"/></title>
    </head>
    <body bgcolor="#CCDDEE">
        <%
            String loginname=null;
            /* DB 类中通过 myLogin ()方法把登录用户的信息保存到 session 对象中,在该页
               面中获取保存在 session 对象中的用户登录名,并把登录名输出到该页面上 */
            ArrayList login=(ArrayList)session.getAttribute("userName");
            if(login==null||login.size()==0){
                loginname="水木清华";
            }else{
                for(int i=login.size()-1;i>=0;i--){
                    UserNameBean nm=(UserNameBean)login.get(i);
                    loginname=nm.getUserName();
                }
            }
        %>
        <table width="100%" align="right" bgcolor="blue">
            <tr height="10" bgcolor="gray" align="center">
                <td><a href="http://localhost:8084/ch04/personMessage
                    /lookMessage.jsp" target="main">个人信息管理</a>
                </td>
                <td><a href="http://localhost:8084/ch04/friendManager
                    /lookFriends.jsp" target="main">通讯录管理</a>
                </td>
                <td><a href="http://localhost:8084/ch04/dateTimeManager
                    /lookDay.jsp" target="main">日程安排管理</a>
                </td>
                <td><a href="http://localhost:8084/ch04/fileManager
                    /lookFile.jsp" target="main">个人文件管理</a>
                </td>
                <td><a href="http://localhost:8084/ch04
                    /login/index.jsp" target="_top">退出主页面</a>
```

```
                </td>
                <td>欢迎<%=loginname%>使用本系统!</td>
            </tr>
        </table>
    </body>
</html>
```

about.jsp 页面代码如下：

```
<%@page contentType="text/html" pageEncoding="UTF-8"%>
<%@taglib prefix="s" uri="/struts-tags" %>
<html>
    <head>
        <meta http-equiv="Content-Type" content="text/html; charset=UTF-8">
        <title><s:text name="个人管理系统"/></title>
    </head>
    <body bgcolor="#AABBCCDD">
    </body>
</html>
```

toop.jsp 页面中用到的 UserNameBean 类的主要功能是通过 DB 中的 myLogin()方法把用户登录名保存在 session 里，并在 JSP 页面中通过调用 session 的方法，获取在 DB 类中 myLogin()方法保存的数据。

UserNameBean.java 的代码如下：

```
package JavaBean;
public class UserNameBean {
    private String userName;
    private String password;
    public UserNameBean(){
    }
    public String getUserName() {
        return userName;
    }
    public void setUserName(String userName) {
        this.userName=userName;
    }
    public String getPassword() {
        return password;
    }
    public void setPassword(String password) {
        this.password=password;
    }
}
```

主页面中包括个人信息管理、通讯录管理、日程安排管理、个人文件管理、退出系统等模块。

4.4.4 个人信息管理功能实现

单击图 4-7 所示页面中的"个人信息管理",出现如图 4-8 所示的个人信息相关页面。

图 4-8 个人信息页面

请参考 toop.jsp 页面代码中的"＜a href="http://localhost:8084/ch04/personMessage/lookMessage.jsp" target="main"＞个人信息管理＜/a＞"。lookMessage.jsp 页面用于获取用户的信息,并把用户信息输出到页面中。

lookMessage.jsp 代码如下:

```
<%@page import="JavaBean.MyMessBean"%>
<%@page import="java.util.ArrayList"%>
<%@page contentType="text/html" pageEncoding="UTF-8"%>
<%@taglib prefix="s" uri="/struts-tags" %>
<html>
    <head>
        <meta http-equiv="Content-Type" content="text/html; charset=UTF-8">
        <title><s:text name="个人信息管理系统"></s:text></title>
    </head>
<body bgcolor="gray">
    <hr noshade/>
    <s:div align="center">
    <table border="0" cellspacing="0" cellpadding="0" width="100%"
        align="center">
        <tr>
            <td width="33%">
                <s:a href="http://localhost:8084/ch04/personMessage
                    /updateMessage.jsp">修改个人信息</s:a>
```

```
            </td>
            <td width="33%">
                <s:text name="查看个人信息"></s:text>
            </td>
            <td width="33%">
                <s:a href="http://localhost:8084/ch04/personMessage
                    /updatePass.jsp">修改个人密码</s:a>
            </td>
        </tr>
    </table>
</s:div>
<hr noshade/>
<table border="5" cellspacing="0" cellpadding="0" bgcolor="#95BDFF"
        width="60%" align="center">
    <%
        /*通过DB类中的myMessage()方法,把登录用户的信息保存到session对象中,在
            该页面中获取保存在session对象中的用户个人信息,并把用户信息输出到页面
            上,如图4-8所示*/
        ArrayList MyMessage=(ArrayList)session.getAttribute("MyMess");
        if(MyMessage==null||MyMessage.size()==0){
            response.sendRedirect("http://localhost:8084/ch04/login/index.jsp");
        }else{
            for(int i=MyMessage.size()-1;i>=0;i--){
                MyMessBean mess=(MyMessBean)MyMessage.get(i);
    %>
            <tr>
                <td height="30">
                    <s:text name="用户姓名"></s:text>
                </td>
                <td><%=mess.getName()%></td>
            </tr>
            <tr>
                <td height="30">
                    <s:text name="用户性别"></s:text>
                </td>
                <td><%=mess.getSex()%></td>
            </tr>
            <tr>
                <td height="30">
                    <s:text name="出生日期"></s:text>
                </td>
                <td><%=mess.getBirth()%></td>
            </tr>
            <tr>
                <td height="30">
```

```html
                    <s:text name="用户民族"></s:text>
                  </td>
                <td><%=mess.getNation()%></td>
              </tr>
              <tr>
                <td height="30">
                    <s:text name="用户学历"></s:text>
                  </td>
                <td><%=mess.getEdu()%></td>
              </tr>
              <tr>
                <td height="30">
                    <s:text name="用户职称"></s:text>
                  </td>
                <td><%=mess.getWork()%></td>
              </tr>
              <tr>
                <td height="30">
                    <s:text name="用户电话"></s:text>
                  </td>
                <td><%=mess.getPhone()%></td>
              </tr>
              <tr>
                <td height="30">
                    <s:text name="用户住址"></s:text>
                  </td>
                <td><%=mess.getPlace()%></td>
              </tr>
              <tr>
                <td height="30">
                    <s:text name="用户邮箱"></s:text>
                  </td>
                <td><%=mess.getEmail()%></td>
              </tr>
        <%
              }
            }
        %>
      </table>
    </body>
</html>
```

lookMessage.jsp 页面中使用 MyMessBean 类保存数据，在该页面上通过 JavaBean 的调用把数据输出到页面上。

MyMessBean.java 的代码如下：

```java
package JavaBean;
public class MyMessBean {
    private String name;
    private String sex;
    private String birth;
    private String nation;
    private String edu;
    private String work;
    private String phone;
    private String place;
    private String email;
    public MyMessBean(){
    }
    public String getName() {
        return name;
    }
    public void setName(String name) {
        this.name=name;
    }
    public String getSex() {
        return sex;
    }
    public void setSex(String sex) {
        this.sex=sex;
    }
    public String getBirth() {
        return birth;
    }
    public void setBirth(String birth) {
        this.birth=birth;
    }
    public String getNation() {
        return nation;
    }
    public void setNation(String nation) {
        this.nation=nation;
    }
    public String getEdu() {
        return edu;
    }
    public void setEdu(String edu) {
        this.edu=edu;
    }
    public String getWork() {
        return work;
```

```
    }
    public void setWork(String work) {
        this.work=work;
    }
    public String getPhone() {
        return phone;
    }
    public void setPhone(String phone) {
        this.phone=phone;
    }
    public String getPlace() {
        return place;
    }
    public void setPlace(String place) {
        this.place=place;
    }
    public String getEmail() {
        return email;
    }
    public void setEmail(String email) {
        this.email=email;
    }
}
```

单击图 4-8 所示页面中的个人信息页面中的"修改个人信息",出现如图 4-9 所示的修改个人信息页面(updateMessage.jsp),可以对个人信息进行修改。

图 4-9 修改个人信息页面

updateMessage.jsp 代码如下：

```jsp
<%@page import="JavaBean.MyMessBean"%>
<%@page import="java.util.ArrayList"%>
<%@page contentType="text/html" pageEncoding="UTF-8"%>
<%@taglib prefix="s" uri="/struts-tags" %>
<html>
    <head>
        <meta http-equiv="Content-Type" content="text/html; charset=UTF-8">
        <title><s:text name="个人信息管理系统"></s:text></title>
    </head>
    <body bgcolor="gray">
        <hr noshade/>
    <s:div align="center">
    <table border="0" cellspacing="0" cellpadding="0" width="100%"
            align="center">
        <tr>
            <td width="33%">
                <s:text name="修改个人信息"></s:text>
            </td>
            <td width="33%">
                <s:a href="http://localhost:8084/ch04/personMessage
                    /lookMessage.jsp">查看个人信息</s:a>
            </td>
            <td width="33%">
                <s:a href="http://localhost:8084/ch04/personMessage
                    /updatePass.jsp">修改个人密码</s:a>
            </td>
        </tr>
    </table>
    </s:div>
    <hr noshade/>
    <s:form action="upMessAction" method="post">
    <table border="5" cellspacing="0" cellpadding="0" bgcolor="#95BDFF"
            width="60%" align="center">
        <%
            ArrayList MyMessage= (ArrayList)session.getAttribute("MyMess");
            if(MyMessage==null||MyMessage.size()==0){
                response.sendRedirect("http://localhost:8084/ch04
                                /login/index.jsp");
            }else{
                for(int i=MyMessage.size()-1;i>=0;i--){
                    MyMessBean mess= (MyMessBean)MyMessage.get(i);
            %>
                <tr>
                    <td height="30">
```

```
            <s:text name="用户姓名"></s:text>
        </td>
        <td><input type="text" name="name"
            value="<%=mess.getName()%>"/>
        </td>
    </tr>
    <tr>
        <td height="30">
            <s:text name="用户性别"></s:text>
        </td>
        <td><input type="text" name="sex"
            value="<%=mess.getSex()%>"/>
        </td>
    </tr>
    <tr>
        <td height="30">
            <s:text name="出生日期"></s:text>
        </td>
        <td><input type="text" name="birth"
            value="<%=mess.getBirth()%>"/>
        </td>
    </tr>
    <tr>
        <td height="30">
            <s:text name="用户民族"></s:text>
        </td>
        <td><input type="text" name="nation"
            value="<%=mess.getNation()%>"/>
        </td>
    </tr>
    <tr>
        <td height="30">
            <s:text name="用户学历"></s:text>
        </td>
        <td><input type="text" name="edu"
            value="<%=mess.getEdu()%>"/>
        </td>
    </tr>
    <tr>
        <td height="30">
            <s:text name="用户职称"></s:text>
        </td>
        <td><input type="text" name="work"
            value="<%=mess.getWork()%>"/>
        </td>
```

```
                </tr>
                <tr>
                    <td height="30">
                        <s:text name="用户电话"></s:text>
                    </td>
                    <td><input type="text" name="phone"
                            value="<%=mess.getPhone()%>"/>
                    </td>
                </tr>
                <tr>
                    <td height="30">
                        <s:text name="用户住址"></s:text>
                    </td>
                    <td><input type="text" name="place"
                            value="<%=mess.getPlace()%>"/>
                    </td>
                </tr>
                <tr>
                    <td height="30">
                        <s:text name="用户邮箱"></s:text>
                    </td>
                    <td><input type="text" name="email"
                            value="<%=mess.getEmail()%>"/>
                    </td>
                </tr>
                <tr>
                    <td colspan="2" align="center">
                    <input type="submit" value="确 定"/>
                          
                    <input type="reset" value="还 原"/>
                    </td>
                </tr>
            <%
                }
            }
            %>
        </table>
    </s:form>
  </body>
</html>
```

updateMessage.jsp 页面对应的业务控制器类为 UpdateMessAction。UpdateMessAction.java 的代码如下：

```
package edu.personManager.Action;
import DBJavaBean.DB;
```

```java
import com.opensymphony.xwork2.ActionSupport;
import javax.servlet.http.HttpServletRequest;
import org.apache.struts2.interceptor.ServletRequestAware;
public class UpdateMessAction extends ActionSupport implements
        ServletRequestAware {
    private String name;
    private String sex;
    private String birth;
    private String nation;
    private String edu;
    private String work;
    private String phone;
    private String place;
    private String email;
    private String userName;
    private HttpServletRequest request;
    private String message=ERROR;
    public String getName() {
        return name;
    }
    public void setName(String name) {
        this.name=name;
    }
    public String getSex() {
        return sex;
    }
    public void setSex(String sex) {
        this.sex=sex;
    }
    public String getBirth() {
        return birth;
    }
    public void setBirth(String birth) {
        this.birth=birth;
    }
    public String getNation() {
        return nation;
    }
    public void setNation(String nation) {
        this.nation=nation;
    }
    public String getEdu() {
        return edu;
    }
    public void setEdu(String edu) {
```

```java
        this.edu=edu;
    }
    public String getWork() {
        return work;
    }
    public void setWork(String work) {
        this.work=work;
    }
    public String getPhone() {
        return phone;
    }
    public void setPhone(String phone) {
        this.phone=phone;
    }
    public String getPlace() {
        return place;
    }
    public void setPlace(String place) {
        this.place=place;
    }
    public String getEmail() {
        return email;
    }
    public void setEmail(String email) {
        this.email=email;
    }
    //实现了接口ServletRequestAware中的方法
    public void setServletRequest(HttpServletRequest hsr) {
        request=hsr;
    }
    public void validate(){
        if(getName()==null||getName().length()==0){
            addFieldError("name","用户姓名不允许为空!");
        }
        if(getSex()==null||getSex().length()==0){
            addFieldError("sex","用户性别不允许为空!");
        }
        if(getBirth()==null||getBirth().length()==0){
            addFieldError("birth","用户生日不允许为空!");
        }else{
            if(getBirth().length()!=10){
                addFieldError("birth","用户生日格式为'yyyy-mm-dd'!");
            }else{
                String an=this.getBirth().substring(4, 5);
                String bn=this.getBirth().substring(7, 8);
```

```
                if(!(an.equals("-"))||!(bn.equals("-"))){
                    addFieldError("birth","用户生日格式为'yyyy-mm-dd'!");
                }
            }
        }
        if(getNation()==null||getNation().length()==0){
            addFieldError("nation","用户民族不允许为空!");
        }
        if(getEdu()==null||getEdu().length()==0){
            addFieldError("edu","用户学历不允许为空!");
        }
        if(getWork()==null||getWork().length()==0){
            addFieldError("work","用户工作不允许为空!");
        }
        if(getPhone()==null||getPhone().length()==0){
            addFieldError("phone","用户电话不允许为空!");
        }
        if(getPlace()==null||getPlace().length()==0){
            addFieldError("place","用户地址不允许为空!");
        }
        if(getEmail()==null||getEmail().length()==0){
            addFieldError("email","用户email不允许为空!");
        }
    }
    public String execute() throws Exception {
        DB mysql=new DB();
        userName=mysql.returnLogin(request);
        String mess=mysql.updateMess(request, userName, this.getName(),
                                    this.getSex(), this.getBirth(),
                                    this.getNation(), this.getEdu(),
                                    this.getWork(), this.getPhone(),
                                    this.getPlace(), this.getEmail());
        if(mess.equals("ok")){
            message=SUCCESS;
        }
        return message;
    }
}
```

单击图 4-9 所示页面中的"修改个人密码",出现如图 4-10 所示的修改个人密码页面(updatePass.jsp),在该页面中可以对登录密码进行修改。

updatePass.jsp 代码如下:

```
<%@page import="java.util.ArrayList"%>
<%@page import="JavaBean.UserNameBean"%>
<%@page contentType="text/html" pageEncoding="UTF-8"%>
```

图 4-10 修改密码页面

```jsp
<%@ taglib prefix="s" uri="/struts-tags" %>
<html>
    <head>
        <meta http-equiv="Content-Type" content="text/html; charset=UTF-8">
        <title><s:text name="个人信息管理系统"></s:text></title>
    </head>
    <body bgcolor="gray">
      <hr noshade/>
      <s:div align="center">
      <table border="0" cellspacing="0" cellpadding="0" width="100%"
            align="center">
        <tr>
            <td width="33%">
                <s:a href="http://localhost:8084/ch04/personMessage
                    /updateMessage.jsp">修改个人信息</s:a>
            </td>
            <td width="33%">
                <s:a href="http://localhost:8084/ch04/personMessage
                    /lookMessage.jsp">查看个人信息</s:a>
            </td>
            <td width="33%">
                <s:text name="修改个人密码"></s:text>
            </td>
        </tr>
      </table>
      </s:div>
      <hr noshade/>
      <s:form action="upPassAction" method="post">
      <table border="5" cellspacing="0" cellpadding="0" bgcolor="#95BDFF"
            width="60%" align="center">
        <%
```

```jsp
            ArrayList login= (ArrayList)session.getAttribute("userName");
            if(login==null||login.size()==0){
                response.sendRedirect("http://localhost:8084/ch04
                            /login/index.jsp");
            }else{
                for(int i=login.size()-1;i>=0;i--){
                    UserNameBean nm= (UserNameBean)login.get(i);
                    %>
                    <tr>
                        <td height="30">
                            <s:text name="用户密码"></s:text>
                        </td>
                    <td><input type="text" name="password1"
                            value="<%=nm.getPassword()%>"/>
                        </td>
                    </tr>
                    <tr>
                        <td height="30">
                            <s:text name="重复密码"></s:text>
                        </td>
                        <td><input type="text" name="password2"
                            value="<%=nm.getPassword()%>"/>
                        </td>
                    </tr>
                    <tr>
                        <td colspan="2" align="center">
                            <input type="submit" value="确 定" size="12"/>
                                  
                            <input type="reset" value="清 除" size="12"/>
                        </td>
                    </tr>
                <%
                }
            }
            %>
        </table>
    </s:form>
    </body>
</html>
```

updatePass.jsp 页面对应的业务控制器类为 UpdatePassAction。UpdatePassAction.java 的代码如下：

```java
package edu.personManager.Action;
import DBJavaBean.DB;
import com.opensymphony.xwork2.ActionSupport;
```

```java
import javax.servlet.http.HttpServletRequest;
import javax.swing.JOptionPane;
import org.apache.struts2.interceptor.ServletRequestAware;

public class UpdatePassAction extends ActionSupport implements
    ServletRequestAware{
    private String password1;
    private String password2;
    private String userName;
    private HttpServletRequest request;
    private String message=ERROR;

    public String getPassword1() {
        return password1;
    }
      public String getPassword1() {
        return password1;
    }
    public void setPassword1(String password1) {
        this.password1=password1;
    }
    public String getPassword2() {
        return password2;
    }
    public void setPassword2(String password2) {
        this.password2=password2;
    }
    public void setServletRequest(HttpServletRequest hsr) {
        request=hsr;
    }
    public void message(String msg){
        int type=JOptionPane.YES_NO_OPTION;
        String title="信息提示";
        JOptionPane.showMessageDialog(null,msg,title,type);
    }
    public void validate(){
        if(!(password1.equals(password2))){
            message("两次密码不同!");
            addFieldError("password2","两次密码不同!");
        }
    }
    public String execute() throws Exception {
        DB mysql=new DB();
        userName=mysql.returnLogin(request);
        String pass=mysql.updatePass(request, userName, this.getPassword1());
```

```
        if(pass.equals("ok")){
            message=SUCCESS;
        }
        return message;
    }
}
```

4.4.5 通讯录管理功能实现

单击图 4-10 所示页面中的"通讯录管理",出现如图 4-11 所示的通讯录信息页面,可以对通讯录进行相关操作。lookFriends.jsp 页面用于获取通讯录的信息,并把通讯录信息输出到主页面中。

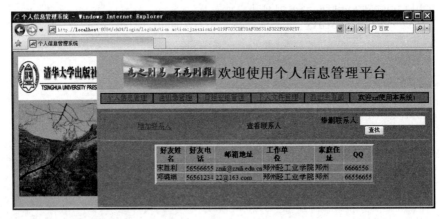

图 4-11 查看联系人页面

lookFriends.jsp 代码如下:

```
<%@page import="JavaBean.MyFriBean"%>
<%@page import="java.util.ArrayList"%>
<%@page contentType="text/html" pageEncoding="UTF-8"%>
<%@taglib  prefix="s" uri="/struts-tags" %>
<html>
    <head>
        <meta http-equiv="Content-Type" content="text/html; charset=UTF-8">
        <title><s:text name="个人信息管理系统"></s:text></title>
    </head>
    <body bgcolor="gray">
      <hr noshade/>
      <s:div align="center">
      <s:form action="findFriAction" method="post">
      <table border="0" cellspacing="0" cellpadding="0" width="100%"
            align="center">
        <tr>
           <td width="33%">
              <s:a href="http://localhost:8084/ch04/friendManager
```

```
                /addFriend.jsp">增加联系人</s:a>
            </td>
            <td width="33%">
                <s:text name="查看联系人"></s:text>
            </td>
            <td width="33%">
                <s:text name="修删联系人:"></s:text>
              <input type="text" name="friendname"/>
              <input type="submit" value="查找"/>
            </td>
        </tr>
    </table>
</s:form>
</s:div>
<hr noshade/>
<table border="5" cellspacing="0" cellpadding="0" bgcolor="#95BDFF"
        width="60%" align="center">
    <tr>
        <th height="30">好友姓名</th>
        <th height="30">好友电话</th>
        <th height="30">邮箱地址</th>
        <th height="30">工作单位</th>
        <th height="30">家庭住址</th>
        <th height="30">QQ</th>
    </tr>
    <%
      ArrayList friends= (ArrayList)session.getAttribute("friends");
      if(friends==null||friends.size()==0){
      %>
      <s:div align="center"><%="您还没有添加联系人!"%></s:div>
      <%
      }else{
          for(int i=friends.size()-1;i>=0;i--){
              MyFriBean ff= (MyFriBean)friends.get(i);
            %>
        <tr>
            <td><%=ff.getName()%></td>
            <td><%=ff.getPhone()%></td>
            <td><%=ff.getEmail()%></td>
            <td><%=ff.getWorkplace()%></td>
            <td><%=ff.getPlace()%></td>
            <td><%=ff.getQQ()%></td>
        </tr>
        <%
      }
```

```
            }
        %>
        </table>
    </body>
</html>
```

lookFriends.jsp 页面对应的业务控制器类为 FindFriAction。FindFriAction.java 的代码如下：

```java
package edu.friendManager.Action;
import DBJavaBean.DB;
import com.opensymphony.xwork2.ActionSupport;
import java.sql.ResultSet;
import javax.servlet.http.HttpServletRequest;
import javax.swing.JOptionPane;
import org.apache.struts2.interceptor.ServletRequestAware;

public class FindFriAction extends ActionSupport implements ServletRequestAware{
    private String friendname;
    private String userName;
    private ResultSet rs=null;
    private String message="error";
    private HttpServletRequest request;
    public String getFriendname() {
        return friendname;
    }
    public void setFriendname(String friendname) {
        this.friendname=friendname;
    }
    public void setServletRequest(HttpServletRequest hsr) {
        request=hsr;
    }
    public void message(String msg){
        int type=JOptionPane.YES_NO_OPTION;
        String title="信息提示";
        JOptionPane.showMessageDialog(null,msg,title,type);
    }
    public void validate(){
        if(this.getFriendname().equals("")||this.getFriendname().length()==0){
            message("联系人姓名不允许为空!");
            addFieldError("friendname","联系人姓名不允许为空!");
        }else{
            try{
                DB mysql=new DB();
                userName=mysql.returnLogin(request);
                rs=mysql.selectFri(request, userName, this.getFriendname());
```

```
                if(!rs.next()){
                    message("联系人姓名不存在!");
                    addFieldError("friendname","联系人姓名不存在!");
                }
            }catch(Exception e){
                e.printStackTrace();
            }
        }
    }
    public String execute() throws Exception {
        DB mysql=new DB();
        userName=mysql.returnLogin(request);
        String fri=mysql.findFri(request, userName, this.getFriendname());
        if(fri.equals("ok")){
            message=SUCCESS;
        }
        return message;
    }
}
```

lookFriends.jsp 页面使用的 JavaBean 是 MyFriBean。
MyFriBean.java 的代码如下：

```
package JavaBean;
public class MyFriBean{
    private String name;
    private String phone;
    private String email;
    private String workplace;
    private String place;
    private String QQ;
    public String getName() {
        return name;
    }
    public void setName(String name) {
        this.name=name;
    }
    public String getPhone() {
        return phone;
    }
    public void setPhone(String phone) {
        this.phone=phone;
    }
    public String getEmail() {
        return email;
    }
```

```java
    public void setEmail(String email) {
        this.email=email;
    }
    public String getWorkplace() {
        return workplace;
    }
    public void setWorkplace(String workplace) {
        this.workplace=workplace;
    }
    public String getPlace() {
        return place;
    }
    public void setPlace(String place) {
        this.place=place;
    }
    public String getQQ() {
        return QQ;
    }
    public void setQQ(String QQ) {
        this.QQ=QQ;
    }
}
```

单击图 4-11 所示的通讯录页面中的"增加联系人",出现如图 4-12 所示的增加联系人信息页面(addFriend.jsp),在该页面中可以增加联系人。

图 4-12 增加联系人页面

addFriend.jsp 代码如下:

```
<%@page contentType="text/html" pageEncoding="UTF-8"%>
<%@taglib prefix="s" uri="/struts-tags" %>
<html>
    <head>
```

```html
        <meta http-equiv="Content-Type" content="text/html; charset=UTF-8">
        <title><s:text name="个人信息管理系统"></s:text></title>
</head>
<body bgcolor="gray">
    <hr noshade/>
    <s:div align="center">
        <s:form action="findFriAction" method="post">
            <table border="0" cellspacing="0" cellpadding="0" width="100%"
                align="center">
            <tr>
                <td width="33%">
                    <s:text name="增加联系人"></s:text>
                </td>
                <td width="33%">
                    <s:a href="http://localhost:8084/ch04/friendManager
                        /lookFriends.jsp">查看联系人</s:a>
                </td>
                <td width="33%">
                    <s:text name="修删联系人:"></s:text>
                    <input type="text" name="friendname"/>
                    <input type="submit" value="查找"/>
                </td>
            </tr>
            </table>
        </s:form>
    </s:div>
    <hr noshade/>
    <form action="addFriAction" method="post">
        <table border="2" cellspacing="0" cellpadding="0" bgcolor="95BDFF"
                width="60%" align="center">
            <tr>
                <td>
                    <s:textfield name="name" label="好友姓名"></s:textfield>
                </td>
            </tr>
            <tr>
                <td>
                    <s:textfield name="phone" label="好友电话"></s:textfield>
                </td>
            </tr>
            <tr>
                <td>
                    <s:textfield name="email" label="邮箱地址"></s:textfield>
                </td>
            </tr>
```

```html
            <tr>
                <td>
                    <s:textfield name="workplace" label="工作单位">
                    </s:textfield>
                </td>
            </tr>
            <tr>
                <td>
                    <s:textfield name="place" label="家庭住址"></s:textfield>
                </td>
            </tr>
            <tr>
                <td>
                    <s:textfield name="QQ" label="QQ"></s:textfield>
                </td>
            </tr>
            <tr>
              <td colspan="2" align="center">
                  <input type="submit" value="确 定" size="12">
                        
                  <input type="reset" value="清 除" size="12">
              </td>
            </tr>
        </table>
    </form>
  </body>
</html>
```

addFriend.jsp 页面对应的业务控制器类为 AddFriAction。AddFriAction.java 的代码如下：

```java
package edu.friendManager.Action;
import DBJavaBean.DB;
import com.opensymphony.xwork2.ActionSupport;
import java.sql.*;
import javax.servlet.http.HttpServletRequest;
import org.apache.struts2.interceptor.ServletRequestAware;

public class AddFriAction extends ActionSupport implements ServletRequestAware{
    private String name;
    private String phone;
    private String email;
    private String workplace;
    private String place;
    private String QQ;
    private ResultSet rs=null;
```

```java
private String message="error";
private HttpServletRequest request;
private String userName=null;
public String getName() {
    return name;
}
public void setName(String name) {
    this.name=name;
}
public String getPhone() {
    return phone;
}
public void setPhone(String phone) {
    this.phone=phone;
}
public String getEmail() {
    return email;
}
public void setEmail(String email) {
    this.email=email;
}
public String getWorkplace() {
    return workplace;
}
public void setWorkplace(String workplace) {
    this.workplace=workplace;
}
public String getPlace() {
    return place;
}
public void setPlace(String place) {
    this.place=place;
}
public String getQQ() {
    return QQ;
}
public void setQQ(String QQ) {
    this.QQ=QQ;
}
public void setServletRequest(HttpServletRequest hsr) {
    request=hsr;
}
public void validate(){
    if(getName()==null||getName().length()==0){
        addFieldError("name","用户姓名不允许为空");
```

```java
        }else{
            try {
                DB mysql=new DB();
                userName=mysql.returnLogin(request);
                rs=mysql.selectFri(request, userName, this.getName());
                if(rs.next()){
                    addFieldError("name","此用户已存在!");
                }
            } catch (SQLException ex) {
                ex.printStackTrace();
            }
        }
        if(getPhone()==null||getPhone().length()==0){
            addFieldError("phone","用户电话不允许为空");
        }
        if(getEmail()==null||getEmail().length()==0){
            addFieldError("email","邮箱地址不允许为空");
        }
        if(getWorkplace()==null||getWorkplace().length()==0){
            addFieldError("workplace","工作单位不允许为空");
        }
        if(getPlace()==null||getPlace().length()==0){
            addFieldError("place","家庭住址不允许为空");
        }
        if(getQQ()==null||getQQ().length()==0){
            addFieldError("QQ","用户 QQ 不允许为空");
        }
    }
    public String execute() throws Exception{
        DB mysql=new DB();
        userName=mysql.returnLogin(request);
        String fri=mysql.insertFri(request, userName, this.getName(),
                        T his.getEmail(), this.getWorkplace(),
                        this.getPlace(), this.getQQ());
        if(fri.equals("ok")){
            message=SUCCESS;
        }
        else if(fri.equals("one")){
            message=INPUT;
        }
        return message;
    }
}
```

在图 4-12 所示页面中的"修删联系人"后,输入数据并单击"查找"按钮,出现如图 4-13 所示的修改和删除联系人页面(findFriend.jsp)。单击"查找"按钮后请求提交到业务控制器类FindFriAction,请参考addFriend.jsp页面中的＜s:form＞中的属性。FindFriAction类代码前面已经给出。请求提交后可以修改联系人。

图4-13 联系人修删页面

findFriend.jsp代码如下：

```
<%@page import="JavaBean.MyFriBean"%>
<%@page import="java.util.ArrayList"%>
<%@page contentType="text/html" pageEncoding="UTF-8"%>
<%@taglib prefix="s" uri="/struts-tags" %>
<html>
    <head>
        <meta http-equiv="Content-Type" content="text/html; charset=UTF-8">
        <title><s:text name="个人信息管理系统"></s:text></title>
    </head>
    <body bgcolor="gray">
      <hr noshade/>
      <s:div align="center">
      <s:form action="findFriAction" method="post">
      <table border="0" cellspacing="0" cellpadding="0" width="100%"
            align="center">
          <tr>
              <td width="33%">
                  <s:a href="http://localhost:8084/ch04/friendManager
                      /addFriend.jsp">增加联系人</s:a>
              </td>
              <td width="33%">
                  <s:a href="http://localhost:8084/ch04/friendManager
                      /lookFriends.jsp">查看联系人</s:a>
              </td>
              <td width="33%">
                  <s:text name="修删联系人:"></s:text>
                  <input type="text" name="friendname"/>
                  <input type="submit" value="查找"/>
              </td>
```

```
                </tr>
            </table>
        </s:form>
    </s:div>
    <hr noshade/>
    <table border="5" cellspacing="0" cellpadding="0" bgcolor="#95BDFF"
            width="60%" align="center">
        <tr>
            <th height="30">用户姓名</th>
            <th height="30">用户电话</th>
            <th height="30">邮箱地址</th>
            <th height="30">工作单位</th>
            <th height="30">家庭住址</th>
            <th height="30">用户 QQ</th>
            <th height="30">用户操作</th>
        </tr>
        <%
            ArrayList friends= (ArrayList)session.getAttribute("findfriend");
            if(friends==null||friends.size()==0){
        %>
        <s:div align="center"><%="您还没有添加联系人!"%></s:div>
        <%
            }else{
                for(int i=friends.size()-1;i>=0;i--){
                    MyFriBean ff= (MyFriBean)friends.get(i);
        %>
        <tr>
            <td><%=ff.getName()%></td>
            <td><%=ff.getPhone()%></td>
            <td><%=ff.getEmail()%></td>
            <td><%=ff.getWorkplace()%></td>
            <td><%=ff.getPlace()%></td>
            <td><%=ff.getQQ()%></td>
            <td>
                <s:a href="http://localhost:8084/ch04/friendManager
                    /updateFriend.jsp">修改</s:a>
                <s:a href="deleteFriAction">删除</s:a>
            </td>
        </tr>
        <%
                }
            }
        %>
    </table>
</body>
</html>
```

findFriend.jsp 页面使用的 JavaBean 类是 MyFriBean,其代码前面已经给出。单击图 4-13 所示页面中联系人后面的"修改",出现如图 4-14 所示的修改联系人页面（updateFriend.jsp），可对好友信息进行修改。

图 4-14　修改联系人页面

updateFriend.jsp 代码如下：

```
<%@page import="JavaBean.MyFriBean"%>
<%@page import="java.util.ArrayList"%>
<%@page contentType="text/html" pageEncoding="UTF-8"%>
<%@taglib  prefix="s" uri="/struts-tags" %>
<html>
   <head>
       <meta http-equiv="Content-Type" content="text/html; charset=UTF-8">
       <title><s:text name="个人信息管理系统"></s:text></title>
   </head>
<body bgcolor="gray">
   <hr noshade/>
   <s:div align="center">
   <s:form action="findFriAction" method="post">
   <table border="0" cellspacing="0" cellpadding="0" width="100%"
           align="center">
       <tr>
           <td width="33%">
               <s:a href="http://localhost:8084/ch04/friendManager
                  /addFriend.jsp">增加联系人</s:a>
           </td>
           <td width="33%">
               <s:a href="http://localhost:8084/ch04/friendManager
                  /lookFriends.jsp">查看联系人</s:a>
           </td>
```

```html
            <td width="33%">
                <s:text name="修删联系人:"></s:text>
                <input type="text" name="friendname"/>
                <input type="submit" value="查找"/>
            </td>
        </tr>
    </table>
</s:form>
</s:div>
<hr noshade/>
<s:form action="upFriAction" method="post">
    <table border="2" cellspacing="0" cellpadding="0" bgcolor="95BDFF"
        width="60%" align="center">
        <%
ArrayList delemess= (ArrayList)session.getAttribute("findfriend");
if(delemess==null||delemess.size()==0){
%>
<s:div align="center"><%="您还没有添加联系人!"%></s:div>
<%
}else{
    for(int i=delemess.size()-1;i>=0;i--){
        MyFriBean ff= (MyFriBean)delemess.get(i);
        %>
    <tr>
        <td><s:text name="用户姓名"></s:text></td>
        <td>
            <input type="text" name="name"
                value="<%=ff.getName()%>"/>
        </td>
    </tr>
    <tr>
        <td><s:text name="用户电话"></s:text></td>
        <td>
            <input type="text" name="phone"
                value="<%=ff.getPhone()%>"/>
        </td>
    </tr>
    <tr>
        <td><s:text name="邮箱地址"></s:text></td>
        <td>
            <input type="text" name="email"
                value="<%=ff.getEmail()%>"/>
        </td>
    </tr>
    <tr>
```

```html
        <td><s:text name="工作单位"></s:text></td>
            <td>
                <input type="text" name="workplace"
                    value="<%=ff.getWorkplace()%>"/>
            </td>
        </tr>
        <tr>
            <td><s:text name="家庭住址"></s:text></td>
            <td>
                <input type="text" name="place"
                    value="<%=ff.getPlace()%>"/>
            </td>
        </tr>
        <tr>
            <td><s:text name="用户QQ"></s:text></td>
            <td>
                <input type="text" name="QQ"
                    value="<%=ff.getQQ()%>"/>
            </td>
        </tr>
        <tr>
          <td colspan="2" align="center">
            <input type="submit" value="确 定"
                size="12">     
            <input type="reset" value="清 除" size="12">
          </td>
        </tr>
        <%
            }
        }
        %>
        </table>
    </s:form>
  </body>
</html>
```

updateFriend.jsp 页面使用的 JavaBean 类是 MyFriBean，其代码前面已经给出。updateFriend.jsp 页面对应的业务控制器类为 UpdateFriAction。该业务控制器对应的代码如下。

UpdateFriAction.java 的代码如下：

```java
package edu.friendManager.Action;
import DBJavaBean.DB;
import com.opensymphony.xwork2.ActionSupport;
import java.sql.ResultSet;
import javax.servlet.http.HttpServletRequest;
```

```java
import org.apache.struts2.interceptor.ServletRequestAware;

public class UpdateFriAction extends ActionSupport implements
   ServletRequestAware{
    private String name;
    private String phone;
    private String email;
    private String workplace;
    private String place;
    private String QQ;
    private String message="error";
    private HttpServletRequest request;
    private ResultSet rs=null;
    private String userName;
    private String friendname;
    public String getName() {
        return name;
    }
    public void setName(String name) {
        this.name=name;
    }
    public String getPhone() {
        return phone;
    }
    public void setPhone(String phone) {
        this.phone=phone;
    }
    public String getEmail() {
        return email;
    }
    public void setEmail(String email) {
        this.email=email;
    }
    public String getWorkplace() {
        return workplace;
    }
    public void setWorkplace(String workplace) {
        this.workplace=workplace;
    }
    public String getPlace() {
        return place;
    }
    public void setPlace(String place) {
        this.place=place;
    }
    public String getQQ() {
        return QQ;
```

```java
    }
    public void setQQ(String QQ) {
        this.QQ=QQ;
    }
    public void setServletRequest(HttpServletRequest hsr) {
        request=hsr;
    }
    public void validate(){
        if(getName()==null||getName().length()==0){
            addFieldError("name","用户姓名不允许为空");
        }
        if(getPhone()==null||getPhone().length()==0){
            addFieldError("phone","用户电话不允许为空");
        }
        if(getEmail()==null||getEmail().length()==0){
            addFieldError("email","邮箱地址不允许为空");
        }
        if(getWorkplace()==null||getWorkplace().length()==0){
            addFieldError("workplace","工作单位不允许为空");
        }
        if(getPlace()==null||getPlace().length()==0){
            addFieldError("place","家庭住址不允许为空");
        }
        if(getQQ()==null||getQQ().length()==0){
            addFieldError("QQ","用户QQ不允许为空");
        }
    }
    public String execute() throws Exception {
        DB mysql=new DB();
        userName=mysql.returnLogin(request);
        friendname=mysql.returnFri(request);
        String fri=mysql.updateFri(request,userName,friendname,this.getName(),
                            this.getPhone(),this.getEmail(),
                            this.getWorkplace(),this.getPlace(),
                            this.getQQ());
        if(fri.equals("ok")){
            message=SUCCESS;
        }
        return message;
    }
}
```

单击图 4-13 所示页面中的"删除",参考 findFriend.jsp 代码中"< s:a href="deleteFriAction">删除</s:a>",请求提交到 DeleteFriAction,该类是一个业务控制器,用于对联系人进行删除。

DeleteFriAction.java 的代码如下:

```java
package edu.friendManager.Action;
import DBJavaBean.DB;
import com.opensymphony.xwork2.ActionSupport;
import javax.servlet.http.HttpServletRequest;
import org.apache.struts2.interceptor.ServletRequestAware;

public class DeleteFriAction extends ActionSupport implements
ServletRequestAware{
    private String message="erroe";
    private String userName;
    private String name;
    private HttpServletRequest request;
    public void setServletRequest(HttpServletRequest hsr) {
        request=hsr;
    }
    public String execute() throws Exception {
        DB mysql=new DB();
        userName=mysql.returnLogin(request);
        name=mysql.returnFri(request);
        String del=mysql.deleteFri(request, userName, name);
        if(del.equals("ok")){
            message=SUCCESS;
        }
        return message;
    }
}
```

4.4.6 日程安排管理功能实现

单击系统主页面中的"日程安排管理",出现如图 4-15 所示的日程信息页面(lookDay.jsp),可以对日程进行相关操作。

图 4-15 查看日程页面

lookDay.jsp 代码如下:

```jsp
<%@page import="JavaBean.MyDayBean"%>
<%@page import="java.util.ArrayList"%>
<%@page contentType="text/html" pageEncoding="UTF-8"%>
<%@taglib prefix="s" uri="/struts-tags" %>
<html>
    <head>
        <meta http-equiv="Content-Type" content="text/html; charset=UTF-8">
        <title><s:text name="个人信息管理系统"></s:text></title>
    </head>
    <body bgcolor="gray">
      <hr noshade/>
      <s:div align="center">
      <s:form action="findDayAction" method="post">
        <table border="0" cellspacing="0" cellpadding="0" width="100%"
              align="center">
          <tr>
             <td width="30%">
                <s:a href="http://localhost:8084/ch04/dateTimeManager
                      /addDay.jsp">增加日程</s:a>
             </td>
             <td width="30%">
                <s:text name="查看日程"></s:text>
             </td>
             <td width="40%">
                <s:text name="日程时间:"></s:text>
                20<input type="text" size="1" name="year"/>年
                <input type="text" size="1" name="month"/>月
                <input type="text" size="1" name="day"/>日
                <input type="submit" value="修删日程"/>
             </td>
          </tr>
        </table>
      </s:form>
      </s:div>
      <hr noshade/>
      <table border="5" cellspacing="0" cellpadding="0" bgcolor="#95BDFF"
              width="60%" align="center">
        <tr>
             <th width="40%">日程时间</th>
             <th width="60%">日程内容</th>
        </tr>
        <%
          ArrayList day=(ArrayList)session.getAttribute("day");
          if(day==null||day.size()==0){
```

```
            %>
            <s:div align="center"><%="您还没有任何日程安排!"%></s:div>
            <%
        }else{
            for(int i=day.size()-1;i>=0;i--){
                MyDayBean dd= (MyDayBean)day.get(i);
            %>
                <tr>
                    <td><%=dd.getDay()%></td>
                    <td><%=dd.getThing()%></td>
                </tr>
                <%
            }
        }
        %>
    </table>
  </body>
</html>
```

lookDay.jsp 页面使用的 JavaBean 类是 MyDayBean。MyDayBean.java 的代码如下：

```
package JavaBean;
import java.sql.ResultSet;
import javax.servlet.http.HttpServletRequest;
import org.apache.struts2.interceptor.ServletRequestAware;

public class MyDayBean implements ServletRequestAware{
    private String Day;
    private String thing;
    private ResultSet rs=null;
    private HttpServletRequest request;
    public String getDay() {
        return Day;
    }
    public void setDay(String Day) {
        this.Day=Day;
    }
    public String getThing() {
        return thing;
    }
    public void setThing(String thing) {
        this.thing=thing;
    }
    public void setServletRequest(HttpServletRequest hsr) {
        request=hsr;
```

            }
    }

单击图 4-15 所示页面中的"增加日程",出现如图 4-16 所示的添加日程页面(addDay. jsp),可以添加日程。

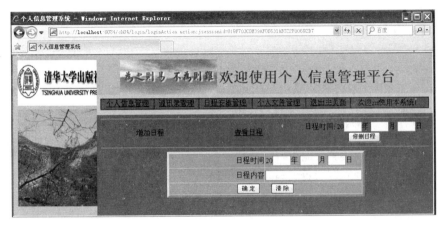

图 4-16 添加日程页面

addDay.jsp 代码如下:

```
<%@page contentType="text/html" pageEncoding="UTF-8"%>
<%@taglib  prefix="s" uri="/struts-tags" %>
<html>
    <head>
        <meta http-equiv="Content-Type" content="text/html; charset=UTF-8">
        <title><s:text name="个人信息管理系统"></s:text></title>
    </head>
<body bgcolor="gray">
    <hr noshade/>
 <s:div align="center">
 <s:form action="findDayAction" method="post">
 <table border="0" cellspacing="0" cellpadding="0" width="100%"
        align="center">
    <tr>
        <td width="30%">
            <s:text name="增加日程"></s:text>
        </td>
        <td width="30%">
            <s:a href="http://localhost:8084/ch04/dateTimeManager
                /lookDay.jsp">查看日程</s:a>
        </td>
        <td width="40%">
            <s:text name="日程时间:"></s:text>
            20<input type="text" size="1" name="year"/>年
         <input type="text" size="1" name="month"/>月
```

```
                <input type="text" size="1" name="day"/>日
                <input type="submit" value="修删日程"/>
            </td>
        </tr>
    </table>
    </s:form>
    </s:div>
    <hr noshade/>
    <s:form action="addDayAction" method="post">
        <table border="5" cellspacing="0" cellpadding="0" bgcolor="#95BDFF"
            width="60%" align="center">
            <tr>
                <td height="30" width="50%" align="right">日程时间</td>
                <td width="50%">
                    20<input type="text" size="1" name="year"/>年
                    <input type="text" size="1" name="month"/>月
                    <input type="text" size="1" name="day"/>日
                </td>
            </tr>
            <tr>
                <td height="30" width="50%" align="right">日程内容</td>
                <td width="50%">
                    <input type="text" size="30" name="thing"/>
                </td>
            </tr>
            <tr>
                <td colspan="2" align="center">
                    <input type="submit" value="确 定" size="12">
                         
                    <input type="reset" value="清 除" size="12">
                </td>
            </tr>
        </table>
    </s:form>
</body>
</html>
```

addDay.jsp 页面对应的业务控制器类为 AddDayAction。
AddDayAction.java 代码如下：

```
package edu.dateTimeManager.Action;
import DBJavaBean.DB;
import com.opensymphony.xwork2.ActionSupport;
import java.sql.ResultSet;
import java.text.SimpleDateFormat;
import java.util.Date;
```

```java
import java.util.StringTokenizer;
import javax.servlet.http.HttpServletRequest;
import javax.swing.JOptionPane;
import org.apache.struts2.interceptor.ServletRequestAware;

public class AddDayAction extends ActionSupport implements ServletRequestAware{
    private String year;
    private String month;
    private String day;
    private String thing;
    private String date;
    private String userName;
    private ResultSet rs=null;
    private String message="error";
    private HttpServletRequest request;
      public String getYear() {
         return year;
    }
    public void setYear(String year) {
         this.year=year;
    }
    public String getMonth() {
         return month;
    }
    public void setMonth(String month) {
         this.month=month;
    }
    public String getDay() {
         return day;
    }
    public void setDay(String day) {
         this.day=day;
    }
    public String getThing() {
         return thing;
    }
    public void setThing(String thing) {
         this.thing=thing;
    }
    public String getTime(){
        String time="";
        SimpleDateFormat ff=new SimpleDateFormat("yyyy-MM-dd");
        Date d=new Date();
        time=ff.format(d);
        return time;
```

```java
    }
    public void message(String msg){
        int type=JOptionPane.YES_NO_CANCEL_OPTION;
        String title="信息提示";
        JOptionPane.showMessageDialog(null, msg, title, type);
    }
    public void setServletRequest(HttpServletRequest hsr) {
        request=hsr;
    }
    public void validate(){
        String mess="";
        boolean Y=true,M=true,D=true;
        boolean DD=false;
        String time=getTime();
        StringTokenizer token=new StringTokenizer(time,"-");
        if(this.getYear()==null||this.getYear().length()==0){
            Y=false;
            mess=mess+" * 年份";
            addFieldError("year","年份不允许为空!");
        }else if(Integer.parseInt("20"+this.getYear())<Integer.parseInt(
                token.nextToken())||this.getYear().length()!=2){
            DD=true;
            addFieldError("year","请正确填写年份!");
        }
        if(this.getMonth()==null||this.getMonth().length()==0){
            M=false;
            mess=mess+" * 月份";
            addFieldError("month","月份不允许为空!");
        }else if(this.getMonth().length()>2||Integer.parseInt(
                this.getMonth())<0||Integer.parseInt(this.getMonth())>12){
            DD=true;
            addFieldError("month","请正确填写月份!");
        }
        if(this.getDay()==null||this.getDay().length()==0){
            D=false;
            mess=mess+" * 日期";
            addFieldError("day","日期不允许为空!");
        }else if(this.getDay().length()>2||Integer.parseInt(
                this.getDay())<0||Integer.parseInt(this.getDay())>31){
            DD=true;
            addFieldError("day","请正确填写日程!");
        }
        if(Y&&M&&D){
            try{
                DB mysql=new DB();
```

```
            userName=mysql.returnLogin(request);
            date="20"+this.getYear()+"-"+this.getMonth()+"-"+this.getDay();
            rs=mysql.selectDay(request, userName, date);
            if(rs.next()){
                message("该日程已有安排!");
                addFieldError("year","该日程已有安排!");
            }
        }catch(Exception e){
            e.printStackTrace();
        }
    }
    if(this.getThing()==null||this.getThing().length()==0){
        mess=mess+" * 日程安排";
        addFieldError("thing","日程安排不允许为空!");
    }
    if(!mess.equals("")){
        mess=mess+"不允许为空!";
        message(mess);
    }
    if(DD){
        message("填写的日程无效!");
    }
}
public String execute() throws Exception{
    DB mysql=new DB();
    userName=mysql.returnLogin(request);
    date="20"+this.getYear()+"-"+this.getMonth()+"-"+this.getDay();
    String dd=mysql.insertDay(request, userName, date, this.getThing());
    if(dd.equals("ok")){
        message=SUCCESS;
    }else if(dd.equals("one")){
        message=INPUT;
    }
    return message;
}
}
```

在图4-16所示页面中的"日程时间"内输入数据后单击"修删日程"按钮,出现如图4-17所示的修改和删除日程页面(findDay.jsp)。单击"修删日程"按钮后请求提交到业务控制器类FindDayAction。请求提交后可以修改日程。

findDay.jsp代码如下:

```
<%@page import="JavaBean.MyDayBean"%>
<%@page import="java.util.ArrayList"%>
<%@page contentType="text/html" pageEncoding="UTF-8"%>
<%@taglib  prefix="s" uri="/struts-tags" %>
```

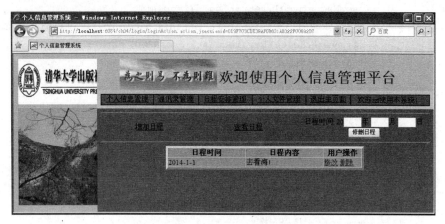

图 4-17 修删日程页面

```
<html>
    <head>
        <meta http-equiv="Content-Type" content="text/html; charset=UTF-8">
        <title><s:text name="个人信息管理系统"></s:text></title>
    </head>
    <body bgcolor="gray">
      <hr noshade/>
      <s:div align="center">
      <s:form action="findDayAction" method="post">
        <table border="0" cellspacing="0" cellpadding="0" width="100%"
            align="center">
          <tr>
            <td width="30%">
                <s:a href="http://localhost:8084/ch04/dateTimeManager
                    /addDay.jsp">增加日程</s:a>
            </td>
            <td width="30%">
                <s:a href="http://localhost:8084/ch04/dateTimeManager
                    /lookDay.jsp">查看日程</s:a>
            </td>
            <td width="40%">
                <s:text name="日程时间:"></s:text>
                20<input type="text" size="1" name="year"/>年
              <input type="text" size="1" name="month"/>月
              <input type="text" size="1" name="day"/>日
                <input type="submit" value="修删日程"/>
            </td>
          </tr>
        </table>
      </s:form>
      </s:div>
```

```
        <hr noshade/>
        <table border="5" cellspacing="0" cellpadding="0" bgcolor="#95BDFF"
              width="60%" align="center">
          <tr>
              <th width="40%">日程时间</th>
              <th width="40%">日程内容</th>
              <th width="20%">用户操作</th>
          </tr>
          <%
            ArrayList day= (ArrayList)session.getAttribute("findday");
            if(day==null||day.size()==0){
            %>
            <s:div align="center"><%="您还没有任何日程安排!"%></s:div>
            <%
          }else{
              for(int i=day.size()-1;i>=0;i--){
                  MyDayBean dd= (MyDayBean)day.get(i);
                  %>
                  <tr>
                      <td><%=dd.getDay()%></td>
                      <td><%=dd.getThing()%></td>
                      <td>
                      <s:a href="http://localhost:8084/ch04
                          /dateTimeManager/updateDay.jsp">修改</s:a>
                      <s:a href="deleteDayAction">删除</s:a>
                      </td>
                  </tr>
                  <%
              }
          }
          %>
        </table>
    </body>
</html>
```

findDay.jsp 页面使用的 JavaBean 类是 MyDayBean,其代码前面已给出。findDay.jsp 页面对应的业务控制器类为 FindDayAction。

FindDayAction.java 的代码如下:

```
package edu.dateTimeManager.Action;
import DBJavaBean.DB;
import com.opensymphony.xwork2.ActionSupport;
import java.sql.ResultSet;
import java.text.SimpleDateFormat;
import java.util.Date;
import java.util.StringTokenizer;
```

```java
import javax.servlet.http.HttpServletRequest;
import javax.swing.JOptionPane;
import org.apache.struts2.interceptor.ServletRequestAware;

public class FindDayAction extends ActionSupport implements ServletRequestAware{
    private String year;
    private String month;
    private String day;
    private String userName;
    private String date;
    private ResultSet rs=null;
    private String message="error";
    private HttpServletRequest request;
    public String getYear() {
        return year;
    }
    public void setYear(String year) {
        this.year=year;
    }
    public String getMonth() {
        return month;
    }
    public void setMonth(String month) {
        this.month=month;
    }
    public String getDay() {
        return day;
    }
    public void setDay(String day) {
        this.day=day;
    }
    public String getTime(){
        String time="";
        SimpleDateFormat ff=new SimpleDateFormat("yyyy-MM-dd");
        Date d=new Date();
        time=ff.format(d);
        return time;
    }
      public void setServletRequest(HttpServletRequest hsr) {
        request=hsr;
    }
    public void message(String msg){
        int type=JOptionPane.YES_NO_CANCEL_OPTION;
        String title="信息提示";
        JOptionPane.showMessageDialog(null, msg, title, type);
```

```
    }
    public void validate(){
        String mess="";
        boolean Y=true,M=true,D=true;
        boolean DD=false;
        String time=getTime();
        StringTokenizer token=new StringTokenizer(time,"-");
        if(this.getYear()==null||this.getYear().length()==0){
            Y=false;
            mess=mess+" * 年份";
            addFieldError("year","年份不允许为空!");
        }else if(Integer.parseInt("20"+this.getYear())<Integer.parseInt(
                token.nextToken())||this.getYear().length()!=2){
            DD=true;
            addFieldError("year","请正确填写年份!");
        }
        if(this.getMonth()==null||this.getMonth().length()==0){
            M=false;
            mess=mess+" * 月份";
            addFieldError("month","月份不允许为空!");
        }else if(this.getMonth().length()>2||Integer.parseInt(
                this.getMonth())<0||Integer.parseInt(this.getMonth())>12){
            DD=true;
            addFieldError("month","请正确填写月份!");
        }
        if(this.getDay()==null||this.getDay().length()==0){
            D=false;
            mess=mess+" * 日期";
            addFieldError("day","日期不允许为空!");
        }else if(this.getDay().length()>2||Integer.parseInt(
                this.getDay())<0||Integer.parseInt(this.getDay())>31){
            DD=true;
            addFieldError("day","请正确填写日程!");
        }
        if(Y&&M&&D){
            try{
                DB mysql=new DB();
                userName=mysql.returnLogin(request);
                date="20"+this.getYear()+"-"+this.getMonth()+"-"+this.getDay();
                rs=mysql.selectDay(request, userName, date);
                if(!rs.next()){
                    message("该日程暂无安排!");
                    addFieldError("year","该日程暂无安排!");
                }
            }catch(Exception e){
```

```
            e.printStackTrace();
        }
    }
    if(!mess.equals("")){
        mess=mess+"不允许为空!";
        message(mess);
    }
    if(DD){
        message("填写的日程无效!");
    }
}
public String execute() throws Exception {
    DB mysql=new DB();
    userName=mysql.returnLogin(request);
    date="20"+this.getYear()+"-"+this.getMonth()+"-"+this.getDay();
    String dd=mysql.findDay(request, userName, date);
    if(dd.equals("ok")){
        message=SUCCESS;
    }
    return message;
}
}
```

单击图 4-17 所示页面中日程后面的"修改",出现如图 4-18 所示的修改日程页面(updateDay.jsp)。可对日程信息进行修改。updateDay.jsp 页面对应的业务控制器类为 UpDayAction。

图 4-18 修改日程页面

updateDay.jsp 代码如下:

```
<%@page import="JavaBean.MyDayBean"%>
<%@page import="java.util.StringTokenizer"%>
<%@page import="java.util.ArrayList"%>
<%@page contentType="text/html" pageEncoding="UTF-8"%>
```

```jsp
<%@taglib prefix="s" uri="/struts-tags"%>
<html>
    <head>
        <meta http-equiv="Content-Type" content="text/html; charset=UTF-8">
        <title><s:text name=""></s:text></title>
    </head>
    <body bgcolor="gray">
        <hr noshade/>
    <s:div align="center">
    <s:form action="findDayAction" method="post">
    <table border="0" cellspacing="0"cellpadding="0"
            width="100%"align="center">
        <tr>
            <td width="30%">
                <s:a href="http://localhost:8084/ch04/dateTimeManager
                    /addDay.jsp">增加日程</s:a>
            </td>
            <td width="30%">
                <s:a href="http://localhost:8084/ch04/dateTimeManager
                    /lookDay.jsp">查看日程</s:a>
            </td>
            <td width="40%">
                <s:text name="日程时间:"></s:text>
                20<input type="text" size="1" name="year"/>年
                <input type="text" size="1" name="month"/>月
                <input type="text" size="1" name="day"/>日
                <input type="submit" value="修删日程"/>
            </td>
        </tr>
    </table>
    </s:form>
    </s:div>
    <hr noshade/>
    <s:form action="upDayAction" method="post">
        <table border="5" cellspacing="0" cellpadding="0" bgcolor="#95BDFF"
            width="60%" align="center">
            <%
                ArrayList day=(ArrayList)session.getAttribute("findday");
                if(day==null||day.size()==0){
            %>
                <s:div align="center"><%="您还没有任何日程安排!"%>
                </s:div>
            <%
            }else{
                for(int i=day.size()-1;i>=0;i--){
```

```jsp
            MyDayBean dd= (MyDayBean)day.get(i);
            StringTokenizer token=new StringTokenizer(dd.getDay().
                        substring(2, dd.getDay().length()),"-");
        %>
        <tr>
            <td height="30" width="50%" align="right">
                        日程时间
            </td>
            <td width="50%">
                20<input type="text" size="1" name="year"
                    value="<%=token.nextToken()%>"/>年
                <input type="text" size="1" name="month"
                    value="<%=token.nextToken()%>"/>月
                <input type="text" size="1" name="day"
                    value="<%=token.nextToken()%>"/>日
            </td>
        </tr>
        <tr>
            <td height="30" width="50%" align="right">日程内容
            </td>
            <td width="50%"><input type="text" size="30"
                name="thing" value="<%=dd.getThing()%>"/>
            </td>
        </tr>
        <tr>
            <td colspan="2" align="center">
                <input type="submit" value="确 定" size="12">
                     
                <input type="reset" value="清 除" size="12">
            </td>
        </tr>
        <%
        }
    }
    %>
        </table>
    </s:form>
  </body>
</html>
```

UpdateDayAction.java 的代码如下：

```java
package edu.dateTimeManager.Action;
import DBJavaBean.DB;
import com.opensymphony.xwork2.ActionSupport;
import java.sql.ResultSet;
```

```java
import java.text.SimpleDateFormat;
import java.util.Date;
import java.util.StringTokenizer;
import javax.servlet.http.HttpServletRequest;
import javax.swing.JOptionPane;
import org.apache.struts2.interceptor.ServletRequestAware;

public class UpdateDayAction extends ActionSupport implements ServletRequestAware{
    private String year;
    private String month;
    private String day;
    private String thing;
    private String message="error";
    private HttpServletRequest request;
    private String userName;
    private String Day;
    private String date;
    private ResultSet rs=null;
    public String getYear() {
        return year;
    }
    public void setYear(String year) {
        this.year=year;
    }
    public String getMonth() {
        return month;
    }
    public void setMonth(String month) {
        this.month=month;
    }
    public String getDay() {
        return day;
    }
    public void setDay(String day) {
        this.day=day;
    }
    public String getThing() {
        return thing;
    }
    public void setThing(String thing) {
        this.thing=thing;
    }
    public void setServletRequest(HttpServletRequest hsr) {
        request=hsr;
    }
```

```java
public String getTime(){
    String time="";
    SimpleDateFormat ff=new SimpleDateFormat("yyyy-MM-dd");
    Date d=new Date();
    time=ff.format(d);
    return time;
}
public void message(String msg){
    int type=JOptionPane.YES_NO_OPTION;
    String title="信息提示";
    JOptionPane.showMessageDialog(null,msg,title,type);
}
public void validate(){
    String mess="";
    boolean DD=false;
    String time=getTime();
    StringTokenizer token=new StringTokenizer(time,"-");
    if(this.getYear()==null||this.getYear().length()==0){
        mess=mess+" * 年份";
        addFieldError("year","年份不允许为空!");
    }else if(Integer.parseInt("20"+this.getYear())
            <Integer.parseInt(
            token.nextToken())||this.getYear().length()!=2){
        DD=true;
        addFieldError("year","请正确填写年份!");
    }
    if(this.getMonth()==null||this.getMonth().length()==0){
        mess=mess+" * 月份";
        addFieldError("month","月份不允许为空!");
    }else if(this.getMonth().length()>2||Integer.parseInt(
            this.getMonth())<0||Integer.parseInt(this.getMonth())>12){
        DD=true;
        addFieldError("month","请正确填写月份!");
    }
    if(this.getDay()==null||this.getDay().length()==0){
        mess=mess+" * 日期";
        addFieldError("day","日期不允许为空!");
    }else if(this.getDay().length()>2||Integer.parseInt(
            this.getDay())<0||Integer.parseInt(this.getDay())>31){
        DD=true;
        addFieldError("day","请正确填写日程!");
    }
    if(this.getThing()==null||this.getThing().length()==0){
        mess=mess+" * 日程安排";
        addFieldError("thing","日程安排不允许为空!");
```

```
        }
        if(!mess.equals("")){
            mess=mess+"不允许为空!";
            message(mess);
        }
        if(DD){
            message("填写的日程无效!");
        }
    }
    public String execute() throws Exception {
        DB mysql=new DB();
        userName=mysql.returnLogin(request);
        Day=mysql.returnDay(request);
        date="20"+this.getYear()+"-"+this.getMonth()+"-"+this.getDay();
        String D=mysql.updateDay(request, userName,Day, date, thing);
        if(D.equals("ok")){
            message=SUCCESS;
        }else if(D.equals("one")){
            message=INPUT;
        }
        return message;
    }
}
```

单击图 4-17 所示页面中的"删除",请求提交到 DeleteFriAction,该类是一个业务控制器 DeleteDayAction,实现日程删除功能。

DeleteDayAction.java 的代码如下:

```
package edu.dateTimeManager.Action;
import DBJavaBean.DB;
import com.opensymphony.xwork2.ActionSupport;
import javax.servlet.http.HttpServletRequest;
import org.apache.struts2.interceptor.ServletRequestAware;

public class DeleteDayAction extends ActionSupport implements
ServletRequestAware{
    private String message="error";
    private String userName;
    private String day;
    private HttpServletRequest request;
        public void setServletRequest(HttpServletRequest hsr) {
        request=hsr;
    }
    public String execute() throws Exception {
        DB mysql=new DB();
        userName=mysql.returnLogin(request);
```

```
        day=mysql.returnDay(request);
        String dd=mysql.deleteDay(request, userName, day);
        if(dd.equals("ok")){
            message=SUCCESS;
        }
        return message;
    }
}
```

4.4.7 个人文件管理功能实现

单击图 4-18 所示页面中的"个人文件管理",出现如图 4-19 所示的文件列表页面(lookFile.jsp)。

图 4-19　个人文件信息页面

单击图 4-19 所示页面中的"上传文件",出现如图 4-20 所示的文件上传页面(fileUp.jsp)。

图 4-20　文件上传页面

在图 4-19 所示页面中的"文件标题"中输入文件名字后单击"下载"按钮,出现如图 4-21 所示的文件下载和删除页面(findFile.jsp),可以下载或删除文件。

图 4-21 文件下载和删除页面

以上页面对应的 JavaBean 以及业务控制器类分别为 MyFileBean.java、AddFileAction.java、FindFileAction.java、DeleteFileAction.java、DownFileAction.java。

4.5 本章小结

本章主要介绍了一个基于 Struts2 的个人信息管理系统的开发过程。通过本章实训的练习，能够在掌握所学理论知识的同时，提高基于 Struts2 的项目开发能力，激发基于 Struts2 的项目开发兴趣，并为集成 Hibernate 框架和 Spring 框架进行项目开发奠定基础。

4.6 习 题

实训题

1. 请根据自己对个人信息的管理经验进一步完善和扩展本章实训项目的功能。
2. 请使用内置验证器对本项目进行校验。
3. 请自己编程实现个人信息管理系统的"个人文件管理"功能。
4. 请为个人信息管理系统添加"个人课程管理"功能。

第 5 章 Hibernate 框架技术入门

在 Java Web 项目开发中,有许多功能模块需要连接数据库系统,实现对数据库表的操作。在 Java 程序设计和 JSP 程序设计技术中,使用 JDBC 技术连接数据库。为了实现与数据库的高效操作,提高 Java Web 的项目性能,可以使用 Hibernate 框架技术。本章主要介绍 Hibernate 的基本内容。

本章主要内容如下所示。
(1) Hibernate 的发展与特点。
(2) Hibernate 的下载与配置。
(3) Hibernate 的工作原理。
(4) Hibernate 的核心组件。
(5) 基于 Struts2＋Hibernate 的登录和注册系统。

5.1 Hibernate 基础知识

Hibernate 是封装了 JDBC 的一种开放源代码的对象/关系映射(Object-Relation Mapping,ORM)框架,使程序员可以使用面向对象的思想来操作数据库。Hibernate 是一种对象/关系映射的解决方案,即将 Java 对象与对象之间的关系映射到数据库表与表之间的关系。

5.1.1 Hibernate 的发展与特点

目前,Hibernate 是 Java 工程师招聘中要求必备的一门技能,也是 Java Web 三大经典框架之一。Christian Bauer 是 Hibernate 之父,现是 Redhat 公司负责开发和维护 Hibernate 的项目经理。

2001 年,Hibernate1 发布,即 Hibernate 的第一个版本;2003 年,Hibernate2 发布,并在当年获得 Jolt2003 大奖(Jolt 大奖素有"软件业界的奥斯卡"之美誉,共设通用类图书、技术类图书、语言和开发环境、框架库和组件、开发者网站等十余个分类大奖),2003 年 Hibernate 被 JBoss 公司收购,成为该公司的子项目之一;2005 年,JBoss 发布 Hibernate3;2006 年,JBoss 公司被 Redhat 公司收购;2012 年 11 月发布 Hibernate 4.1.8。

Hibernate 是封装了 JDBC 与 ORM 技术的数据持久性解决方案。在 Java 世界中,Hibernate 是众多 ORM 软件中获得关注最多、使用最广泛的框架。它成功地实现了透明持久化,以面向对象的 HQL 语句封装 SQL 语句,为开发人员提供了一个简单灵活且面向对象的数据访问接口。Hibernate 是一个开源软件,开发人员可以很方便地获得软件源代码。当遇到问题时,程序员可以深入到源代码中查看究竟,甚至修改 Hibernate 内部错误并将修改方案提供给 JBoss 组织,从而帮助 Hibernate 框架技术改进。

Hibernate 自发布以来受到业界的欢迎,目前许多 Java 程序员学习和使用它来开发商业应用软件。另外,网络上有大量介绍和讨论 Hibernate 应用的文章,JBoss 网站也提供了一个完善的社区,所以一旦在使用中遇到问题,开发者可以轻松地在网络上搜索到相应的解决方法,这又进一步吸引了更多的程序员来学习 Hibernate,吸引更多的公司采用 Hibernate 开发软件。

Hibernate 为使用者考虑得十分周全,对于一个普通的程序员来说,只需学习不到 10 个类的用法就可以开发基于 Hibernate 框架的应用系统,实际使用起来十分方便。

Hibernate 提供了透明持久化功能,支持第三方框架,即能与其他框架进行整合,如 Struts、Spring 等,不但提供面向对象的 HQL,而且支持传统的 SQL 语句。

在基于 MVC 设计模式的 Java Web 应用中,Hibernate 可以作为应用的数据访问层或持久层。它具有以下特点。

(1) Hibernate 是一个开放源代码的对象关系映射框架,它对 JDBC 进行了非常轻量级的对象封装,使得 Java 工程师可以随心所欲地使用面向对象编程思维来操作数据库。Hibernate 可以应用在任何使用 JDBC 的场合,既可以在 Java 的客户端程序使用,也可以在 Servlet、JSP 的 Java Web 应用中使用,最具革命意义的是,Hibernate 可以在 Java EE 框架中取代 CMP,完成数据持久化的重任。

(2) Hibernate 的目标是成为 Java 中处理数据持久性问题的一种完整解决方案。它协调应用程序与关系型数据库的交互,把开发者解放出来专注于项目的业务逻辑问题。

(3) Hibernate 是一种非强迫性的解决方案。开发者在写业务逻辑和持久化类时,不会被要求遵循许多 Hibernate 特定的规则和设计模式。这样,Hibernate 就可以与大多数新的和现有的应用程序进行集成,而不需要对应用程序的其余部分做破坏性的改动。

5.1.2 Hibernate 软件包的下载和配置

由于许多软件公司现在主要使用的是 Hibernate3 版本,本书的实例和项目也使用 Hibernate3 版本。本书使用的是 Hibernate 3.6.0。Hibernate 4.1.8 于 2012 年 11 月发布,如需使用 Hibernate4 进行 Java Web 项目开发,可以在其官方网站下载。

1. 软件包下载

由于 Hibernate 先被 JBoss 公司收购,后来 JBoss 被 Redhat 公司收购,所以 Hibernate 可以在以下 3 个网站下载:www.redhat.com、www.jboss.org、www.hibernate.org。可根据需要在上述 3 个网站下载要使用的 Hibernate 版本。下载页面如图 5-1 所示。

单击图 5-1 所示页面的上边或者右侧的 Download,出现如图 5-2 所示的下载页面。

单击图 5-2 所示页面中的 release bundles,出现如图 5-3 所示的 Hibernate 不同版本下载选择页面。

单击图 5-3 所示页面中需要下载的 Hibernate 版本,如 Hibernate 4.1.8,出现如图 5-4 所示的页面,单击 4.1.8.Final,出现如图 5-5 所示页面后,在该页面中选择 hibernate-release-4.1.8.Final.zip 进行下载。

2. Hibernate4 软件包中主要文件

解压缩 hibernate-release-4.1.8.Final.zip 文件后得到一个名为 hibernate-release-4.1.8.Final 的文件夹,该文件夹结构如图 5-6 所示。

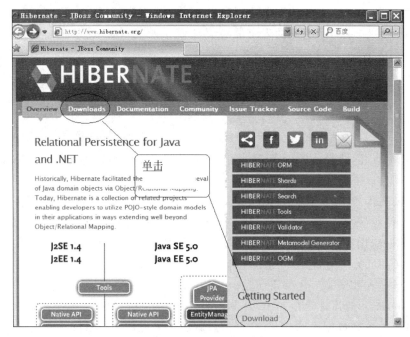

图 5-1　Hibernate4 下载页面

图 5-2　Release bundle 下载页面

图 5-3 Hibernate 版本选择页面

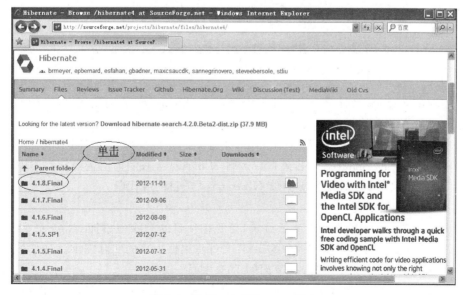

图 5-4 选择一个 Hibernate 版本

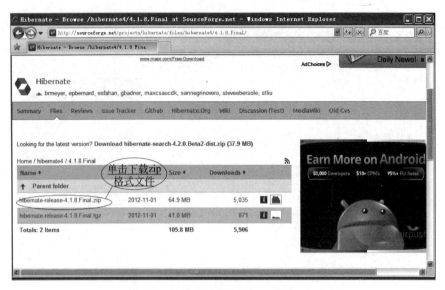

图 5-5　下载 Hibernate4 文件

图 5-6　Hibernate4 文件夹结构

（1）documentation 文件夹：该文件夹中存放了 Hibernate4 的相关文档，包括 Hibernate4 的参考文档和 API 文档等。

（2）lib 文件夹：该文件夹存放 Hibernate4 框架的核心类库以及 Hibernate4 的第三方类库。该文件夹下的 required 子目录存放运行 Hibernate4 项目时必需的核心类库。

（3）project 文件夹：该文件夹存放 Hibernate4 项目的源代码。

3. Hibernate 的配置

Hibernate 的 lib 文件夹有 4 个子目录，需要在项目的类库中添加 required 和 jpa 子目录下面的所有 JAR 文件，其他目录中的 JAR 文件可根据项目的实际需求添加。例如，使用连接池需要添加"lib\optional\c3p0"下面的 JAR 文件。

由于 NetBeans 7.2 和 MyEclipse 10.6 中都集成有 Hibernate，所以可以使用工具中自带的 Hibernate。集成的 Hibernate 版本一般不是 Hibernate 的最新版本。NetBeans 7.2 集成的是 Hibernate 3.2.5；MyEclipse 10.6 集成有 Hibernate 3.1、3.2、3.3 和 4.1。由于

Hibernate 各版本之间存在一些细节差异,有可能在配置文件和映射文件中存在差异导致项目无法运行。使用本书进行项目开发时建议使用 Hibernate 3.6.0。在 NetBeans 7.2、MyEclipse 10.6 和 Eclipse 4.2 中配置 Hibernate 与第 1 章中介绍的配置 Struts 2.3.4 的方法相似,这里不再介绍。

5.1.3 Hibernate 的工作原理

Hibernate 的工作原理如图 5-7 所示。

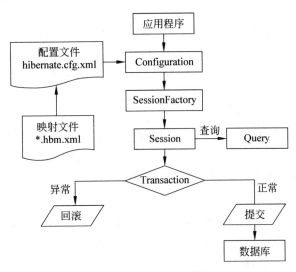

图 5-7 Hibernate 的工作原理

Hibernate 的工作过程如下。

首先,Configuration 读取 Hibernate 的配置文件 hibernate.cfg.xml 和映射文件 *.hbm.xml 中的信息,即加载配置文件和映射文件,并通过 Hibernate 配置文件生成一个多线程的 SessionFactory 对象;然后,多线程 SessionFactory 对象生成一个线程 Session 对象;Session 对象生成 Query 对象或者 Transaction 对象;可通过 Session 对象的 get()、load()、save()、update()、delete() 和 saveOrUpdate() 等方法对 PO 进行加载、保存、更新、删除等;也可利用 Query 对象执行查询操作;如果没有异常,Transaction 对象将提交这些数据到数据库中。

5.2 Hibernate 的核心组件

在基于 Hibernate 框架的项目开发时,非常关键的一点就是要使用 Hibernate 的核心类和接口,即核心组件。Hibernate 的核心组件位于业务层和持久化层之间。Hibernate 除核心组件外还包括 Hibernate 配置文件(hibernate.cfg.xml 或 hibernate.properties)、映射文件(xxx.hbm.xml)和持久化类(Persistent Objects,PO)。

1. Configuration

Configuration 负责配置并启动 Hibernate,创建 SessionFactory 对象。在 Hibernate 的启动过程中,Configuration 类的实例首先定位映射文件位置、读取配置,然后创建

SessionFactory 对象。

2. SessionFactory

SessionFactroy 负责初始化 Hibernate。它充当数据存储源的代理,并负责创建 Session 对象,这里用到了工厂模式。需要注意的是 SessionFactory 并不是轻量级的,因为一般情况下,一个项目通常只需要一个 SessionFactory 就可以了,当需要多次操作数据库时,可以为每个数据库指定一个 SessionFactory 线程对象。SessionFactroy 是产生 Session 实例的工厂。

3. Session

Session 负责执行持久化对象的操作,它用 get()、load()、save()、update() 和 delete() 等方法来对 PO 进行加载、保存、更新及删除等操作。但需要注意的是 Session 对象是非线程安全的。同时,Hibernate 的 Session 不同于 JSP 应用中的 HttpSession。这里使用的 Session 术语,其实指的是 Hibernate 中的 Session。

4. Transaction

Transaction 负责事务相关的操作,用来管理 Hibernate 事务,它的主要方法有 commit() 和 rollback(),可以使用 Session 的 beginTransaction() 方法生成。它是可选的,开发人员也可以设计编写自己的底层事务处理代码。

5. Query

Query 负责执行各种数据库查询。它可以使用 HQL 语言对 PO 进行查询操作。Query 对象可以使用 Session 的 createQuery() 方法生成。

6. Hibernate 的配置文件

Hibernate 的配置文件主要用来配置数据库连接参数,例如,数据库的驱动程序、URL、用户名和密码、数据库方言等。它有两种格式:hibernate.cfg.xml 和 hibernate.properties。两者的配置内容基本相同,但前者比后者使用方便一些,如 hibernate.cfg.xml 可以在其 <mapping> 子元素中定义用到的 xxx.hbm.xml 映射文件列表,而使用 hibernate.properties 时则需要在程序中以编码方式指明映射文件。一般情况下,hibernate.cfg.xml 是 Hibernate 的默认配置文件。

7. 映射文件

映射文件(xxx.hbm.xml)用来把 PO 与数据库中的表、PO 之间的关系与表之间的关系以及 PO 的属性与表字段一一映射起来,它是 Hibernate 的核心文件。

8. 持久化对象

持久化对象 PO 可以是普通的 JavaBean,唯一特殊的是它们与 Session 相关联。PO 在 Hibernate 中存在三种状态:临时状态(Transient)、持久化状态(Persistent)和脱管状态(Detached)。当一个 JavaBean 对象在内存中孤立存在不与数据库中的数据有任何关联关系时,那么这个 JavaBean 对象就称为临时对象(Transient Object);当它与一个 Session 相关联时,就变成持久化对象(Persistent Object);在这个 Session 被关闭的同时,这个对象也会脱离持久化状态,变成脱管对象(Detached Object),这时可以被应用程序的任何层自由使用,例如,可用做与表示层(V)打交道的数据传输对象。

5.3 基于 Struts2 和 Hibernate 的登录和注册系统

下面使用 Struts 2.3.4 和 Hibernate 3.6.0 开发一个实现登录和注册功能的项目,该项目的文件结构如图 5-8 所示。将项目中使用到的 Struts 2.3.4、Hibernate 3.6.0 以及 MySQL 5.0 驱动的 JAR 文件添加到项目 ch05 的"库"中,如图 5-9 所示。

图 5-8 项目文件结构图

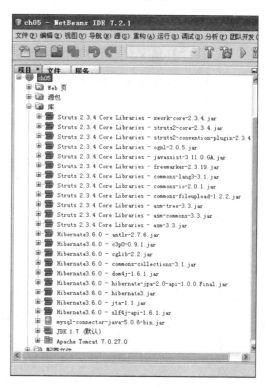

图 5-9 项目所需的 JAR 文件

1. 项目介绍

本项目实现用户登录和用户注册功能。有一个登录页面(login.jsp),代码如例 5-1 所示,登录页面对应的业务控制器为 LoginAction,该 Action 中覆盖了 validate()方法,使用手工验证对登录页面进行验证,该业务控制器类代码如例 5-4 所示,如果输入的用户名和密码正确,进入登录成功页面(success.jsp),代码如例 5-2 所示;如果用户没有注册需先注册,注册页面(register.jsp)代码如例 5-3 所示,该注册页面对应的业务控制器为 RegisterAction,代码如例 5-5 所示,注册成功后返回登录页面。还需要配置 web.xml,代码如例 1-3 所示;在 struts.xml 中配置 Action,代码如例 5-6 所示。

该项目使用 MySQL 数据库。数据库名为 test,有一张名为 info 的表,表的字段名称、类型以及长度如图 5-10 所示。

在基于 Struts2+Hibernate 的项目开发中,连接数据库时需要 Hibernate 的配置文件 hibernate.cfg.xml 或者 hibernate.properties,本项目使用的是 hibernate.cfg.xml,代码如例 5-7 所示,配置文件主要用于加载数据库的驱动以及与数据库建立连接等,该配置文件一

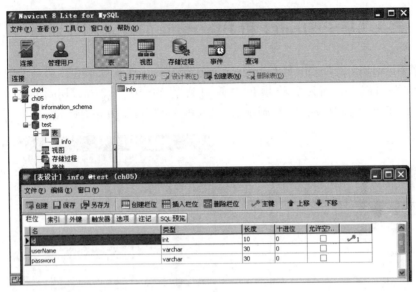

图 5-10　数据库 info 表的结构

般与 struts.xml 文件放在同一位置,参考图 5-8。该配置文件在项目运行时需要加载,本项目编写一个加载该配置文件的类 HibernateSessionFactory,该类封装了配置文件的加载方法,即加载配置文件的 JavaBean,代码如例 5-8 所示。

另外,在使用 Hibernate 时,项目所用数据库中的每一张表都对应一个持久化对象 PO。通过 PO 对象把页面中的数据保存起来并把数据存到数据库中。为了简化开发,本项目的登录页面和注册页面都使用同一张表,登录页面和注册页面中的数据都保存在同一个 PO 对象 UserInfoPO 中,代码如例 5-9 所示。每个 PO 对象一般都会对应一个映射文件 UserInfoPO.hbm.xml,代码如例 5-10 所示,映射文件配置 PO 对象与数据库中表之间的映射关系;映射文件一般都放在和其对应的 PO 对象所在包中。为了封装登录和注册功能对数据的操作,即登录和注册页面要实现的业务逻辑,本项目编写一个名为 LoginRegisterInfo 的 JavaBean,该类提供了登录和注册业务处理,可以实现登录和注册的持久化业务操作,代码如例 5-11 所示。

图 5-11　登录页面

以上文件所在的路径可参考项目文件结构图 5-8。

2. 在 web.xml 中配置核心控制器 FilterDispatcher

参考 1.3.1 节中的例 1-3。

3. 编写视图组件（JSP 页面）

登录页面（login.jsp）运行效果如图 5-11 所示，代码如例 5-1 所示。

【例 5-1】 登录页面（login.jsp）。

```jsp
<%@page contentType="text/html" pageEncoding="UTF-8"%>
<%@taglib  prefix="s" uri="/struts-tags" %>
<html>
    <head>
        <meta http-equiv="Content-Type" content="text/html; charset=UTF-8">
        <title><s:text name="基于SH的登录和注册系统"></s:text></title>
    </head>
    <body bgcolor="#CCCCFF">
        <s:form action="login" method="post">
            <br><br><br><br><br><br>
            <table border="1" align="center" bgcolor="#AABBCCDD">
                <tr>
                    <td>
                        <s:textfield name="userName" label="用户名字"
                            size="16"/>
                    </td>
                </tr>
                <tr>
                    <td>
                        <s:password name="password" label="用户密码"
                            size="18"/>
                    </td>
                </tr>
                <tr>
                    <td colspan="2" align="center">
                        <s:submit value="登录"/>
                    </td>
                </tr>
                <tr>
                    <td colspan="2" align="center">
                        <s:a href="http://localhost:8084/ch05/register.jsp">注册
                        </s:a>
                    </td>
                </tr>
            </table>
        </s:form>
    </body>
</html>
```

登录成功页面(success.jsp),代码如例 5-2 所示。

【例 5-2】 登录成功页面(success.jsp)。

```
<%@page contentType="text/html" pageEncoding="UTF-8"%>
<%@taglib prefix="s" uri="/struts-tags" %>
<html>
    <head>
        <meta http-equiv="Content-Type" content="text/html; charset=UTF-8">
        <title><s:text name="基于 SH 的登录和注册系统"></s:text></title>
    </head>
    <body bgcolor="#CCCCFF">
        <hr>
        <table border="0" align="center" bgcolor="#AABBCCDD">
            <tr>
                <td>
                    欢迎${userName},登录成功!
                </td>
            </tr>
        </table>
        <hr>
    </body>
</html>
```

注册页面(register.jsp)如图 5-12 所示,代码如例 5-3 所示。

图 5-12 注册页面

【例 5-3】 注册页面(register.jsp)。

```
<%@page contentType="text/html" pageEncoding="UTF-8"%>
<%@taglib prefix="s" uri="/struts-tags" %>
<html>
    <head>
        <meta http-equiv="Content-Type" content="text/html; charset=UTF-8">
        <title><s:text name="基于 SH 的登录和注册系统"></s:text></title>
    </head>
    <body bgcolor="#CCCCFF">
```

```html
<s:form action="register" method="post">
    <br><br><br><br><br><br>
    <table border="1" align="center" bgcolor="#AABBCCDD">
        <tr>
            <td>
                <s:textfield name="userName" label="用户名字"
                    size="16"/>
            </td>
        </tr>
        <tr>
            <td>
                <s:password name="password1" label="用户密码"
                    size="18"/>
            </td>
        </tr>
        <tr>
            <td>
                <s:password name="password2" label="再次输入密码"
                    size="18"/>
            </td>
        </tr>
        <tr>
            <td colspan="2" align="center">
                <input type="submit" value="提交"/>  
                       
                <input type="reset" value="清空"/>
            </td>
        </tr>
        <tr>
            <td colspan="2" align="center">
                <s:a href="http://localhost:8084/ch05/login.jsp">返回
                </s:a>
            </td>
        </tr>
    </table>
</s:form>
</body>
</html>
```

4. 编写业务控制器 Action

登录页面对应的业务控制器是 LoginAction，代码如例 5-4 所示。

【例 5-4】 登录页面对应的业务控制器（LoginAction.java）。

```java
package loginRegisterAction;
import com.opensymphony.xwork2.ActionSupport;
import loginRegisterDao.LoginRegisterInfo;
```

```java
import PO.UserInfoPO;
import java.util.List;

public class LoginAction extends ActionSupport{
    private String userName;
    private String password;
    private String message="error";
    private List list;
    public String getUserName() {
        return userName;
    }
    public void setUserName(String userName) {
        this.userName=userName;
    }
    public String getPassword() {
        return password;
    }
    public void setPassword(String password) {
        this.password=password;
    }
    public void validate(){
        if(this.getUserName()==null||this.getUserName().length()==0){
            addFieldError("userName","用户名不能为空!");
        }else{
            LoginRegisterInfo info=new LoginRegisterInfo();
            list=info.queryInfo("userName", this.getUserName());
            if(list.size()==0){
                addFieldError("userName","该用户尚未注册!");
            }else{
                UserInfoPO ui=new UserInfoPO();
                int count=0;
                for(int i=0;i<list.size();i++){
                    count++;
                    ui= (UserInfoPO)list.get(i);
                    if(this.getUserName().equals(ui.getUserName())){
                        if(ui.getPassword().equals(this.getPassword())){
                            message=SUCCESS;
                        }else{
                            addFieldError("password","登录密码不正确!");
                        }
                    }
                }
            }
        }
    }
```

```java
    public String execute() throws Exception{
        return message;
    }
}
```

【例 5-5】 注册页面对应的业务控制器(RegisterAction.java)。

```java
package loginRegisterAction;
import PO.UserInfoPO;
import com.opensymphony.xwork2.ActionSupport;
import loginRegisterDao.LoginRegisterInfo;
import java.util.List;

public class RegisterAction extends ActionSupport{
    private String userName;
    private String password1;
    private String password2;
    private String mess="error";
    private List list;
    public String getUserName() {
        return userName;
    }
    public void setUserName(String userName) {
        this.userName=userName;
    }
    public String getPassword1() {
        return password1;
    }
    public void setPassword1(String password1) {
        this.password1=password1;
    }
    public String getPassword2() {
        return password2;
    }
    public void setPassword2(String password2) {
        this.password2=password2;
    }
    public void validate(){
        if(this.getUserName()==null||this.getUserName().length()==0){
            addFieldError("userName","用户名不能为空!");
        }else{
            LoginRegisterInfo info=new LoginRegisterInfo();
            list=info.queryInfo("userName", this.getUserName());
            UserInfoPO ui=new UserInfoPO();
            for(int i=0;i<list.size();i++){
                ui= (UserInfoPO)list.get(i);
```

```java
                if(ui.getUserName().equals(this.getUserName())){
                    addFieldError("userName","用户名已存在!");
                }
            }
        }
        if(this.getPassword1()==null||this.getPassword1().length()==0){
            addFieldError("password1","登录密码不允许为空!");
        }else if(this.getPassword2()==null||this.getPassword2().length()==0){
            addFieldError("password2","重复密码不允许为空!");
        }else if(!this.getPassword1().equals(this.getPassword2())){
            addFieldError("password2","两次密码不一致!");
        }
    }
    public UserInfoPO userInfo(){
        UserInfoPO info=new UserInfoPO();
        info.setUserName(this.getUserName());
        info.setPassword(this.getPassword1());
        return info;
    }
    public String execute() throws Exception{
        LoginRegisterInfo lr=new LoginRegisterInfo();
        String ri=lr.saveInfo(userInfo());
        if(ri.equals("success")){
            mess="success";
        }
        return mess;
    }
}
```

5. 修改 struts.xml 配置 Action

修改配置文件 struts.xml，代码如例 5-6 所示。

【例 5-6】 在 struts.xml 配置 Action(struts.xml)。

```xml
<!DOCTYPE struts PUBLIC
"-//Apache Software Foundation//DTD Struts Configuration 2.0//EN"
"http://struts.apache.org/dtds/struts-2.0.dtd">
<struts>
    <include file="example.xml"/>
    <package name="default" extends="struts-default">
        <action name="register" class="loginRegisterAction.RegisterAction">
            <result name="success">/login.jsp</result>
            <result name="input">/register.jsp</result>
            <result name="error">/register.jsp</result>
        </action>
        <action name="login" class="loginRegisterAction.LoginAction">
            <result name="success">/success.jsp</result>
```

```xml
            <result name="input">/login.jsp</result>
            <result name="error">/login.jsp</result>
        </action>
    </package>
</struts>
```

6. Hibernate 的配置文件

使用 Hibernate 需要通过 Hibernate 的配置文件加载数据库驱动以及与数据库建立连接,配置文件为 hibernate.cfg.xml,代码如例 5-7 所示。

【例 5-7】 Hibernate 的配置文件(hibernate.cfg.xml)。

```xml
<!--表明解析本 XML 文件的 DTD 文档位置,DTD 是 Document Type Definition 的缩写,即文档类
型的定义,XML 解析器使用 DTD 文档来检查 XML 文件的合法性。Hibernate 版本不一样配置文件中
的 DTD 信息会有一定的差异,例如,将 Hibernate 3.2 配置文件的 DTD 信息放在 Hibernate 3.6.0
的配置文件中使用时将导致找不到配置文件信息的异常或者其他异常。Hibernate 配置文件可以
在下载 Hibernate 的文件夹中找到。切记,使用与版本相对应的 DTD 信息-->
<!DOCTYPE hibernate-configuration PUBLIC "-//Hibernate/Hibernate Configuration
    DTD 3.0//EN" "http://hibernate.sourceforge.net/hibernate-configuration-
    3.0.dtd">
<!--Hibernate 配置文件的根元素,其他元素要在其中使用-->
<hibernate-configuration>
<!--指定初始化 Hibernate 参数的元素。其中,指定 Hibernate 初始化参数,表明以下的配置针
    对 session-factory。SessionFactory 是 Hibernate 中的一个接口,这个接口主要负责保存
    Hibernate 的配置信息以及对 Session 的操作-->
    <session-factory>
        <!--指定连接数据库所用的驱动,本例中的 com.mysql.jdbc.Driver 是 MySQL 的驱动
            名称-->
        <property name="connection.driver_class">com.mysql.jdbc.Driver</property>
        <!--设置数据库连接使用的 URL,"jdbc:mysql://localhost:port/test",其中,
            localhost 表示 MySQL 服务器名称,即连接地址,此处为本机。port 代表 MySQL 服务器
            的端口号,默认是 3306。test 是数据库名,即要连接的数据库名-->
        <property name="connection.url">jdbc:mysql://localhost:3306/test
        </property>
        <!--指定数据库的用户名(登录名)-->
        <property name="connection.username">root</property>
        <!--指定数据库的连接密码-->
        <property name="connection.password">root</property>
        <!--指定数据库的方言,每种数据库都有对应的方言-->
        <property name="dialect">org.hibernate.dialect.MySQLInnoDBDialect
        </property>
        <!--加入映射文件,可以加入多个映射文件-->
        <mapping resource="PO/UserInfoPO.hbm.xml"/>
    </session-factory>
</hibernate-configuration>
```

7. 加载 Hibernate 配置文件的类

加载 Hibernate.cfg.xml 文件的类为 HibernateSessionFactory,代码如例 5-8 所示。

【例 5-8】 加载 Hibernate.cfg.xml 文件的类（HibernateSessionFactory.java）。

```java
package addHibernateFile;
import org.hibernate.Session;
import org.hibernate.SessionFactory;
import org.hibernate.cfg.Configuration;

public class HibernateSessionFactory {
    private SessionFactory sessionFactory;
    public HibernateSessionFactory(){
    }
    public SessionFactory config(){
        try{
            Configuration configuration=new Configuration();
            Configuration configure=configuration.configure("hibernate.cfg.xml");
            return configure.buildSessionFactory();
        }catch(Exception e){
            e.getMessage();
            return null;
        }
    }
    public Session getSession(){
        sessionFactory=config();
        return sessionFactory.openSession();
    }
}
```

8. PO 对象以及对应的映射文件

PO 对象的类为 UserInfoPO，代码如例 5-9 所示。

【例 5-9】 PO 对象的类（UserInfoPO.java）。

```java
package PO;

public class UserInfoPO {
    private int id;
    private String userName;
    private String password;
    public int getId() {
        return this.id;
    }
    public void setId(int id) {
        this.id=id;
    }
    public String getUserName() {
        return this.userName;
    }
```

```java
    public void setUserName(String userName) {
        this.userName=userName;
    }
    public String getPassword() {
        return this.password;
    }
    public void setPassword(String password) {
        this.password=password;
    }
}
```

PO 对应的映射文件是 UserInfoPO.hbm.xml,代码如例 5-10 所示。

【例 5-10】 PO 对应的映射文件(UserInfoPO.hbm.xml)。

```xml
<?xml version="1.0" encoding="UTF-8"?>
<!--映射文件的DTD-->
<!DOCTYPE hibernate-mapping PUBLIC "-//Hibernate/Hibernate Mapping DTD 3.0//EN" "http://hibernate.sourceforge.net/hibernate-mapping-3.0.dtd">
<!--映射文件的根元素-->
<hibernate-mapping>
        <!--配置PO对象与数据库中表对应的关系使用class元素,name配置PO对象对应的
            类,table配置该PO对象在数据库中对应的表名,catalog配置表对应的数据库名-->
        <class name="PO.UserInfoPO" table="info" catalog="test">
        <!--id元素配置PO对象与数据库中表对应的id字段,name配置PO对象对应的属性,
            type指定PO对象属性的类型,column元素配置表中和PO对象属性对应的字段,把PO
            对象中的指定属性存到表中指定字段中,generator元素将主键自动加入序列-->
        <id name="id" type="int">
            <column name="id"/>
            <generator class="assigned"/>
        </id>
        <!--property元素配置PO对象中的某个属性对应的表中某个字段名,即实现一一对应
            的映射关系。name指定PO对象的属性,type指定属性的类型-->
        <property name="userName" type="string">
            <!--column元素配置对应的表中字段,name指定对应表的字段,length指定字段
                长度,not-null设置字段是否为空-->
            <column name="userName" length="30" not-null="true"/>
        </property>
        <property name="password" type="string">
            <column name="password" length="30" not-null="true"/>
        </property>
    </class>
</hibernate-mapping>
```

9. 封装对数据操作的类

将登录和注册业务功能封装到类 LoginRegisterInfo(JavaBean)中,代码如例 5-11 所示。

【例 5-11】 数据库操作类(LoginRegisterInfo.java)。

```java
package loginRegisterDao;
import addHibernateFile.HibernateSessionFactory;
import PO.UserInfoPO;
import java.util.List;
import javax.swing.JOptionPane;
import org.hibernate.Query;
import org.hibernate.Session;
import org.hibernate.Transaction;

public class LoginRegisterInfo {
    private Session session;
    private Transaction transaction;
    private Query query;
    HibernateSessionFactory getSession;
    public LoginRegisterInfo(){
    }
    public String  saveInfo(UserInfoPO info){
        String mess="error";
        getSession=new HibernateSessionFactory();
        session=getSession.getSession();
        try{
            transaction=session.beginTransaction();
            session.save(info);
            transaction.commit();
            mess="success";
            return mess;
        }catch(Exception e){
            message("RegisterInfo.error:"+e);
            e.printStackTrace();
            return null;
        }
    }
    public List queryInfo(String type,Object value){
        getSession=new HibernateSessionFactory();
        session=getSession.getSession();
        try{
            String hqlsql="from UserInfoPO as u where u.userName=?";
            query=session.createQuery(hqlsql);
            query.setParameter(0, value);
            List list=query.list();
            transaction=session.beginTransaction();
            transaction.commit();
            return list;
        }catch(Exception e){
```

```
            message("LoginRegisterInfo 类中有异常,异常为:"+e);
            e.printStackTrace();
            return null;
        }
    }
    public void message(String mess){
        int type=JOptionPane.YES_NO_OPTION;
        String title="提示信息";
        JOptionPane.showMessageDialog(null, mess, title, type);
    }
}
```

10. 项目部署和运行

项目部署运行后,登录页面效果如图 5-11 所示。如果用户没有注册,单击图 5-11 所示页面中的"注册"按钮,将出现如图 5-12 所示的注册页面,在图 5-12 所示页面中注册,数据如图 5-13 中所示,注册完后单击"提交"按钮,如果注册成功将返回登录页面,如图 5-14 所示,并把用户名返回到用户名文本区域中,在图 5-14 所示页面中输入密码后单击"登录"按钮,如果登录成功就进入登录成功页面,如图 5-15 所示。

图 5-13 输入数据注册用户

图 5-14 用户注册成功返回登录页面

图 5-15 登录成功页面

5.4 本章小结

本章主要介绍了 Hibernate 框架的基础知识,通过本章的学习应对 Hibernate 框架有一定的了解,在后续两章中将对 Hibernate 进行深入介绍。通过本章的学习应了解和掌握以下内容。

(1) Hibernate 的发展与特点。
(2) Hibernate 的下载和配置。
(3) Hibernate 的工作原理。
(4) Hibernate 的核心组件。
(5) 基于 Struts2＋Hibernate 的 Java Web 项目开发。

5.5 习 题

5.5.1 选择题

1. Hibernate1 版本发布于( )。
   A. 2001 年　　　　B. 2003 年　　　　C. 2006 年　　　　D. 2011 年
2. Hibernate 中存放类库的子目录是( )。
   A. documentation　B. lib　　　　　　C. project　　　　D. apps
3. Hibernate 中用于加载配置文件的是( )。
   A. Configuration　B. SessionFactory　C. Session　　　　D. Transaction

5.5.2 填空题

1. Hibernate 是封装了_____和_____的持久层解决方案。
2. Hibernate 的配置文件格式有_____和_____。
3. Hibernate 中映射文件的格式是_____。
4. Hibernate 中 PO 对象的三种状态是_____、_____和_____。

5.5.3 简答题

1. 简述 Hibernate 的特点。

2. 简述 Hibernate 的工作原理。

5.5.4 实训题

1. 完成一个简单的 Struts2＋Hibernate 的 Java Web 项目。
2. 把 5.3 节实例中的手工验证改为内置验证器验证。
3. 把例 5-8 中的类 HibernateSessionFactory 改为静态方法加载 Hibernate 配置文件的方式。

第 6 章 Hibernate 核心组件详解

本章将在第 5 章的基础上详细介绍 Hibernate 框架的核心组件部分。
本章主要内容如下所示。
(1) Hibernate 的配置文件。
(2) Hibernate 的 PO 对象。
(3) Hibernate 的映射文件。
(4) Hibernate 的核心类和接口。
(5) Hibernate 常用方法应用实例。

6.1 Hibernate 的配置文件

Hibernate 框架的配置文件主要用来为程序配置连接数据库的参数,例如,数据库的驱动程序名、URL、用户名和密码等。Hibernate 的基本配置文件有两种形式:hibernate.cfg.xml 和 hibernate.properties。hibernate.cfg.xml 包含了 Hibernate 与数据库的基本连接信息,在 Hibernate 工作的初始阶段,这些信息被先后加载到 Configuration 和 SessionFactory 实例中;该文件还包含了 Hibernate 的基本映射信息,即系统中每一个类和与其对应的数据库表之间的关联信息,在 Hibernate 工作的初始阶段,这些信息通过 hibernate.cfg.xml 的 mapping 元素被加载到 Configuration 和 SessionFactory 实例中。这两种文件包含了 Hibernate 运行期间用到的所有参数。两者的配置内容基本相同,但前者的使用稍微方便一些,例如,在 hibernate.cfg.xml 中可以定义要用到的 xxx.hbm.xml 映射文件,而使用 hibernate.properties 则需要在程序中以编码方式指明映射文件。hibernate.cfg.xml 是 Hibernate 的默认配置文件。

下面分别对这两种格式的 Hibernate 配置文件进行介绍,在 Hibernate 下载文件夹的 "\project\etc"中有这两种配置文件的模板。

6.1.1 hibernate.cfg.xml

hibernate.cfg.xml 配置文件定义了连接各种数据库所需要的参数,而且还定义了程序中用到的映射文件。一般把它作为 Hibernate 的默认配置文件。

在第 5 章中已经对 hibernate.cfg.xml 配置文件有了简单的了解,下面详细介绍 hibernate.cfg.xml 配置文件的结构,如例 6-1 所示。

【例 6-1】 hibernate.cfg.xml 配置文件的基本结构(hibernate.cfg.xml)。

```
<?xml version='1.0' encoding='utf-8'?>
<!--表明解析本 XML 文件的 DTD 文档位置,DTD 是 Document Type Definition 的缩写,即文档类
    型的定义,XML 解析器使用 DTD 文档来检查 XML 文件的合法性。Hibernate 版本不一样配置文件
    中的 DTD 信息会有一定的差异,例如,把 Hibernate 3.2 配置文件的 DTD 信息放在 Hibernate 3.6.0
```

的配置文件中使用时将导致找不到配置文件信息的异常或者其他异常。Hibernate 配置文件可以在下载 Hibernate 的文件夹中找到。切记,使用与版本相对应的 DTD 信息-->
```
<!DOCTYPE hibernate-configuration PUBLIC "-//Hibernate/Hibernate Configuration
    DTD 3.0//EN" "http://hibernate.sourceforge.net/hibernate-configuration-3.0.
    dtd">
```
<!--Hibernate 配置文件的根元素,其他元素要包含在其中-->
```
<hibernate-configuration>
```
<!--指定初始化 Hibernate 参数的元素,其中指定 Hibernate 初始化参数,表明以下的配置针对 session-factory。SessionFactory 是 Hibernate 中的一个接口,这个接口主要负责保存 Hibernate 的配置信息以及对 Session 的操作-->
```
    <session-factory>
```
    <!--指定连接数据库所用的驱动,本例中的 com.mysql.jdbc.Driver 是 MySQL 驱动-->
```
        <property name="connection.driver_class">com.mysql.jdbc.Driver</property>
```
    <!--设置数据库的连接 url:jdbc:mysql://localhost:port/test,其中,localhost 表示 mysql 服务器为本机,port 代表 mysql 服务器的端口号,默认是 3306。test 是数据库名,即要连接的数据库名-->
```
        <property name="connection.url">jdbc:mysql://localhost:3306/test
        </property>
```
    <!--指定数据库的用户名(登录名)-->
```
        <property name="connection.username">root</property>
```
    <!--指定数据库的连接密码-->
```
        <property name="connection.password">root</property>
```
    <!--指定连接池里最大连接个数,使用连接池需要加载所用的连接池的 JAR 文件,JAR 文件在 Hibernate 文件夹下的"lib\optional\c3p0"中-->
```
        <property name="hibernate.c3p0.max_size">20</property>
```
    <!--指定连接池里最小连接个数-->
```
        <property name="hibernate.c3p0.min_size">1</property>
```
    <!--指定连接池里连接的超时时长,即最大时间-->
```
        <property name="hibernate.c3p0.timeout">5000</property>
```
    <!--指定数据库的方言,每种数据库都有对应的方言-->
```
        <property name="dialect">org.hibernate.dialect.MySQLInnoDBDialect
        </property>
```
    <!--根据需要自动创建数据表-->
```
        <property name="hbm2ddl.auto">update</property>
```
    <!--设置是否将 Hibernate 发送给数据库的 SQL 显示出来,这是非常有用的功能。在调试 Hibernate 的时候,让 Hibernate 打印 SQL 语句,有助于迅速发现并解决问题-->
```
        <property name="show_sql">true</property>
```
    <!--开启二级缓存-->
```
        <property name="hibernate.cache.use_second_level_cache">true</property>
```
    <!--指定缓存产品所需的类-->
```
        <propertyname="hibernate.cache.provider_class">
            org.hibernate.cache.EhCacheProvider
        </property>
```
    <!--启用查询缓存-->
```
        <property name="hibernate.cache.use_query_cache">true</property>
```

```xml
<property name="hibernate.jdbc.fetch_size">30</property>
<property name="hibernate.jdbc.batch_size">5</property>
<!--加入映射文件。可以加入多个映射文件-->
<mapping resource="aaa(路径)/xxx.hbm.xml"/>
    </session-factory>
</hibernate-configuration>
```

上述 hibernate.cfg.xml 文件包含一个根元素＜hibernate-configuration＞,该元素有一个子元素＜session-factory＞。＜session-factory＞元素有两个子元素：＜property＞和＜mapping＞。＜property＞元素用来指定数据库连接参数,其属性 name 用来指定数据库连接参数的名字,这些参数名字由 Hibernate 框架定义,都代表特定意义,例如,connection.driver_class 表示加载驱动,元素后面的 com.mysql.jdbc.Driver 代表加载的驱动,数据库不一样后面加载的驱动名字不一样。＜mapping＞元素用来指定要用到的映射文件,其属性 resource 用来指定要用到映射文件的路径和映射文件的名字。

其中,hibernate.jdbc.fetch_size 和 hibernate.jdbc.batch_size 的设置非常重要,与 Hibernate 的增、删、改、查性能紧密相关。

hibernate.jdbc.fetch_size 设定 JDBC 的 Statement(处理查询数据结果集的接口)读取数据时每次从数据库中取出的记录条数。例如,一次查询 1 万条记录,对于 Oracle 的 JDBC 驱动来说,是不会一次性把 1 万条记录取出来的,而只会取出 FetchSize 条记录,当遍历完了这些记录后,再去数据库取出 FetchSize 条记录,因此大大减少了无谓的内存消耗。当然,FetchSize 设得越大,读数据库的次数越少,速度越快;而 FetchSize 设得越小,读数据库的次数越多,速度越慢。

Oracle 数据库的 JDBC 驱动默认 FetchSize＝10,这是一个非常保守的设定。根据测试,当 FetchSize＝50 时,性能会提升一倍之多;当 FetchSize＝100 时,性能还能继续提升 20%;如果 FetchSize 继续增大,性能提升就不显著了。因此,建议使用 Oracle 时将 FetchSize 设定为 50。但不是所有数据库都支持 FetchSize 特性,例如,MySQL 就不支持。MySQL 的表现就如同前面说的最坏的情况,总是一下就把 1 万条记录完全取出来,内存消耗非常惊人。

BatchSize 设定对数据库进行批量删除、批量更新和批量插入时的批次大小,有点类似于设置 Buffer 缓冲区大小的意思。BatchSize 越大,批量操作向数据库发送 SQL 的次数越少,速度就越快。很多人做 Hibernate 和 JDBC 的插入性能测试会惊奇地发现,Hibernate 的速度至少是 JDBC 的两倍,就是因为 Hibernate 使用了 BatchSize,而他们使用的 JDBC 没有使用 BatchSize。

6.1.2 hibernate.properties

hibernate.properties 配置文件是 Hibernate 框架提供的另一种配置数据库参数的形式,文件使用♯为注释,去掉♯就可以使用里面的设置,该文件给出了配置数据库的方法和对常用数据库的配置。例 6-2 是 Hibernate3.6.0 文件夹"\project\etc"下的 Hibernate 框架提供的模板。

【例 6-2】 Hibernate 的 hibernate.properties 配置文件模板(hibernate.properties)。

```
###Query Language,使用查询语言
#该配置的含义是在 Hibernate 应用程序里面输入 yes 的时候,Hibernate 就会把字
#符'Y'插入数据库中,当输入 no 的时候,就会把字符'N'插入数据库中。
hibernate.query.substitutions yes 'Y', no 'N'
##select the classic query parser
#hibernate.query.factory_class,即 Hibernate 使用的工厂模式
org.hibernate.hql.internal.classic.ClassicQueryTranslatorFactory
##JNDI Datasource 数据库的配置
hibernate.connection.datasource jdbc/test
hibernate.connection.username db2
hibernate.connection.password db2
##HypersonicSQL 数据库的配置
hibernate.dialect org.hibernate.dialect.HSQLDialect
hibernate.connection.driver_class org.hsqldb.jdbcDriver
hibernate.connection.username sa
hibernate.connection.password
hibernate.connection.url jdbc:hsqldb:./build/db/hsqldb/hibernate
hibernate.connection.url jdbc:hsqldb:hsql://localhost
hibernate.connection.url jdbc:hsqldb:test
##H2 (www.h2database.com)数据库的配置
hibernate.dialect org.hibernate.dialect.H2Dialect
hibernate.connection.driver_class org.h2.Driver
hibernate.connection.username sa
hibernate.connection.password
hibernate.connection.url jdbc:h2:mem:./build/db/h2/hibernate
hibernate.connection.url jdbc:h2:testdb/h2test
hibernate.connection.url jdbc:h2:mem:imdb1
hibernate.connection.url jdbc:h2:tcp://dbserv:8084/sample;
hibernate.connection.url jdbc:h2:ssl://secureserv:8085/sample;
hibernate.connection.url jdbc:h2:ssl://secureserv/testdb;cipher=AES
##MySQL 数据库的配置
hibernate.dialect org.hibernate.dialect.MySQLDialect
hibernate.dialect org.hibernate.dialect.MySQLInnoDBDialect
hibernate.dialect org.hibernate.dialect.MySQLMyISAMDialect
hibernate.connection.driver_class com.mysql.jdbc.Driver
hibernate.connection.url jdbc:mysql:///test
hibernate.connection.username gavin
hibernate.connection.password
##Oracle 数据库的配置
hibernate.dialect org.hibernate.dialect.Oracle8iDialect
hibernate.dialect org.hibernate.dialect.Oracle9iDialect
hibernate.dialect org.hibernate.dialect.Oracle10gDialect
hibernate.connection.driver_class oracle.jdbc.driver.OracleDriver
hibernate.connection.username ora
hibernate.connection.password ora
```

```
hibernate.connection.url jdbc:oracle:thin:@localhost:1521:orcl
hibernate.connection.url jdbc:oracle:thin:@localhost:1522:XE
##PostgreSQL 数据库配置
hibernate.dialect org.hibernate.dialect.PostgreSQLDialect
hibernate.connection.driver_class org.postgresql.Driver
hibernate.connection.url jdbc:postgresql:template1
hibernate.connection.username pg
hibernate.connection.password
##DB2 数据库的配置
hibernate.dialect org.hibernate.dialect.DB2Dialect
hibernate.connection.driver_class com.ibm.db2.jcc.DB2Driver
hibernate.connection.driver_class COM.ibm.db2.jdbc.app.DB2Driver
hibernate.connection.url jdbc:db2://localhost:50000/somename
hibernate.connection.url jdbc:db2:somename
hibernate.connection.username db2
hibernate.connection.password db2
    ⋮
##MS SQL Server 数据库的配置
hibernate.dialect org.hibernate.dialect.SQLServerDialect
hibernate.connection.username sa
hibernate.connection.password sa
###Hibernate Connection Pool 连接池
hibernate.connection.pool_size 1
###C3P0 Connection Pool
hibernate.c3p0.max_size 2
hibernate.c3p0.min_size 2
hibernate.c3p0.timeout 5000
hibernate.c3p0.max_statements 100
hibernate.c3p0.idle_test_period 3000
hibernate.c3p0.acquire_increment 2
hibernate.c3p0.validate false
hibernate.show_sql true
hibernate.default_batch_fetch_size 8
hibernate.jdbc.fetch_size 25
```

在例 6-2 中只是列出了 hibernate.properties 配置文件的一部分配置参数，如需使用其他配置参数，请参考 Hibernate 框架提供的 hibernate.properties 模板。

6.2 Hibernate 的 PO 对象

6.2.1 Hibernate 的 PO 对象基础知识

在 Hibernate 的应用程序中，每一个数据库中的表都对应一个持久化对象 PO。PO 可以看成是与数据库表相映射的 Java 对象。最简单的 PO 对应数据库中某个表中的一条记录，多个记录可以对应 PO 的集合。PO 中不应该包含任何对数据库的操作。

PO 类即持久化类,其实就是一个普通 JavaBean,只要声明时遵循一定的规则就是一个 PO。例 6-3 就是一个持久化类。

【**例 6-3**】 持久化类(UserInfoPO.java)。

```java
package PO;

public class UserInfoPO{
    private int id;
    private String userName;
    private String password;
    public int getId() {
        return this.id;
    }
    public void setId(int id){
        this.id=id;
    }
    public String getUserName(){
        return this.userName;
    }
    public void setUserName(String userName){
        this.userName=userName;
    }
    public String getPassword(){
        return this.password;
    }
    public void setPassword(String password){
        this.password=password;
    }
}
```

该 PO 对应数据库中的 info 表。info 表有 3 个字段:id(int 类型)、userName(varchar 类型)和 password(varchar 类型)。可以看出该 PO 主要是为 info 表中的字段定义访问方法,每一个字段对应一对 getter 和 setter 方法。定义 PO 应遵循以下 3 个规则。

1. 为持久化字段声明私有属性且提供 getter 和 settter 方法

Hibernate 框架中持久化 JavaBean 属性的常用形式为 getter 和 setter。属性不需要声明为 public。Hibernate 可以对 default、protected 或 private 的 getter 和 setter 方法的属性一视同仁地执行持久化。

2. 实现一个无参构造方法

所有的持久化类都必须有一个默认的构造方法,这样 Hibernate 就可以使用 newInstance()来实例化它们。在 Hibernate 中,为了运行期间代理的生成,建议构造方法至少是包内可见的。

3. 提供一个标识符属性

标识符属性如例 6-3 中的 id,用于映射数据库表的主键字段。这个属性可以叫任何名字,其类型可以是任何的原始类型、原始类型的包装类型、java.lang.String 或者是 java

.util.Date)。

标识符属性是可选的。可以不用管它,让 Hibernate 内部来追踪对象的识别。

实际上,一些功能只对那些声明了标识符属性的类起作用,例如,脱管对象的重新关联(级联更新或级联合并),session.saveOrUpdate()和sssion.merge()。

建议为持久化类声明命名一致的标识符属性,并使用可以为空(也就是说不是原始类型)的类型。

6.2.2 Hibernate 的 PO 对象状态

Hibernate 的 PO 对象有三种状态:临时状态(又称为临时态)、持久状态(又称为持久态)和脱管状态(又称为脱管态)。处于持久态的对象也称为 PO,临时对象和脱管对象也称为 VO(Value Object)。

1. 临时态

由 new 命令开辟内存空间刚生成的 Java 对象就处于临时态。

例如:

```
UserInfoPO ui=new UserInfoPO();
```

如果没有变量对该对象进行引用,它将被 Java 虚拟机回收。

临时对象在内存中是孤立存在的,它是携带信息的载体,不和数据库的数据有任何关联关系。在 Hibernate 中,可通过 Session 的 save()或 saveOrUpdate()方法将临时对象与数据库关联起来,并将数据插入数据库中,此时该临时对象转变成持久化对象。

2. 持久态

处于该状态的对象在数据库中具有对应的记录,并拥有一个持久化标识。如果使用 Hibernate 的 delete()方法,对应的持久对象就变成临时对象,因数据库中的对应数据已被删除,该对象不再与数据库的记录关联。

当一个 Session 执行 close()或 clear()之后,持久对象会变成脱管对象,此时该对象虽然具有数据库识别值,但它已不在 Hibernate 持久层的管理之下。

持久对象具有如下特点。

(1) 和 Session 实例关联。

(2) 在数据库中有与之关联的记录。

3. 脱管态

当与某持久对象关联的 Session 被关闭后,该持久对象转变为脱管对象。当脱管对象被重新关联到 Session 上时,将再次转变成持久对象。

脱管对象拥有数据库的标识值,可通过 update()、saveOrUpdate()等方法,转变成持久对象。

脱管对象具有如下特点。

(1) 本质上与临时对象相同,在没有任何变量引用它时,JVM 会在适当的时候将它回收。

(2) 比临时对象多了一个数据库记录标识值。

4. Session 中改变 PO 对象状态的常用方法

通过 get()或 load()方法得到的 PO 对象都处于持久态,但如果执行了 delete(),该 PO

对象就处于临时态(表示和 Session 脱离关联);因执行 delete()而变成临时态的 PO 对象可以通过调用 save()或 saveOrUpdate()变成持久态;当把 Session 关闭时,Session 缓存中的持久态 PO 对象也变成脱管态;因关闭 Session 而变成脱管态的 PO 对象可以通过调用 lock()、save()、update()变成持久态;持久态 PO 对象可以通过调用 delete()变成临时状态。

5. save()和 update()区别

save()的作用是保存一个新的对象;update()可以把一个脱管状态的对象(一定要和一个记录对应)更新到数据库。

6. update()和 saveOrUpdate()区别

saveOrUpdate()基本上就是合并了 save()和 update()。

6.3 Hibernate 的映射文件

Hibernate 的映射文件把一个 PO 对象与一个表映射起来。每一个表对应一个扩展名为 hbm.xml 的映射文件。映射文件也称为映射文档,用于向 Hibernate 提供关于将对象持久化到关系型数据库表中的信息。

持久化对象的映射定义可全部存储在同一个映射文件中,也可将每个对象的映射定义存储在独立的文件中。后一种方法较好,因为将大量持久化类的映射定义存储在一个文件中比较麻烦,建议采用每个类一个文件的方法来组织映射文档。映射文件的命名规则是使用持久化类的类名,并使用扩展名 hbm.xml。

映射文件需要在 hibernate.cfg.xml 中注册,最好与功能相关对象类放在同一目录中,这样修改起来很方便。例 6-4 是 UserInfoPO 类对应的映射文件。

【例 6-4】 PO 对象的映射文件(UserInfoPO.hbm.xml)。

```
<?xml version="1.0" encoding="UTF-8"?>
<!--映射文件的DTD-->
<!DOCTYPE hibernate-mapping PUBLIC "-//Hibernate/Hibernate Mapping DTD 3.0//EN" "
http://hibernate.sourceforge.net/hibernate-mapping-3.0.dtd">
<!--映射文件的根元素-->
<hibernate-mapping>
    <!--配置PO对象与数据库中表的对应关系使用class元素,name配置PO对象对应的类,
    table配置该PO对象在数据库中对应的表名,catalog配置表对应的数据库名-->
    <class name="PO.UserInfoPO" table="info" catalog="test">
    <!--id元素配置PO对象与数据库中表的id字段,name配置PO对象对应的属性,type指定
        PO对象该属性的类型,column元素配置表和PO对象属性对应的字段,即把PO对象中的指定
        属性存到表中指定字段中,generator元素将主键自动加入序列-->
        <id name="id" type="int">
            <column name="id"/>
            <generator class="assigned"/>
        </id>
    <!--property元素配置PO对象中的某个属性对应表中的某个字段,即实现一一对应的
        映射关系。name指定PO对象的属性,type指定属性的类型-->
```

```xml
        <property name="userName" type="string">
            <!--column 元素配置对应的表中字段,length 指定字段长度,not-null 设置字段
            是否为空-->
            <column name="userName" length="30" not-null="true"/>
        </property>
        <property name="password" type="string">
            <column name="password" length="30" not-null="true"/>
        </property>
    </class>
</hibernate-mapping>
```

下面分别介绍 Hibernate 映射文件中的元素。

1. 根元素

<hibernate-mapping>,每一个.hbm.xml 文件都有一个唯一的根元素,包含可选的属性 package。

package:指定一个包前缀。

2. <class>定义类

该元素是根元素的子元素,用以定义一个持久化类与数据表的映射关系,在映射文件中可以有多对该元素。下面所列是包含的一些常用属性。

(1) name:指定持久化类。

(2) table:指定对应的数据库表名。

(3) catalog:指定对应的数据库。

(4) batch-size:指定一个用于根据标识符抓取实例时使用的 batch-size(批次抓取数量)。

3. <id>定义主键

Hibernate 使用 OID(对象标识符)来标识对象的唯一性,OID 是关系数据库中主键在 Java 对象模型中的等价物,在运行时,Hibernate 根据 OID 来维持 Java 对象和数据库表中记录的对应关系。常用属性如下所示。

(1) name:持久化类的标识符属性名字。

(2) type:标识久化类的属性类型。

(3) column:数据库表的主键字段的名字。

(4) access:Hibernate 用来访问属性值的策略。

如果表使用联合主键,那么可以映射类的多个属性为标识符属性。<composite-id>元素接受<key-property>属性映射和<key-many-to-one>属性映射作为子元素。

例如,下面的代码段定义了两个字段作为联合主键:

```xml
<composite-id>
    <key-property name="name" />
    <key-property name="password" />
</composite-id>
```

4. ＜generator＞设置主键生成方式

该元素的作用是指定主键的生成器,通过一个 class 属性指定生成器对应的类(通常与＜id＞元素结合使用)。

```
<id name="id" column="id" type="integer">
    <!--assigned 是 Hibernate 主键生成器的实现算法之一,由 Hibernate 根据底层数据库自
        行判断采用 identity 或 sequence 作为主键生成方式-->
    <generator class="increment" />
</id>
```

Hibernate 提供的内置生成器如下所示。

1) increment(递增)

increment 是 org. hibernate. id. IncrementGenerator 类的快捷名字,用于为 long、short 或者 int 类型生成唯一标识。只有在没有其他进程往同一张表中插入数据时才能使用,在集群下不要使用。

2) identity(标识)

identity 是 org. hibernate. id. IdentityGenerator 类的快捷名字,为 DB2、MySQL、MS SQL Server 和 Sybase 等数据库的内置标识字段提供支持。返回的标识符是 long、short 或者 int 类型的。

3) sequence(序列)

sequence 是 org. hibernate. id. SequenceGenerator 类的快捷名字,为 DB2、Oracle 等数据库的内置序列(sequence)提供支持。返回的标识符是 long、short 或者 int 类型的。

4) seqhilo(序列高/低位)

seqhilo 是 org. hibernate. id. SequenceHiLoGenerator 类的快捷名字,使用一个高/低位算法来高效地生成 long、short 或者 int 类型的标识符,需要指定一个数据库 sequence 的名字。

5) uuid. hex

uuid. hex 是 org. hibernate. id. UUIDHexGenerator 类的快捷名字,使用一个 128 位的 UUID 算法生成字符串类型的标识符,在同一个网络中是唯一的(使用了 IP 地址)。UUID 被编码为一个 32 位十六进制的字符串,包含 IP 地址、JVM 的启动时间(精确到 14s)、系统时间和一个计数器值(在 JVM 中是唯一的)。

6) assigned

assigned 是 org. hibernate. id. Assigned 类的快捷名字,可让应用程序在执行 save() 方法之前为对象分配一个标识符。如果需要为应用程序分配一个标识符(而非由 Hibernate 来生成它们),可以使用 assigned 生成器。

7) foreign

foreign 是 org. hibernate. id. ForeignGenerator 类的快捷名字。它使用另外一个相关对象的标识符,和＜one-to-one＞元素一起使用。

5. ＜property＞定义属性

用于持久化类的属性与数据库表字段之间的映射,包含如下属性。

(1) name:持久化类的属性名。

(2) column：数据库表的字段名。

(3) insert：表明用于 insert 的 SQL 语句中是否包含这个被映射的字段，默认为 true。

(4) update：表明用于 update 的 SQL 语句中是否包含这个被映射的字段，默认为 true。

(5) lazy：指定实例变量第一次被访问时，这个属性是否延迟抓取，默认为 false。

(6) type：Hibernate 映射类型的名字。如果没有指定类型，Hibernate 会使用反射来得到这个名字的属性，以此来猜测正确的 Hibernate 类型。type 属性可以指定为以下几种类型之一。

① Hibernate 基础类型，如 integer、string、character、date、timestamp、float、binary、serializable、object 和 blob 等。

② 一个 Java 类。该类属于一种默认基础类型，如 int、float、char、java.lang.String、java.util.Date、java.lang.Integer 和 java.sql.Clob 等。

③ 一个可以序列化的 Java 类。

④ 一个自定义的类。

6.4 Hibernate 的 Configuration 类

Configuration 类的主要作用是解析 Hibernate 的配置文件和映射文件中的信息，即负责管理 Hibernate 的配置信息。Hibernate 运行时需要获取一些底层实现的基本信息，如数据库驱动程序类、数据库的 URL、数据库登录名、数据库登录密码等，这些信息定义在 Hibernate 的配置文件中。通过 Configuration 对象的 buildSessionFactory() 方法可创建 SessionFactory 对象，因此 Configuration 对象一般只有在获取 SessionFactory 对象时使用。当获取了 SessionFactory 对象之后，由于配置信息已经由 Hibernate 维护并绑定在返回的 SessionFactory 中，该 Configuration 将不再有价值。

加载 Hibernate 配置文件的格式如下：

```
Configuration cfg=new Configuration().configure("hibernate.cfg.xml");
```

一般情况下，hibernate.cfg.xml 和 struts.xml 文件都放在源包中，且都在默认包中。在使用 NetBeans 7.2 开发项目时，这两个配置文件放在项目的\src\java 文件夹中；在使用 MyEclipse 10.6 开发项目时，它们放在项目的\src 文件夹中。如果找到该文件，configure() 方法会首先访问＜session-factory＞元素，并获取该元素的 name 属性；如果 name 属性值非空，将用这个配置的值来覆盖 hibernate.propperties 中 hibernate.session_factory_name 的配置值。从这里可以看出，hibernate.cfg.xml 中的配置信息可以覆盖 hibernate.properties 的配置信息。

Configuration 的 configure() 方法还支持带参数的访问方式，可以用参数指定配置文件的位置，从而不使用默认路径中设置的 hibernate.cfg.xml。

例如：

```
File file=new File("c:\\cfg\\hibernate.cfg.xml");
Configuration config=new Configuration().configure(file);
```

6.5 Hibernate 的 SessionFactory 接口

SessionFactroy 接口负责初始化 Hibernate。它充当数据存储源的代理,并负责创建 Session 对象。可以通过 Configuration 实例构建 SessionFactory 对象。

例如:

```
Configuration cfg=new Configuration().configure("hibernate.cfg.xml");
SessionFactory sf=cfg.buildSessionFactory();
```

Configuration 对象会根据当前的配置信息生成 SessionFactory 对象。SessionFactory 对象一旦构造完毕,即被赋予特定的配置信息,即以后配置的改变不会影响到已创建的 SessionFactory 对象。如果要把改动后的配置信息赋给 SessionFactory 对象,需要从新的 Configuration 对象生成新的 SessionFactory 对象。

SessionFactory 是线程安全的,可以被多个线程调用。因为构造 SessionFactory 很消耗资源,所以多数情况下一个应用中只初始化一个 SessionFactory,为不同的线程提供 Session。

当客户端发送一个请求线程时,SessionFactory 生成一个 Session 对象来处理客户请求。

6.6 Hibernate 的 Session 接口

6.6.1 Session 接口的基础知识

Session 接口是 Hibernate 中的核心接口,不同于 Java Web 应用中的 HttpSession 接口,虽然都可以将其翻译为"会话"。在 Hibernate 框架中除非特别说明,否则 Session 均指 Hibernate 中的 Session 对象。

Session 对象是 Hibernate 技术的核心,持久化对象的生命周期、事务的管理及持久化对象(PO)的增加、修改和删除都是通过 Session 对象来完成的。Hibernate 在操作数据库之前必须先取得 Session 对象,相当于 JDBC 在操作数据库之前必须先取得 Connection 对象一样。Session 对象不是线程安全的,一个 Session 对象最好只由一个单一线程来使用。同时该对象的生命周期要比 SessionFactory 短,其生命通常在完成数据库的一个短暂的系列操作之后结束。一个应用系统中可以自始至终只使用一个 SessionFactory 对象。Session 对象通过 SessionFactory 对象的 getCurrentSession()或者 openSession()方法获取,代码如下:

```
Configuration cfg=new Configuration().configure("hibernate.cfg.xml");
SessionFactory sf=cfg.buildSessionFactory();
Session session=sf.openSession();
```

获取 Session 对象后,Hibernate 内部并不会立即获取操作数据库的 java.sql.Connection 对象,而是等待 Session 对象真正需要对数据库进行增、删、改、查等操作时,才会从数据库连接池中获取 Connection 对象。关闭 Session 对象时,则是将 Connection 对象返回到连接池中,而不是直接关闭 Connection 对象。

SessionFactory 对象是线程安全的,允许多个线程同时存取该对象而不存在数据共享冲突的问题。然而 Session 对象不是线程安全的,如果试图让多个线程共享一个 Session 对象,将会发生数据共享混乱的问题。

一个持久化类从定义上与普通的 JavaBean 类没有任何区别,但是它与 Session 关联起来后,就具有了持久化的能力。当然,这种持久化操作是受 Session 控制的,即通过 Session 对象装载、保存、创建或查询持久化对象(PO)。Session 类提供有 save()、delete()、update() 和 load() 等方法,来分别完成对持久化对象的保存、删除、修改、加载等操作。Session 类的方法按用途可分为以下 5 类。

(1) 获取持久化对象:get() 和 load() 等方法。

(2) 持久化对象的保存、更新和删除:save()、update()、saveOrUpdate() 和 delete() 等方法。

(3) createQuery() 方法:用来从 Session 生成 Query 对象,此方法将在讲解 Query 接口的章节中介绍。

(4) beginTransaction() 方法:用来从 Session 生成 Transaction 对象,此方法将在讲解 Transaction 接口的章节中介绍。

(5) 管理 Session 的方法:isOpen()、flush()、clear()、evict() 和 close() 等方法,其中 isOpen() 方法用来检查 Session 是否仍然打开;flush() 方法用来清理 Session 缓存,并把缓存中的 SQL 语句发送出去;clear() 方法用来清除 Session 中的所有缓存对象;evict() 方法用于清除 Session 缓存中的某个对象;close() 方法用来关闭 Session。

下面通过实例分别介绍 Session 类的主要方法。实例中的持久化对象为 UserInfoPO 类,代码见例 6-3,其映射文件为 UserInfoPO.hbm.xml,代码见例 6-4。

6.6.2 通过方法获取持久化对象

获取持久化对象的方法主要有两种:get() 方法和 load() 方法,它们都是通过主键 id 来获取 PO 对象的。

1. 使用 get() 方法获取 PO 对象

通常可以使用 get() 方法获取 PO 对象。

例如:

```
public Object get(Class className, Serializable id)
```

className 是类名,id 是对象的主键值,如果 id 的类型是 int,可通过 new Integer(id) 的方法生成一个 Integer 对象。

以下代码获取一个主键 id 为 66 的 UserInfoPO 对象:

```
UserInfoPO ui=(UserInfoPO)session.get(UserInfoPO.class, new Integer(66));
```

get() 方法获取 PO 对象时执行过程如下。

(1) 首先通过 id 在 Session 缓存中查找对象,如果存在与 id 主键值对应的对象,直接将其返回。

（2）如果在 Session 缓存中没有查询到对应的对象，则在二级缓存中查找，找到后将其返回。

（3）如果在前两个步骤的操作中都没有找到该对象，则从数据库加载拥有此 id 的对象。

从以上步骤中可以看出，get()方法并不总是发送 SQL 语句查询数据库，只有缓存中无此数据时，才向数据库发送 SQL 语句以取得数据。

2. 使用 load()方法获取 PO 对象

load()方法和 get()方法一样都可以通过主键 id 的值从数据库中加载一个持久化对象。

例如：

```
UserInfoPO ui=(UserInfoPO)session.load(UserInfoPO.class, new Integer(66));
```

get()方法和 load()方法的区别如下。

（1）在立即加载 PO 对象时（当 Hibernate 从数据库中取得数据组装好一个对象后，会立即再从数据库取得数据组装与此对象相关联的对象），如果对象存在，get()方法和 load()方法没有区别，它们都可取得已初始化的对象；但当对象不存在且是立即加载时，get()方法返回 null，而 load()方法则弹出一个异常。因此使用 load()方法时，要确认查询的主键 id 一定是存在的，从这一点来讲，它没有 get()方法方便。

（2）在延迟加载对象时（在 Hibernate 从数据库中取得数据组装好一个对象后，不会立即再从数据库取得数据组装与此对象相关联的对象，而是等到需要时，才会从数据库取得数据组装关联对象），get()方法仍然使用立即加载的方式发送 SQL 语句，并得到已初始化的对象，而 load()方法则根本不发送 SQL 语句，它返回一个代理对象，这个对象直到被访问使用时才被初始化。

6.6.3 操作持久化对象的常用方法

1. 使用 save()方法操作 PO 对象

Session 中的 save()方法可将一个 PO 对象的属性取出放入 PreparedStatement（具有预编译功能的 SQL 类）语句中，然后向数据库表中插入一条记录（或者多条记录，如果有级联关系）。下面的代码把一个新建 UserInfoPO 对象持久化到数据库中，即把一条记录插入到数据库表中。

例如：

```
UserInfoPO ui=new UserInfoPO();
ui.setId(66);                                    //为对象设定一个 id 值
Session session=sf.openSession();                //打开 Session
Transaction tx=session.beginTransaction();       //开启事务
session.save(ui);                                //调用 save()方法保存数据
tx.commit();                                     //提交事务
session.close();                                 //关闭 Session
```

上述代码等价于：

insert into info(id, userName, password) values(66,"李想","123456A")

当 Session 保存一个 PO 对象的时候，按照以下的步骤进行操作。
(1) 根据映射文件中的配置为主键 id 设置生成算法，为 info 指定一个 id。
(2) 将 UserInfoPO 对象存入 Session 对象的内部缓存中。
(3) 当事务提交时，清理 Session 对象缓存中的数据，并将新对象通过 Hibernate 框架生成的 insert into 语句持久化到数据库中。

如果需要为新对象强制指定一个 id 值，可以调用 Session 的 save(Object obj, Serializable id) 重载方法，例如，在上述代码中，虽然映射文件配置时已经设置了使用 assigned 算法生成 id 值，但也可以使用 save(Object, id) 方法强制改变 id 值。

例如：

```
session.save(ui, new Integer(66));
```

这种方法在使用代理主键时一般不推荐，除非确实需要指定特别的 id 时才使用。

在调用 save() 方法时，并不立即执行 SQL 语句，而是等到清理完缓存后才执行。如果在调用 save() 方法后又修改了 UserInfoPO 的属性，则 Hibernate 框架将会发送一条 insert into 语句和一条 update 语句来完成持久化操作。

例如：

```
UserInfoPO ui=new UserInfoPO();
ui.setId(66);                                    //为对象设定一个 id
ui.setUserName("李想");                          //设定用户名的值
Session session=sf.openSession();                //打开 Session
Transaction tx=session.beginTransaction();       //开启事务
session.save(ui);
ui.setUserName("李想");                          //修改用户名的值
tx.commit();                                     //提交事务
session.close();                                 //关闭 Session
```

上述代码等价于：

insert into info(id,userName, password) values(66,"李想","123456A")
update info set userName=？password=？where id=？

因此，最好是在对象状态稳定（也即属性不会再变化）时再调用 save() 方法，这样可以少执行一条 update 语句。

调用 save() 方法将临时对象保存到数据库中，对象的临时状态将变为持久化状态。当对象在持久化状态时，会一直位于 Session 的缓存中，对它的任何操作在事务提交时都将同步保存到数据库中，因此，对一个已经持久化的对象调用 save() 或 update() 方法是没有意义的。

例如：

```
UserInfoPO ui=new UserInfoPO();
ui.setId(66);                                       //为对象设定一个 id
ui.setUserName("李想");                              //设定用户名的值
Session session=sf.openSession();                   //打开 Session
Transaction tx=session.beginTransaction();          //开启事务
session.save(ui);
ui.setUserName("李想");                              //修改用户名的值
session.save(user);                                 //无效
session.update(user);                               //无效
tx.commit();                                        //提交事务
session.close();                                    //关闭 Session
```

上述代码仍等价于：

```
insert into info(id,userName, password) values(66,"李想","123456A")
update info set userName=?password=? where id=?
```

2. 使用 update()方法操作 PO 对象

Session 的 update()方法可以用来更新脱管对象到持久化对象。

例如：

```
UserInfoPO ui=new UserInfoPO();
Session session=sf.openSession();                   //打开 Session
Transaction tx=session.beginTransaction();          //开启事务
ui= (UserInfoPO)session.get(UserInfoPO.class, new Integer(66));
ui.setUserName("李想");
session.update(ui);                                 //更新脱管对象
tx.commit();                                        //提交事务
session.close();                                    //关闭 Session
```

使用 update()方法时，Hibernate 框架并不立即发送 SQL 语句，而是将对象的更新操作积累起来，在事务提交时由 flush()清理缓存，并发送一条 SQL 语句完成全部更新操作。

3. 使用 saveOrUpdate()方法操作 PO 对象

在实际 Java Web 项目开发中，Java Web 程序员往往并不会注意一个对象是脱管对象还是临时对象，但是对脱管对象使用 save()方法是不对的，对临时对象使用 update()方法也是不对的。为了解决这个问题，便产生了 saveOrUpdate()方法。

saveOrUpdate()方法兼具 save()和 update()方法的功能，对于传入的对象，saveOrUpdate()方法首先判断该对象是脱管对象还是临时对象，然后调用合适的方法。

saveOrUpdate()方法的应用如下：

```
UserInfoPO ui=new UserInfoPO();
ui.setId(66);                                       //为对象设定一个 id 值
Session session=sf.openSession();                   //打开 Session
Transaction tx=session.beginTransaction();          //开启事务
session.saveOrUpdate(ui);                           //使用方法保存数据
```

```
tx.commit();                                      //提交事务
session.close();                                  //关闭 Session
```

4. 使用 delete()方法操作 PO 对象

Session 的 delete()方法负责删除一个对象(包括持久对象和脱管对象)。
例如:

```
UserInfoPO ui=new UserInfoPO();
Session session=sf.openSession();                 //打开 Session
Transaction tx=session.beginTransaction();        //开启事务
ui=(UserInfoPO)session.get(UserInfoPO.class, new Integer(66));
session.delete(ui);                               //删除持久对象
tx.commit();                                      //提交事务
session.close();                                  //关闭 Session
```

上述代码等价于:

```
select * from info where id=66
delete from info where id=66
```

使用 delete()方法删除对象时,会出现一些性能上的问题。例如,从以上代码中可以看出,当删除一个对象时,先调用 get()方法加载这个对象,然后调用 delete()方法删除对象,但此方法发送了一条 select 语句和一条 delete 语句。实际上 select 语句是不必要的,这种情况在批量删除时尤其明显。为了解决批量删除时产生的性能问题,常用的办法是使用批量删除操作。

例如:

```
UserInfoPO ui=new UserInfoPO();
Session session=sf.openSession();                 //打开 Session
Transaction tx=session.beginTransaction();        //开启事务
Query q=session.createQuery("delete from UserInfoPO");  //使用 HQL 语句进行删除
q.executeUpdate();                                //删除对象
tx.commit();                                      //提交事务
session.close();                                  //关闭 Session
```

上述代码等价于:

```
delete from info
```

只用一条语句就完成了批量删除的操作。但这样处理也有问题,批量删除后的数据还会存在缓存中,因此程序查询时可能得到脏数据(数据库中已不存在的数据),因此在使用批量删除时,要综合考虑性能和数据一致性的问题。

6.7 Hibernate 的 Transaction 接口

Transaction 接口是对事务实现的一个抽象,这些实现包括 JDBC 事务或者 JTA(Java Transaction,API,Java 事务 API)等。JTA 允许应用程序执行分布式事务处理(在两个或多个网络计算机资源上访问并且更新数据)。

Hibernate 框架中的事务通过配置 hibernate.cfg.xml 文件选择使用 JDBC 或者是 JTA 事务控制。在 Transaction 接口中主要定义了 commit()和 rollback()两个方法,前者是提交事务的方法;后者是回滚事务的方法。此外还提供了 wasCommitted()方法。

Transaction 的运行与 Session 接口相关,可调用 Session 的 beginTransaction()方法生成一个 Transaction 实例。

例如:

```
Transaction tx=session.beginTransaction();
```

一个 Session 实例可以与多个 Transaction 实例相关联,但一个特定的 Session 实例在任何时候必须与至少一个未提交的 Transaction 实例相关联。

Transaction 接口的常用方法如下。

(1) commit():提交相关联的 Session 实例。

(2) rollback():撤销事务操作。

(3) wasCommitted():检查事务是否提交。

Transaction 的应用如例 6-5 所示。

【例 6-5】 Transaction 的应用(Test.java)。

```
import org.hibernate.*;
import org.hibernate.cfg.*;

public class Test {
    try{
        SessionFactory sf=
        new Configuration.configure("hibernate.cfg.xml").buildSessionFacory();
        Session session=sf.openSession();
        Transaction tx=session.beginTransaction();
        Query query=session.createQuery("from UserInfoPO u where u.age>?");
        query.setInteger(0,20);
        List list=query.list();
        for (int i=0;i<list.size();i++) {
            UserInfoPO ui=(UserInfoPO)list.get(i);
            System.out.prinln(ui.getUserName());
        }
        tx.commit();
        session.close();
    } catch (HibernateException e) {
        e.printStackTrace();
    }
}
```

6.8 Hibernate 的 Query 接口

6.8.1 Query 接口的基本知识

使用 Query 的对象可以方便地查询数据库中的数据，它主要使用 HQL 或者本地 SQL（Native SQL）查询数据。Query 对象不仅能查询数据，还可以绑定参数、限制查询记录数量，并实现批量删除和批量更新等。

例如：

```
Configuration cfg=new Configuration().configure("hibernate.cfg.xml");
SessionFactory sf=cfg.buildSessionFactory();
Session session=sf.getCurrentSession();
Transaction tx=session.beginTransaction();
Query query=session.createQuery("from UserInfoPO");
List list=query.list();
tx.commit();
```

上面代码表示 Query 对象通过 Session 对象的 createQuery()方法创建，方法的参数值"from UserInfoPO"是 HQL 语句，表示要读取所有 UserInfoPO 类型的对象，即读取 UserInfoPO 表中的所有记录，把每条记录封装成 UserInfoPO 对象后保存到 List 对象中并返回 List 对象。

Query 对象在 Session 对象关闭之前有效，否则会抛出 SessionException 类型的异常。因为 Session 对象就像 JDBC 中的 Connection 对象，即数据库的一次连接。关闭 Connection 对象，Statement 对象就不能再使用，所以关闭 Session 后就不能再使用 Query 对象。

6.8.2 Query 接口的常用方法

Query 接口的常用方法如下所示。

（1）setXxx()方法：用于设置 HQL 中问号或变量的值。

（2）list()方法：返回查询结果，并把查询结果转换成 List 对象。

（3）excuteUpdate()方法：执行更新或删除语句。

1. 使用 setXxx()方法为 HQL 语句设置参数

Query 接口中 setXxx()方法主要用来为 HQL 中的问号"?"和变量设置参数，根据参数的数据类型，常用的 setXxx()方法如下。

（1）setBinary()：设置类型为 binary 的参数。

（2）setBoolean()：设置类型为 boolean 的参数。

（3）setByte()：设置类型为 byte 的参数。

（4）setCharacter()：设置类型为 char 的参数。

（5）setDate()：设置类型为 Date 的参数。

（6）setDouble()：设置类型为 double 的参数。

（7）setFloat()：设置类型为 float 的参数。

(8) setInteger()：设置类型为 int 的参数。

(9) setLong()：设置类型为 long 的参数。

(10) setString()：设置类型为 String 的参数。

上述方法都有两种使用方式。

(1) setString(int position,String value)：用于设置 HQL 中"?"的值；其中 position 代表"?"在 HQL 中的位置，value 是要为"?"设置的值。

例如：

```
Query query=session.createQuery("from UserInfoPO u where u.age>? and u.userName like ?");
query.setInteger(0,22);                //设置第一个问号的值为 22
query.setString(1,"%志%");              //设置第二个问号的值为"%志%"
```

(2) setString(String paraName,String value)：用于设置 HQL 中":"后所跟变量的值；其中 paraName 代表 HQL 中":"后所跟变量,value 为要为该变量设置的值。

例如：

```
Query query=session.createQuery("from UserInfoPO u where u.age>:minAge and u
.userName like: userName");
query.setInteger("minAge",22);          //设置变量 minAge 的值
query.setString("userName","%志%");     //设置变量 userName 的值
```

在 HQL 中使用变量代替问号"?"，然后用 setXxx()方法为该变量设值。

2. 使用 list()方法获取查询结果

Query 中的 list()方法用于获取查询结果,并将查询结果转变成一个 List 接口的实例。

例如：

```
Query query=session.createQuery("from UserInfoPO u where u.age>?");
query.setInteger(0,20);
List list=query.list();
for (int i=0;i<list.size();i++) {
    ui=(UserInfoPO)list.get(i);
    System.out.prinln(ui.getUserName());
}
```

3. 使用 excuteUpdate()方法更新或者删除数据

Query 的 excuteUpdate()方法用于批量更新或批量删除操作。

例如：

```
Query q=session.createQuery("delete from UserInfoPO");
q.executeUpdate();                      //删除对象
```

上述代码等价于：

```
delete from info
```

4. 使用命名查询(namedQuery)

可以不将 HQL 语句写在程序中，而是写入映射文件 *.hbm.xml 中,这样便于在需要

时修改 HQL。这时需要在映射文件(*.hbm.xml)中使用＜query＞标记,把 HQL 语句放入＜![CDATA[]]＞之中,如例 6-6 所示。

【例 6-6】 UserInfoPO 类对应的映射文件(UserInfoPO.hbm.xml)。

```xml
<?xml version="1.0" encoding="UTF-8"?>
<!DOCTYPE hibernate-mapping PUBLIC "-//Hibernate/Hibernate Mapping DTD 3.0//EN" "
http://hibernate.sourceforge.net/hibernate-mapping-3.0.dtd">
<hibernate-mapping>
    <class name="PO.UserInfoPO" table="info" catalog="test">
        <id name="id" type="int">
            <column name="id"/>
            <generator class="assigned"/>
        </id>
        <property name="userName" type="string">
            <column name="userName" length="30" not-null="true"/>
        </property>
        <property name="password" type="string">
            <column name="password" length="30" not-null="true"/>
        </property>
    </class>
    <!--设置命名查询,query 元素中的 name 指定要被调用的 HQL 名字-->
    <query name="queryUser_byAgeAndName">
        <![CDATA[ from UserInfoPO u where u.age>:minAge and u.userName
        like:userName ]]>
    </query>
</hibernate-mapping>
```

＜query＞标记的 name 属性用来设定查询外部 HQL 时的名称,使用命名查询的实例代码如下:

```
Query query=session.getUserNamedQuery("queryUser_byAgeAndName");
query.setInteger("minAge",22);
query.setString("userName","%志%");
List list=query.list();
for(int i=0;i<list.size();i++){
    ui=(UserInfoPO)list.get(i);
    System.out.println(ui.getUserName());
}
```

6.9 基于 Struts2＋Hibernate 的学生信息管理系统

本节使用 Struts2＋Hibernate 开发一个学生信息管理系统,通过对学生信息的增、删、改、查训练来熟悉 Hibernate 中常用方法的使用。项目使用 MySQL 数据库,数据库名为 student,表为 stuinfo,表的字段以及数据类型如图 6-1 所示。

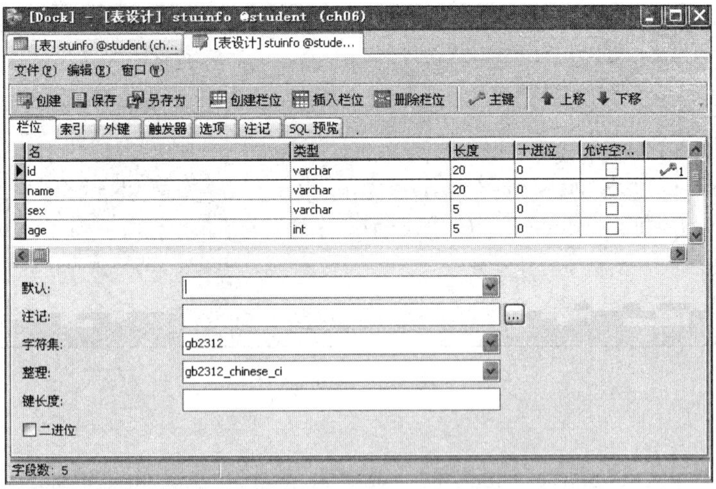

图 6-1 表 stuinfo 的字段以及数据类型

6.9.1 项目介绍、主页面以及查看学生信息功能的实现

项目的文件结构如图 6-2 所示。

首先,项目有一个主页面(index.jsp),代码如例 6-7 所示,页面运行效果如图 6-3 所示,单击图 6-3 所示页面中的"点此进入"后请求提交给业务控制器类 LookMessageAction,该控制器主要用于实现查看学生信息功能,代码如例 6-8 所示;业务控制器调用 JavaBean 进行处理;该项目封装了类 StudentDao,该类中封装了对数据库操作的方法,StudentDao 类的代码如例 6-9 所示;需要加载 Hibernate 的配置文件 hibernate.cfg.xml,代码如例 6-10 所示;项目提供一个加载配置文件的类 HibernateSessionFactory,代码如例 6-11 所示;LookMessageAction 业务请求处理成功后把查询到的学生信息通过内置对象 session 保存起来并返回到查看学生信息页面(lookMessage.jsp),页面跳转关系请参考例 6-15,代码如例 6-12 所示,页面效果如图 6-4 所示,该页面用 PO 对象获取保存在内置对象 session 中的数据并显示,PO 对象 Stuinfo 的代码如例 6-13 所示;该 PO 对象对应的映射文件(Stuinfo.hbm.xml)的代码如例 6-14 所示;该项目配置 Action(struts.xml)的代码如例 6-15 所示。

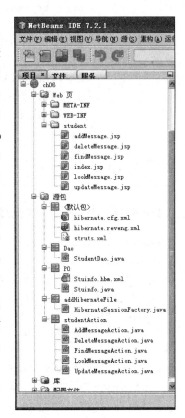

图 6-2 项目文件结构图

【例 6-7】 主页面(index.jsp)。

```
<%@page contentType="text/html" pageEncoding=
"UTF-8"%>
<%@taglib prefix="s" uri="/struts-tags" %>
<html>
```

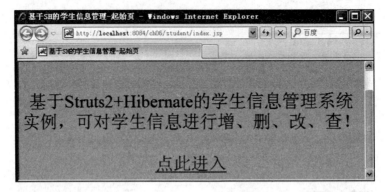

图 6-3 主页面

```
<head>
    <meta http-equiv="Content-Type" content="text/html; charset=UTF-8">
    <title><s:text name="基于 SH 的学生信息管理-起始页"></s:text></title>
</head>
<body bgcolor="#CCCCFF">
    <s:div align="center">
        <br/><br/><br/><br/><br/>
        <font color="black" size="6">基于 Struts2+Hibernate 的学生信息管理
          系统实例,可对学生信息进行增、删、改、查!
        </font>
        <br/><br/><br/>
        <s:a href="lookMessageAction">
            <font color="blue" size="6">点此进入</font>
        </s:a>
    </s:div>
</body>
</html>
```

【例 6-8】 查看学生信息功能控制器(LookMessageAction.java)。

```java
package studentAction;
import Dao.StudentDao;
import com.opensymphony.xwork2.ActionSupport;
import java.util.List;
import javax.servlet.http.HttpServletRequest;
import org.apache.struts2.ServletActionContext;

public class LookMessageAction extends ActionSupport{
    private HttpServletRequest request;
    private String message="input";
    public String execute() throws Exception{
        request=ServletActionContext.getRequest();
        StudentDao dao=new StudentDao();        //实例化
        List list=dao.findAllInfo();            //调用 StudentDao 类中的 findAllInfo()方法
```

```java
            request.getSession().setAttribute("count", list.size());
                                                            //向 session 对象传值
            request.getSession().setAttribute("allInfo", list);
            message="success";
            return message;
        }
    }
```

【例 6-9】 封装数据库操作的类(StudentDao.java)。

```java
package Dao;
import addHibernateFile.HibernateSessionFactory;
import PO.Stuinfo;
import java.util.List;
import javax.swing.JOptionPane;
import org.hibernate.Query;
import org.hibernate.Session;
import org.hibernate.Transaction;

public class StudentDao {
    private Transaction transaction;
    private Session session;
    private Query query;
    public StudentDao(){
    }
    public boolean saveInfo(Stuinfo info){
        try{
            session=HibernateSessionFactory.getSession();
            transaction=session.beginTransaction();
            session.save(info);
            transaction.commit();
            session.close();
            return true;
        }catch(Exception e){
            message("saveInfo.error:"+e);
            e.printStackTrace();
            return false;
        }
    }
    public List findInfo(String type,Object value){
        session=HibernateSessionFactory.getSession();
        try{
            transaction=session.beginTransaction();
            //HQL 语句
            String queryString="from Stuinfo as model where model."+type+"=?";
            query=session.createQuery(queryString);
```

```java
                query.setParameter(0, value);
                List list=query.list();
                transaction.commit();
                session.close();
                return list;
            }catch(Exception e){
                message("findInfo.error:"+e);
                e.printStackTrace();
                return null;
            }
        }
        public List findAllInfo(){
            session=HibernateSessionFactory.getSession();
            try{
                transaction=session.beginTransaction();
                String queryString="from Stuinfo";
                query=session.createQuery(queryString);
                List list=query.list();
                transaction.commit();
                session.close();
                return list;
            }catch(Exception e){
                message("findInfo.error:"+e);
                e.printStackTrace();
                return null;
            }
        }
        public boolean deleteInfo(String id){
            try{
                session=HibernateSessionFactory.getSession();
                transaction=session.beginTransaction();
                Stuinfo info=new Stuinfo();
                info= (Stuinfo)session.get(Stuinfo.class, id);
                session.delete(info);
                transaction.commit();
                session.close();
                return true;
            }catch(Exception e){
                message("deleteInfo.error:"+e);
                e.printStackTrace();
                return false;
            }
        }
        public boolean updateInfo(Stuinfo info){
            try{
```

```
            session=HibernateSessionFactory.getSession();
            transaction=session.beginTransaction();
            session.update(info);
            transaction.commit();
            session.close();
            return true;
        }catch(Exception e){
            message("updateInfo.error:"+e);
            return false;
        }
    }
    public void message(String mess){
        int type=JOptionPane.YES_NO_OPTION;
        String title="提示信息";
        JOptionPane.showMessageDialog(null, mess, title, type);
    }
}
```

【例 6-10】 Hibernate 的配置文件(hibernate.cfg.xml)。

```
<?xml version="1.0" encoding="UTF-8"?>
<!DOCTYPE hibernate-configuration PUBLIC
    "-//Hibernate/Hibernate Configuration DTD 3.0//EN"
    "http://hibernate.sourceforge.net/hibernate-configuration-3.0.dtd">
<hibernate-configuration>
  <session-factory>
    <property name="hibernate.dialect">org.hibernate.dialect.MySQLDialect
    </property>
    <property name="hibernate.connection.driver_class">com.mysql.jdbc.Driver
    </property>
    <property name="hibernate.connection.url">jdbc:mysql://localhost:3306/student
    </property>
    <property name="hibernate.connection.username">root</property>
    <property name="hibernate.connection.password">root</property>
    <mapping resource="PO/Stuinfo.hbm.xml"/>
  </session-factory>
</hibernate-configuration>
```

【例 6-11】 加载配置文件的类(HibernateSessionFactory.java)。

```
package addHibernateFile;
import javax.swing.JOptionPane;
import org.hibernate.Session;
import org.hibernate.SessionFactory;
import org.hibernate.cfg.Configuration;

public class HibernateSessionFactory {
```

```
    private static SessionFactory sessionFactory;
    private static Configuration configuration=new Configuration();
    public HibernateSessionFactory(){
    }
    static{
        try{
            Configuration configure=configuration.configure("hibernate.cfg.xml");
            sessionFactory=configure.buildSessionFactory();
        }catch(Exception e){
            message("生成 SessionFactoyr 失败："+e);
        }
    }
    public static Session getSession(){
        return sessionFactory.openSession();
    }
    public static void message(String mess){
        int type=JOptionPane.YES_NO_OPTION;
        String title="提示信息";
        JOptionPane.showMessageDialog(null, mess, title, type);
    }
}
```

【例 6-12】 查看学生信息页面(lookMessage.jsp)。

```
<%@page contentType="text/html" pageEncoding="UTF-8"
        import="java.util.ArrayList,PO.Stuinfo"%>
<%@taglib prefix="s" uri="/struts-tags" %>
<html>
    <head>
        <meta http-equiv="Content-Type" content="text/html; charset=UTF-8">
        <title><s:text name="学生信息管理系统-查看"></s:text></title>
    </head>
    <body bgcolor="pink">
        <s:div align="center">
        <hr color="red"/>
        <br>
        <table align="center" width="80%">
            <tr>
                <td width="25%">
                    查看学生信息
                </td>
                <td width="25%">
                  <s:a href="http://localhost:8084/ch06/student/addMessage.jsp">
                        添加学生信息
                    </s:a>
                </td>
```

```
            <td width="25%">
                <s:a href="http://localhost:8084/ch06/student/findMessage.jsp">
                        修改学生信息
                </s:a>
            </td>
            <td width="25%">
                <s:a href="http://localhost:8084/ch06/student/deleteMessage.
                  jsp">
                        删除学生信息
                </s:a>
            </td>
        </tr>
</table>
<br/>
<hr color="red"/>
<br/><br/><br/>
<span>你要查询的数据表中共有
        <%=request.getSession().getAttribute("count")%>人</span>
</s:div>
<table align="center" width="80%" border="5">
    <tr>
            <th>记录条数</th>
            <th>学号</th>
            <th>姓名</th>
            <th>性别</th>
            <th>年龄</th>
            <th>体重</th>
    </tr>
    <%
        ArrayList list=(ArrayList)session.getAttribute("allInfo");
        if(list.isEmpty()){
            %>
            <tr>
                    <td align="center"><span>暂无学生信息!</span></td>
            </tr>
            <%
        }else{
            for(int i=0;i<list.size();i++){
                Stuinfo info=(Stuinfo)list.get(i);
                %>
                <tr>
                        <td align="center"><%=i+1%></td>
                        <td><%=info.getId()%></td>
                        <td><%=info.getName()%></td>
                        <td><%=info.getSex()%></td>
```

```
                <td><%=info.getAge()%></td>
                <td><%=info.getWeight()%></td>
              </tr>
             <%
              }
             }
            %>
        </table>
    </body>
</html>
```

查看学生信息页面效果如图 6-4 所示。

图 6-4 查看学生信息页面

【例 6-13】 PO 对象 Stuinfo(Stuinfo.java)。

```
package PO;

public class Stuinfo  implements java.io.Serializable {
    private String id;
    private String name;
    private String sex;
    private int age;
    private float weight;
    public String getId() {
        return this.id;
    }
    public void setId(String id) {
        this.id=id;
    }
    public String getName() {
        return this.name;
    }
    public void setName(String name) {
        this.name=name;
    }
```

```java
    public String getSex() {
        return this.sex;
    }

    public void setSex(String sex) {
        this.sex=sex;
    }
    public int getAge() {
        return this.age;
    }
    public void setAge(int age) {
        this.age=age;
    }
    public float getWeight() {
        return this.weight;
    }
    public void setWeight(float weight) {
        this.weight=weight;
    }
}
```

【例 6-14】 PO 对象对应的映射文件(Stuinfo.hbm.xml)。

```xml
<?xml version="1.0"?>
<!DOCTYPE hibernate-mapping PUBLIC
    "-//Hibernate/Hibernate Mapping DTD 3.0//EN"
    "http://hibernate.sourceforge.net/hibernate-mapping-3.0.dtd">
<hibernate-mapping>
    <class name="PO.Stuinfo" table="stuinfo" catalog="student">
        <id name="id" type="string">
            <column name="id" length="20" />
            <generator class="assigned" />
        </id>
        <property name="name" type="string">
            <column name="name" length="20" not-null="true" />
        </property>
        <property name="sex" type="string">
            <column name="sex" length="5" not-null="true" />
        </property>
        <property name="age" type="int">
            <column name="age" not-null="true" />
        </property>
        <property name="weight" type="float">
            <column name="weight" precision="10" scale="0" not-null="true" />
        </property>
    </class>
```

```
    </hibernate-mapping>
```

【例 6-15】 Struts2 的配置文件(struts.xml)。

```xml
<!DOCTYPE struts PUBLIC
    "-//Apache Software Foundation//DTD Struts Configuration 2.0//EN"
    "http://struts.apache.org/dtds/struts-2.0.dtd">
<struts>
    <include file="example.xml"/>
    <!--Configuration for the default package.-->
    <package name="default" extends="struts-default">
        <action name="lookMessageAction"
            class="studentAction.LookMessageAction">
            <result name="success">/student/lookMessage.jsp</result>
            <result name="input">/student/index.jsp</result>
        </action>
        <action name="addMessageAction"
             class="studentAction.AddMessageAction">
            <result name="success" type="chain">lookMessageAction</result>
            <result name="input">/student/addMessage.jsp</result>
        </action>
        <action name="findMessageAction"
             class="studentAction.FindMessageAction">
            <result name="success">/student/updateMessage.jsp</result>
            <result name="input">/student/findMessage.jsp</result>
        </action>
        <action name="updateMessageAction"
             class="studentAction.UpdateMessageAction">
            <result name="success" type="chain">lookMessageAction</result>
            <result name="input">/student/updateMessage.jsp</result>
        </action>
        <action name="deleteMessageAction"
             class="studentAction.DeleteMessageAction">
            <result name="success" type="chain">lookMessageAction</result>
            <result name="input">/student/deleteMessage.jsp</result>
        </action>
    </package>
</struts>
```

6.9.2 添加学生信息功能的实现

在图 6-4 所示页面中单击"添加学生信息",出现如图 6-5 所示的添加学生页面(addMessage.jsp),代码如例 6-16 所示,该页面的业务控制器类为 AddMessageAction,代码如例 6-17 所示,该功能使用到的 PO 为 Stuinfo,代码前面已经介绍,需要的映射文件以及对数据的处理方法请参考前述代码。

图 6-5 添加学生页面

【例 6-16】 添加学生页面(addMessage.jsp)。

```
<%@page contentType="text/html" pageEncoding="UTF-8"%>
<%@taglib prefix="s" uri="/struts-tags" %>
<html>
    <head>
        <meta http-equiv="Content-Type" content="text/html; charset=UTF-8">
        <title><s:text name="学生信息管理系统-增加"></s:text></title>
    </head>
    <body bgcolor="pink">
        <s:div align="center">
            <hr color="red"/>
        <br/>
        <table align="center" width="80%">
            <tr>
                <td width="25%">
                    <s:a href="http://localhost:8084/ch06/student/lookMessage.jsp">
                        查看学生信息
                    </s:a>
                </td>
                <td width="25%">
                    添加学生信息
                </td>
                <td width="25%">
                    <s:a href="http://localhost:8084/ch06/student/findMessage.jsp">
```

```
                    修改学生信息
                </s:a>
            </td>
            <td width="25%">
                <s:a href="http://localhost:8084/ch06/student/deleteMessage.
                    jsp">
                    删除学生信息
                </s:a>
            </td>
        </tr>
    </table>
    <br/>
    <hr color="red"/>
    <center><font color="red" size="6">添加学生信息</font></center>
    </s:div>
    <s:form action="addMessageAction" method="post">
        <table align="center" width="30%" bgcolor="gray" border="5">
            <tr>
                <td>
                    <s:textfield name="id" label="学号" maxLength="16"/>
                </td>
                <td>
                    <s:textfield name="name" label="姓名">
                    </s:textfield>
                </td>
                <td>
                    <s:select name="sex" label="性别" list="{'男','女'}"/>
                </td>
                <td>
                    <s:textfield name="age" label="年龄"/>
                </td>
                <td>
                    <s:textfield name="weight" label="体重"/>
                </td>
                <td colspan="2">
                    <s:submit value="提交"/>
                    <s:reset value="清除"/>
                </td>
            </tr>
        </table>
    </s:form>
</body>
</html>
```

【例6-17】 添加学生页面对应的业务控制器（AddMessageAction.java）。

```java
package studentAction;
import Dao.StudentDao;
import PO.Stuinfo;
import com.opensymphony.xwork2.ActionSupport;
import java.util.List;
import javax.swing.JOptionPane;

public class AddMessageAction extends ActionSupport{
    private String id;
    private String name;
    private String sex;
    private int age;
    private float weight;
    private String message="input";
    public String getId() {
        return id;
    }
    public void setId(String id) {
        this.id=id;
    }
    public String getName() {
        return name;
    }
    public void setName(String name) {
        this.name=name;
    }
    public String getSex() {
        return sex;
    }
    public void setSex(String sex) {
        this.sex=sex;
    }
    public int getAge() {
        return age;
    }
    public void setAge(int age) {
        this.age=age;
    }
    public float getWeight() {
        return weight;
    }
    public void setWeight(float weight) {
        this.weight=weight;
    }
```

```java
public void validate(){
    if(this.getId()==null||this.getId().length()==0){
        addFieldError("id","学号不允许为空!");
    }else{
        StudentDao dao=new StudentDao();
        List list=dao.findInfo("id", this.getId());
        if(!list.isEmpty()){
            addFieldError("id","学号已存在!");
        }
    }
    if(this.getName()==null||this.getName().length()==0){
        addFieldError("name","姓名不允许为空!");
    }
    if(this.getAge()>130){
        addFieldError("age","请认真核实年龄!");
    }
    if(this.getWeight()>500){
        addFieldError("weight","请认真核实体重!");
    }
}
public String execute() throws Exception{
    StudentDao dao=new StudentDao();
    boolean save=dao.saveInfo(info());
    if(save){
        message=SUCCESS;
    }
    return message;
}
public Stuinfo info(){
    Stuinfo info=new Stuinfo();
    info.setId(this.getId());
    info.setName(this.getName());
    info.setSex(this.getSex());
    info.setAge(this.getAge());
    info.setWeight(this.getWeight());
    return info;
}
public void message(String mess){
    int type=JOptionPane.YES_NO_OPTION;
    String title="提示信息";
    JOptionPane.showMessageDialog(null, mess, title, type);
}
}
```

6.9.3 修改学生信息功能的实现

在图 6-5 所示页面中单击"修改学生信息",出现如图 6-6 所示的"修改学生信息"页面

(findMessage.jsp),代码如例 6-18 所示,该页面的业务控制器类为 FindMessageAction,代码如例 6-19 所示。

图 6-6　修改学生信息页面

【例 6-18】　修改学生页面(findMessage.jsp)。

```jsp
<%@page contentType="text/html" pageEncoding="UTF-8"%>
<%@page import="java.util.ArrayList,PO.Stuinfo"%>
<%@taglib  prefix="s" uri="/struts-tags" %>
<html>
    <head>
        <meta http-equiv="Content-Type" content="text/html; charset=UTF-8">
        <title><s:text name="学生信息管理系统-查找"></s:text></title>
    </head>
    <body bgcolor="pink">
        <s:div align="center">
            <hr color="red"/>
            <br/>
            <table align="center" width="80%">
                <tr>
                    <td width="25%">
                        <s:a href="http://localhost:8084/ch06/student/lookMessage.jsp">
                            查看学生信息
                        </s:a>
                    </td>
                    <td width="25%">
                        <s:a href="http://localhost:8084/ch06/student/addMessage.jsp">
                            添加学生信息
                        </s:a>
                    </td>
                    <td width="25%">
                        修改学生信息
                    </td>
                    <td width="25%">
```

```html
                    <s:a href="http://localhost:8084/ch06/student/deleteMessage.
                        jsp">
                            删除学生信息
                    </s:a>
                </td>
            </tr>
        </table>
        <br/>
        <hr color="red"/>
        <br/><br/><br/>
        <font size="5">修改学生信息</font>
    </s:div>
    <s:form action="findMessageAction" method="post">
        <table align="center" width="40%" border="5">
            <tr>
                <td>
                    请选择要修改学生的学号：
                </td>
                <td>
                    <select name="id">
                        <%
                            ArrayList list=(ArrayList)session.getAttribute(
                                                            "allInfo");
                            if(list.isEmpty()){
                        %>
                            <option value="null">null</option>
                        <%
                            }else{
                            for(int i=0;i<list.size();i++){
                                Stuinfo info=(Stuinfo)list.get(i);
                        %>
                                <option value="<%=info.getId()%>">
                                    <%=info.getId()%/>
                        <%
                                }
                            }
                        %>
                    </select>
                </td>
                <td>
                    <s:submit value="确定"></s:submit>
                </td>
            </tr>
        </table>
    </s:form>
```

```
      </body>
</html>
```

【例 6-19】 修改学生信息对应的业务控制器(FindMessageAction.java)。

```java
package studentAction;
import Dao.StudentDao;
import com.opensymphony.xwork2.ActionSupport;
import java.util.List;
import javax.servlet.http.HttpServletRequest;
import javax.swing.JOptionPane;
import org.apache.struts2.ServletActionContext;

public class FindMessageAction extends ActionSupport{
    private String id;
    private HttpServletRequest request;
    private String message="input";
    public String getId() {
        return id;
    }
    public void setId(String id) {
        this.id=id;
    }
    public void validate(){
        if(this.getId().equals("null")){
            message("暂无学生信息!");
            addFieldError("id","暂无学生信息!");
        }
    }
    public String execute() throws Exception{
        request=ServletActionContext.getRequest();
        StudentDao dao=new StudentDao();
        List list=dao.findInfo("id", this.getId());
        request.getSession().setAttribute("oneInfo", list);
        message=SUCCESS
        return message;
    }
    public void message(String mess){
        int type=JOptionPane.YES_NO_OPTION;
        String title="提示信息";
        JOptionPane.showMessageDialog(null, mess, title, type);
    }
}
```

在图 6-6 所示页面中选择需要修改的学号后单击"确定"按钮,业务控制由 FindMessage-Action 来完成,如果业务控制器 FindMessageAction 返回结果为 success,页面跳转到如

图 6-7 所示的修改学生信息页面(updateMessage.jsp),代码如例 6-20 所示,该页面对应的业务控制器类为 UpdateMessageAction。

图 6-7 修改信息页面

【例 6-20】 修改学生信息页面(updateMessage.jsp)。

```
<%@page import="PO.Stuinfo"%>
<%@page import="java.util.ArrayList"%>
<%@page contentType="text/html" pageEncoding="UTF-8"%>
<%@taglib prefix="s" uri="/struts-tags" %>
<html>
    <head>
        <meta http-equiv="Content-Type" content="text/html; charset=UTF-8">
        <title><s:text name="学生信息管理系统-修改"/></title>
    </head>
    <body bgcolor="pink">
        <s:div align="center">
            <hr color="red"/>
        <br/>
        <table align="center" width="80%">
            <tr>
                <td width="25%">
                  <s:a href="http://localhost:8084/ch06/student/lookMessage.jsp">
                        查看学生信息
                    </s:a>
                </td>
                <td width="25%">
                  <s:a href="http://localhost:8084/ch06/student/addMessage.jsp">
```

```
                    添加学生信息
                </s:a>
            </td>
            <td width="25%">
                    修改学生信息
            </td>
            <td width="25%">
                <s:a href="http://localhost:8084/ch06/student/deleteMessage.
                    jsp">
                    删除学生信息
                </s:a>
            </td>
        </tr>
    </table>
<br/>
<hr color="red"/>
<br/><br/><br/>
<font size="5">修改学生信息</font>
</s:div>
<s:form action="updateMessageAction" method="post">
    <table align="center" width="30%" bgcolor="gray" border="5">
        <%
        ArrayList list=(ArrayList)session.getAttribute("oneInfo");
        Stuinfo info=(Stuinfo)list.get(0);
        %>
            <tr>
                <td>
                    学号
                </td>
                <td>
                    <input name="id" value="<%=info.getId()%>"
                        readonly="readonly"/>
                </td>
            </tr>
            <tr>
                <td>
                    姓名
                </td>
                <td>
                    <input name="name"
                        value="<%=info.getName()%>"/>
                </td>
            </tr>
            <tr>
                <td>
```

```html
                    性别
                </td>
                <td>
                    <input name="sex" value="<%=info.getSex()%>"/>
                </td>
            </tr>
            <tr>
                <td>
                    年龄
                </td>
                <td>
                    <input name="age" value="<%=info.getAge()%>"/>
                </td>
            </tr>
            <tr>
                <td>
                    体重
                </td>
                <td>
                    <input name="weight"
                        value="<%=info.getWeight()%>"/>
                </td>
            </tr>
            <tr>
                <td colspan="2">
                    <s:submit value="提交"></s:submit>
                </td>
            </tr>
            <tr>
                <td align="center" colspan="2">
                  <s:a href="http://localhost:8084/ch06/student/
                        findMessage.jsp">返回</s:a>
                </td>
            </tr>
        </table>
    </s:form>
  </body>
</html>
```

【例6-21】 修改学生信息页面对应的业务控制器（UpdateMessageAction.java）。

```java
package studentAction;
import Dao.StudentDao;
import PO.Stuinfo;
import com.opensymphony.xwork2.ActionSupport;
import javax.swing.JOptionPane;
```

```java
public class UpdateMessageAction extends ActionSupport{
    private String id;
    private String name;
    private String sex;
    private int age;
    private float weight;
    private String message="input";
    public String getId() {
        return id;
    }
    public void setId(String id) {
        this.id=id;
    }
    public String getName() {
        return name;
    }
    public void setName(String name) {
        this.name=name;
    }
    public String getSex() {
        return sex;
    }
    public void setSex(String sex) {
        this.sex=sex;
    }
    public int getAge() {
        return age;
    }
    public void setAge(int age) {
        this.age=age;
    }
    public float getWeight() {
        return weight;
    }
    public void setWeight(float weight) {
        this.weight=weight;
    }
    public void validate(){
        if(this.getName()==null||this.getName().length()==0){
            addFieldError("name","姓名不允许为空!");
        }
        if(this.getAge()>130){
            addFieldError("age","请认真核实年龄!");
        }
```

```
        if(this.getWeight()>500){
            addFieldError("weight","请认真核实体重!");
        }
    }
    public String execute() throws Exception{
        StudentDao dao=new StudentDao();
        boolean update=dao.updateInfo(info());
        if(update){
            message=SUCCESS;
        }
        return message;
    }
    public Stuinfo info(){
        Stuinfo info=new Stuinfo();
        info.setId(this.getId());
        info.setName(this.getName());
        info.setSex(this.getSex());
        info.setAge(this.getAge());
        info.setWeight(this.getWeight());
        return info;
    }
    public void message(String mess){
        int type=JOptionPane.YES_NO_OPTION;
        String title="提示信息";
        JOptionPane.showMessageDialog(null, mess, title, type);
    }
}
```

6.9.4 删除学生信息功能的实现

在图 6-7 所示页面中单击"删除学生信息",出现如图 6-8 所示的"删除学生信息"页面(deleteMessage),代码如例 6-22 所示,该页面的业务控制器类为 DeleteMessageAction,代码如例 6-23 所示。

图 6-8 删除学生信息页面

【例 6-22】 删除学生信息页面(deleteMessage)。

```jsp
<%@page import="PO.Stuinfo"%>
<%@page import="java.util.ArrayList"%>
<%@page contentType="text/html" pageEncoding="UTF-8"%>
<%@taglib prefix="s" uri="/struts-tags" %>
<html>
    <head>
        <meta http-equiv="Content-Type" content="text/html; charset=UTF-8">
        <title><s:text name="学生信息管理系统-删除"/></title>
    </head>
    <body bgcolor="pink">
        <s:div align="center">
            <hr color="red"/>
            <br/>
            <table align="center" width="80%">
                <tr>
                    <td width="25%">
                        <s:a href="http://localhost:8084/ch06/student/lookMessage.jsp">
                            查看学生信息
                        </s:a>
                    </td>
                    <td width="25%">
                        <s:a href="http://localhost:8084/ch06/student/addMessage.jsp">
                            添加学生信息
                        </s:a>
                    </td>
                    <td width="25%">
                        <s:a href="http://localhost:8084/ch06/student/findMessage.jsp">
                            修改学生信息
                        </s:a>
                    </td>
                    <td width="25%">
                        删除学生信息
                    </td>
                </tr>
            </table>
            <br/>
            <hr color="red"/>
            <br/><br/><br/>
            <font size="5">删除学生信息</font>
        </s:div>
        <s:form action="deleteMessageAction" method="post">
            <table align="center" width="40%" border="5">
                <tr>
                    <td>
```

```
                    请选择要删除学生的学号：
                </td>
                <td>
                    <select name="id">
                        <%
                            ArrayList list=(ArrayList)session.getAttribute(
                                                    "allInfo");
                            if(list.isEmpty()){
                                %>
                                <option value="null">null</option>
                                <%
                            }else{
                                for(int i=0;i<list.size();i++){
                                    Stuinfo info=(Stuinfo)list.get(i);
                                    %>
                                    <option value="<%=info.getId()%>">
                                        <%=info.getId()%></option>
                                    <%
                                }
                            }
                        %>
                    </select>
                </td>
                <td>
                    <s:submit value="确定"/>
                </td>
            </tr>
        </table>
    </s:form>
</body>
</html>
```

【例 6-23】 删除页面对应的业务控制器（DeleteMessageAction.java）。

```
package studentAction;
import Dao.StudentDao;
import com.opensymphony.xwork2.ActionSupport;
import javax.swing.JOptionPane;

public class DeleteMessageAction extends ActionSupport{
    private String id;
    private String message;
    public String getId() {
        return id;
    }
    public void setId(String id) {
```

```
            this.id=id;
    }
    public void validate(){
        if(this.getId().equals("null")){
            message("暂无学生信息!");
            addFieldError("id","暂无学生信息!");
        }
    }
    public String execute() throws Exception{
        StudentDao dao=new StudentDao();
        boolean del=dao.deleteInfo(this.getId());
        if(del){
            message=SUCCESS;
        }
        return message;
    }
    public void message(String mess){
        int type=JOptionPane.YES_NO_OPTION;
        String title="提示信息";
        JOptionPane.showMessageDialog(null, mess, title, type);
    }
}
```

6.10 本章小结

本章详细介绍了 Hibernate 框架的核心组件,通过本章的学习应对 Hibernate 框架有了较深入的认识。应掌握以下内容。

(1) Hibernate 的配置文件。
(2) Hibernate 的 PO 对象。
(3) Hibernate 的映射文件。
(4) Hibernate 的核心类和接口。

6.11 习 题

6.11.1 选择题

1. Hibernate 的默认配置文件是(   )。
   A. hibernate.cfg.xml   B. hibernate.properties
   C. hibernate.hbm.xml   D. hibernate.xml
2. Hibernate 的 Configuration 类主要用来加载(   )。
   A. hibernate.cfg.xml   B. hibernate.properties
   C. hibernate.hbm.xml   D. hibernate.xml

3. Hibernate 中的 SessionFactory 对象是（　　）。
   A. 非线程安全的　　　　　　　　B. 线程安全的
   C. 不是线程对象　　　　　　　　D. PO 对象

6.11.2　填空题

1. Hibernate 的基本配置文件有两种形式：_____ 和 _____。
2. Hibernate 的每个表对应一个扩展名为 _____ 的映射文件。
3. Hibernate 中获取持久化对象的方法主要有 _____ 和 _____。

6.11.3　简答题

1. 简述 Hibernate 配置文件的作用。
2. 简述 Hibernate 的 Configuration 类的作用。
3. 简述 Hibernate 的 Session 的作用。

6.11.4　实训题

1. 使用 Struts2＋Hibernate 框架开发一个应用程序获取持久化对象（PO）。
2. 使用 Struts2＋Hibernate 框架开发一个应用程序，使用 Hibernate 常用方法操作 PO。

第 7 章　Hibernate 的高级组件

前面章节介绍了 Hibernate 的基础知识,通过使用这些知识可以构建基于 Hibernate 框架的应用程序。本章将进一步介绍 Hibernate 框架的一些高级技术,利用这些技术可以构建复杂、高效的、基于 Hibernate 框架的应用程序。通过对这些高级组件功能的学习,将帮助人们进一步了解和使用 Hibernate 框架。

本章主要内容如下所示。

(1) 使用关联关系操纵对象。

(2) Hibernate 框架中的数据查询方式。

(3) Hibernate 中的事务管理。

(4) Hibernate 中的 Cache 管理。

7.1　利用关联关系操纵对象

数据对象之间的关联关系有一对一、一对多和多对多等几种形式。在数据库操作中,数据对象之间的关联关系使用 JDBC 处理很困难。本节讲解如何在 Hibernate 中处理这些对象之间的关联关系。如果直接使用 JDBC 执行这种级联操作会非常烦琐。Hibernate 通过把实体对象之间的关联关系在映射文件中声明,比较简便地解决了这类关联关系操作问题。

7.1.1　一对一关联关系

一对一关联关系在实际生活中是比较常见的,例如,学生(Student)与学生证(Card)的关系,通过学生证可以找到学生。一对一关联关系在 Hibernate 中的实现有两种方式:主键关联和外键关联。

1. 主键关联

主键关联的重点是:关联的两个表共享一个主键值。例如,一个单位在网上的一个系统中注册为会员。则会员(单位)有一个登录账号,单位注册为会员的数据保存在表 company 中,每个会员的登录账号保存在表 login 中;一个会员只有一个登录账号,一个登录账号只属于一个会员,两个表之间是一对一的对应关系。

两个表 company 和 login 对应的 PO 分别为 Company 与 Login,表之间是一对一关系。它们共用一个主键值 id,这个主键可由 company 表或 login 表生成。问题是如何让另一张表引用已经生成的主键值呢?例如,company 表填入了主键 id 的值,login 表如何引用它?这需要在 Hibernate 的映射文件中使用主键的 foreign 生成机制。

为了表示 Company 与 Login 之间的一对一关联关系,在 Company 与 Login 的映射文件 Company.hbm.xml 和 Login.hbm.xml 中都要使用<one-to-one>标记,代码如例 7-1 和例 7-2 所示。

【例7-1】 Company 类的映射文件(Company.hbm.xml)。

```xml
<?xml version="1.0"?>
<!DOCTYPE hibernate-mapping PUBLIC
    "-//Hibernate/Hibernate Mapping DTD 3.0//EN"
    "http://hibernate.sourceforge.net/hibernate-mapping-3.0.dtd">
<hibernate-mapping package="PO">
  <class name="Company" table="company">
    <id column="ID" name="id" type="integer">
        <generator class="identity"/>
    </id>
    <property name="companyname" column="COMPANYNAME" type="string"/>
    <property name="linkman"  column="LINKMAN" type="string"/>
    <property name="telephone" column="TELEPHONE" type="string"/>
    <property name="email" column="EMAIL" type="string"/>
    <!--映射 Company 与 Login 的一对一主键关联-->
    <one-to-one name="login" cascade="all" class="PO.Login" lazy="false"
        fetch="join" outer-join="true"/>
  </class>
</hibernate-mapping>
```

<class>元素的 lazy 属性设定为 true,表示延迟加载,如果 lazy 的值设置为 false,则表示立即加载。下面对立即加载和延迟加载这两个概念进行说明。

(1) 立即加载：表示 Hibernate 在从数据库中取得数据组装好一个对象(比如会员 1)后,会立即再从数据库取得数据组装此对象所关联的对象(例如,登录账号 1)。

(2) 延迟加载：表示 Hibernate 在从数据库中取得数据组装好一个对象(比如会员 1)后,不会立即再从数据库取得数据组装此对象所关联的对象(例如,登录账号 1),而是等到需要时,才从数据库取得数据组装此关联对象。

<one-to-one>元素的 cascade 属性表明操作是否从父对象级联到被关联的对象,它的取值可以是如下几种。

① none：在保存、删除或修改对象时,不对其附属对象(关联对象)进行级联操作。这是默认设置。

② save-update：在保存、更新当前对象时,级联保存、更新附属对象(临时对象、游离对象)。

③ delete：在删除当前对象时,级联删除附属对象。

④ all：所有情况下均进行级联操作,即包括 save-update 和 delete 操作。

⑤ delete-orphan：删除和当前对象解除关系的附属对象。

<one-to-one>元素 fetch 属性的可选值是 join 和 select,默认值是 select。当 fetch 属性设定为 join 时,表示连接抓取(Join fetching)：Hibernate 通过在 select 语句中使用 Outerjoin(外连接)来获得对象的关联实例或者关联集合。当 fetch 属性设定为 select 时,表示查询抓取(Select Fetching)：需要另外发送一条 select 语句抓取当前对象的关联实体或

集合。

例 7-2 中＜one-to-one＞元素的 cascade 属性设置为 all，表示增加、删除及修改 Company 对象时，都会级联增加、删除和修改对应的 Login 对象。

【例 7-2】 Login 类的映射文件（Login.hbm.xml）。

```xml
<?xml version="1.0" encoding="UTF-8"?>
<!DOCTYPE hibernate-mapping PUBLIC
    "-//Hibernate/Hibernate Mapping DTD 3.0//EN"
    "http://hibernate.sourceforge.net/hibernate-mapping-3.0.dtd">
<hibernate-mapping package="PO">
  <class name="Login" table="login">
    <id column="ID" name="id" type="integer">
        <generator class="foreign">
            <param name="property">company</param>
        </generator>
    </id>
    <property name="loginname" column="LOGINNAME" type="string"/>
    <property name="loginpwd" column="LOGINPWD" type="string"/>
    <!--映射 Company 与 Login 的一对一关联-->
    <one-to-one name="company" class="PO.Company" constrained="true" />
  </class>
</hibernate-mapping>
```

在例 7-2 中，Login.hbm.xml 的主键 id 使用外键生成机制（foreign），引用表 company 的主键作为 login 表的主键值。company 在该映射文件的＜one-to-one＞元素中进行了定义，它是 Company 对象的代号。＜one-to-one＞元素的属性 constrained="true"表示 login 引用了 company 的主键作为外键。

2. 外键关联

外键关联的要点是：两个表各自有不同的主键，但其中一个表有一个外键引用另一个表的主键。例如，客户（Client）和客户地址（Address）是外键关联的一对一关系，它们在数据库中对应的表分别是 client 表和 address 表。Client 类的映射文件中的 address 为外键引用 Address 类的对应表中的主键，乍一看是多对一的关系，但是在 Client 类对应的映射文件设置一多对应的属性 address 时，设置 address 为 unique 的值为 true，即这个外键是唯一的，即成为一对一关系。

Address 的映射文件 Address.hmb.xml 的代码如例 7-3 所示。但 Client 的映射文件 Client.hbm.xml 的代码如例 7-4 所示。

【例 7-3】 Address 类的映射文件（Address.hbm.xml）。

```xml
<?xml version="1.0"?>
<!DOCTYPE hibernate-mapping PUBLIC
    "-//Hibernate/Hibernate Mapping DTD 3.0//EN"
    "http://hibernate.sourceforge.net/hibernate-mapping-3.0.dtd">
```

```xml
<hibernate-mapping package="PO">
  <class name="Address" table="address">
    <id column="ID" name="id" type="integer">
        <generator class="identity"/>
    </id>
    <property name="province" column="PROVINCE" type="string"/>
    <property name="city" column="CITY" type="string"/>
    <property name="street" column="STREET" type="string"/>
    <property name="zipcode" column="ZIPCODE" type="string"/>
    <!--映射 Client 与 Address 的一对一外键关联-->
    <one-to-one name="client" class="PO.Client" property-ref="address"/>
  </class>
</hibernate-mapping>
```

【例 7-4】 Client 类的映射文件(Client.hbm.xml)。

```xml
<?xml version="1.0"?>
<!DOCTYPE hibernate-mapping PUBLIC
    "-//Hibernate/Hibernate Mapping DTD 3.0//EN"
    "http://hibernate.sourceforge.net/hibernate-mapping-3.0.dtd">
<hibernate-mapping package="PO">
  <class name="Client" table="client">
    <id column="ID" name="id" type="integer">
        <!--不再是 foreign 了-->
        <generator class="identity"/>
    </id>
    <property column="CLIENTNAME" name="clientname" type="string"/>
    <property column="PHONE" name="phone" type="string"/>
    <property column="EMAIL" name="email" type="string"/>
    <!--映射 Client 到 Address 的一对一外键关联,唯一的多对一,实际上变成一对一关系-->
    <many-to-one name="address" class="PO.Address" column="address"
        cascade="all" lazy="false" unique="true"/>
  </class>
</hibernate-mapping>
```

在例 7-4 中,＜many-to-one＞元素的 name 属性声明外键关联对象的代号,class 属性声明该外键关联对象的类,column 属性声明该外键在数据表中对应的字段名,unique 属性表示使用 DDL 为外键字段生成一个唯一约束。

以外键关联对象的一对一关系,其本质上已经变成了一对多的双向关联,应直接按照一对多和多对一的要求编写它们的映射文件。当＜many-to-one＞元素的 unique 属性设定为 true 时,多对一的关系实际上变成了一对一的关系。

7.1.2 一对一关联关系的应用实例

本实例练习一对一关联关系的主键关联和外键关联。

1. 项目介绍

该项目中会员(Company)有一个登录账号(Login),其中 PO 对象 Company 和 Login

对应的表 company 与 login 使用一对一主键关联，Company 类的代码如例 7-5 所示，该 PO 对象对应的映射文件 Company.hbm.xml 代码如例 7-1 所示，Login 类的代码如例 7-6 所示，该 PO 对象对应的映射文件 Login.hbm.xml 代码如例 7-2 所示，表 company 的字段以及数据类型如图 7-1 所示，login 表的字段以及数据类型如图 7-2 所示；客户（Client）和客户地址（Address）使用一对一的外键关联，Address 类的代码如例 7-7 所示，该 PO 对象对应的映射文件 Address.hbm.xml 代码如例 7-3 所示，Client 类的代码如例 7-8 所示，该 PO 对象对应的映射文件 Client.hbm.xml 代码如例 7-4 所示，表 address 的字段以及数据类型如图 7-3 所示，client 表的字段以及数据类型如图 7-4 所示；该项目连接的是 MySQL 数据库，数据库名为 onetoone，需要的 Hibernate 配置文件为 hibernate.cfg.xml，代码如例 7-9 所示；加载该配置文件的类为 HibernateSessionFactory，代码如例 7-10 所示；为了实现对数据的操作，封装一个类 OneOneDAO，代码如例 7-11 所示；为了对上述关联关系进行测试编写

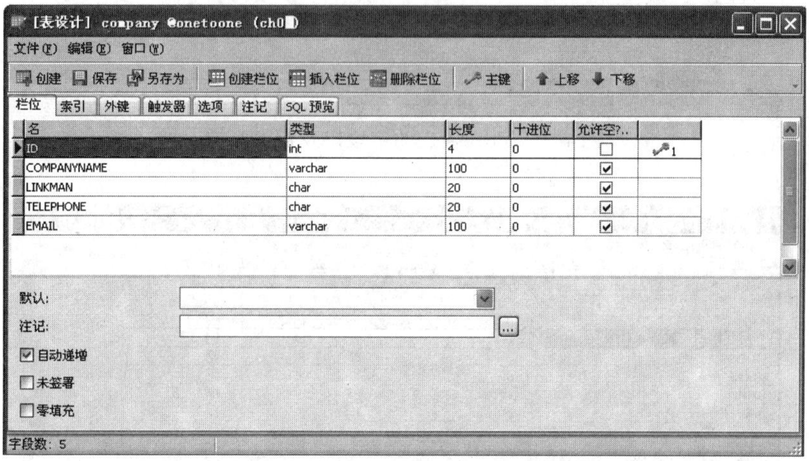

图 7-1　company 表的字段以及数据类型

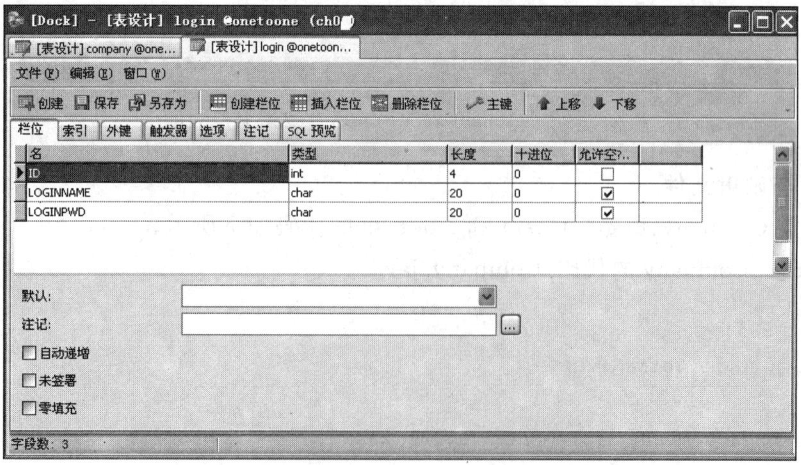

图 7-2　login 表的字段以及数据类型

了一个测试类 TestBean,代码如例 7-12 所示,编写一个 JSP 页面对以上关联关系进行测试;项目提供一个 JSP 页面(index.jsp)把测试数据显示出来,代码如例 7-13 所示,在该页面中调用 TestBean 类。项目的文件结构如图 7-5 所示。

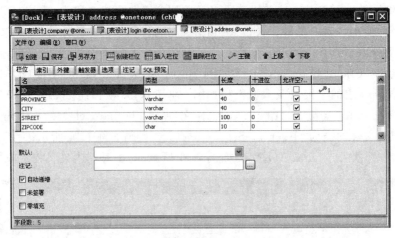

图 7-3　address 表的字段以及数据类型

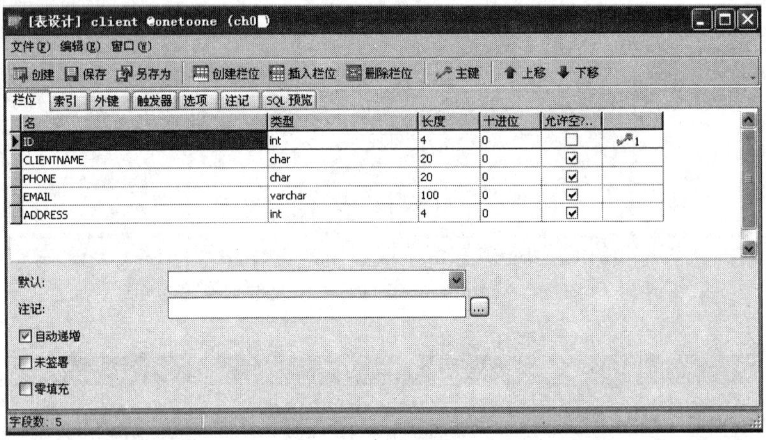

图 7-4　client 表的字段以及数据类型

2. PO 与映射文件

PO 对象 Company、Login、Client 和 Address 的代码如下所示。

【例 7-5】 Company 的代码(Company.java)。

```
package PO;
import java.io.Serializable;

public class Company implements Serializable{
    private Integer id;
    private String companyname;                //单位名称
    private String linkman;                    //单位联系人
    private String telephone;                  //联系电话
```

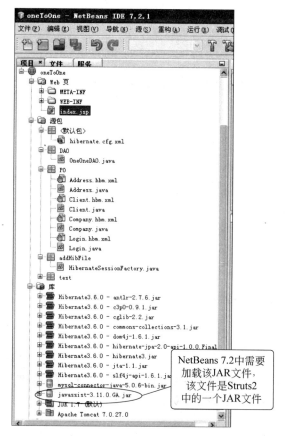

图 7-5　项目文件结构图

```
private String email;                              //邮箱
private Login login;                               //关联另外一个类,即 Login 类
public Integer getId() {
    return id;
}
public void setId(Integer id) {
    this.id=id;
}
public String getCompanyname() {
    return companyname;
}
public void setCompanyname(String companyname) {
    this.companyname=companyname;
}
public String getLinkman() {
    return linkman;
}
public void setLinkman(String linkman) {
    this.linkman=linkman;
```

```java
    }
    public String getTelephone() {
        return telephone;
    }
    public void setTelephone(String telephone) {
        this.telephone=telephone;
    }
    public String getEmail() {
        return email;
    }
    public void setEmail(String email) {
        this.email=email;
    }
    public Login getLogin() {
        return login;
    }
    public void setLogin(Login login) {
        this.login=login;
    }
}
```

PO 对象 Company 对应的映射文件 Company.hbm.xml 代码如例 7-1 所示,参考 7.1.1 节。

【例 7-6】 Login 的代码(Login.java)。

```java
package PO;
import java.io.Serializable;

public class Login implements Serializable{
    private Integer id;
    private String loginname;                       //登录账号
    private String loginpwd;                        //登录密码
    private Company company;                        //关联另外一个类 Company
    public Integer getId() {
        return id;
    }
    public void setId(Integer id) {
        this.id=id;
    }
    public String getLoginname() {
        return loginname;
    }
    public void setLoginname(String loginname) {
        this.loginname=loginname;
    }
    public String getLoginpwd() {
        return loginpwd;
```

```java
    }
    public void setLoginpwd(String loginpwd) {
        this.loginpwd=loginpwd;
    }
    public Company getCompany() {
        return company;
    }
    public void setCompany(Company company) {
        this.company=company;
    }
}
```

PO 对象 Login 对应的映射文件 Login.hbm.xml 代码如例 7-2 所示。

【例 7-7】 Address 类的代码(Address.java)。

```java
package PO;
import java.io.Serializable;

public class Address implements Serializable{
    private Integer id;
    private String province;            //省份
    private String city;                //城市
    private String street;              //街道
    private String zipcode;             //邮编
    private Client client;              //关联另外一个类
    public Integer getId() {
        return id;
    }
    public void setId(Integer id) {
        this.id=id;
    }
    public String getProvince() {
        return province;
    }
    public void setProvince(String province) {
        this.province=province;
    }
    public String getCity() {
        return city;
    }
    public void setCity(String city) {
        this.city=city;
    }
    public String getStreet() {
        return street;
    }
```

```java
    public void setStreet(String street) {
        this.street=street;
    }
    public String getZipcode() {
        return zipcode;
    }
    public void setZipcode(String zipcode) {
        this.zipcode=zipcode;
    }
    public Client getClient() {
        return client;
    }
    public void setClient(Client client) {
        this.client=client;
    }
}
```

PO 对象 Address 对应的映射文件 Address.hbm.xml 代码如例 7-3 所示。

【例 7-8】 Client 类的代码(Client.java)。

```java
package PO;
import java.io.Serializable;

public class Client implements Serializable{
    private Integer id;
    private String clientname;              //客户名称
    private String phone;                   //客户电话
    private String email;                   //客户邮箱
    private Address address;                //关联另外一个 PO
    public Integer getId() {
        return id;
    }
    public void setId(Integer id) {
        this.id=id;
    }
    public String getClientname() {
        return clientname;
    }
    public void setClientname(String clientname) {
        this.clientname=clientname;
    }
    public String getPhone() {
        return phone;
    }
    public void setPhone(String phone) {
        this.phone=phone;
```

```
    }
    public String getEmail() {
        return email;
    }
    public void setEmail(String email) {
        this.email=email;
    }
    public Address getAddress() {
        return address;
    }
    public void setAddress(Address address) {
        this.address=address;
    }
}
```

PO 对象 Client 对应的映射文件 Client.hbm.xml 代码如例 7-4 所示。

3. Hibernate 的配置文件

【例 7-9】 配置文件(hibernate.cfg.xml)。

```xml
<?xml version="1.0"?>
<!DOCTYPE hibernate-configuration PUBLIC
    "-//Hibernate/Hibernate Configuration DTD 3.0//EN"
    "http://hibernate.sourceforge.net/hibernate-configuration-3.0.dtd">
<hibernate-configuration>
    <session-factory>
        <property name="connection.driver_class">com.mysql.jdbc.Driver
        </property>
        <property name="dialect">
            org.hibernate.dialect.MySQLDialect
        </property>
        <property name="connection.url">
            <!--onetoone 为数据库名-->
            jdbc:mysql://localhost:3306/onetoone?
            useUnicode=true&amp;characterEncoding=gb2312
        </property>
        <property name="connection.username">root</property>
        <property name="connection.password">root</property>
        <mapping resource="PO/Address.hbm.xml"/>
        <mapping resource="PO/Client.hbm.xml"/>
        <mapping resource="PO/Company.hbm.xml"/>
        <mapping resource="PO/Login.hbm.xml"/>
    </session-factory>
</hibernate-configuration>
```

4. 加载 Hibernate 配置文件的类

加载配置文件的类为 HibernateSessionFactory。

【例 7-10】 HibernateSessionFactory 类（HibernateSessionFactory.java）。

```java
package addHibFile;
import org.hibernate.HibernateException;
import org.hibernate.Session;
import org.hibernate.SessionFactory;
import org.hibernate.cfg.Configuration;

public class HibernateSessionFactory {
    private HibernateSessionFactory() {
    }
    private static String CONFIG_FILE_LOCATION="/hibernate.cfg.xml";
    /*它的作用是为每一个使用该变量的线程都提供一个变量值的副本,使每一个线程都可以独
      立地改变自己的副本,而不会和其他线程的副本冲突。从线程的角度看,就好像每一个线程
      都完全拥有该变量。这样可以实现子线程的安全*/
    private static final ThreadLocal threadLocal=new ThreadLocal();
    private static final Configuration cfg=new Configuration();
    private static SessionFactory sessionFactory;
    public static Session currentSession() throws HibernateException {
        Session session=(Session) threadLocal.get();
        if (session==null) {
            if (sessionFactory==null) {
                try {
                    cfg.configure(CONFIG_FILE_LOCATION);
                    sessionFactory=cfg.buildSessionFactory();
                }
                catch (Exception e) {
                    System.err.println("生成 SessionFactory 失败!");
                    e.printStackTrace();
                }
            }
            session=sessionFactory.openSession();
            threadLocal.set(session);
        }
        return session;
    }
    public static void closeSession() throws HibernateException {
        Session session=(Session) threadLocal.get();
        threadLocal.set(null);
        if (session !=null) {
            session.close();
        }
    }
}
```

5. 封装对数据操作的类

为了实现对数据库的操作封装一个类，即 OneOneDAO 的 JavaBean，该类中封装了前述需要实现的功能。

【例 7-11】 OneOneDAO 类（OneOneDAO.java）。

```java
package DAO;
import addHibFile.HibernateSessionFactory;
import org.hibernate.*;
import PO.*;

public class OneOneDAO {
    //添加会员的方法
    public void addCompany(Company company) {
        Session session=HibernateSessionFactory.currentSession();
        Transaction ts=null;
        try{
            ts=session.beginTransaction();
            session.save(company);
            ts.commit();
        }catch(Exception ex){
            ts.rollback();
            System.out.println("【系统错误】在 OneOneDAO 的 addCompany 方法中出错：");
            ex.printStackTrace();
        }finally{
            HibernateSessionFactory.closeSession();
        }
    }
    //获取会员信息
    public Company loadCompany(Integer id) {
        Session session=HibernateSessionFactory.currentSession();
        Transaction ts=null;
        Company company=null;
        try{
            ts=session.beginTransaction();
            company=(Company)session.get(Company.class,id);
            ts.commit();
        }catch(Exception ex){
            ts.rollback();
            System.out.println("【系统错误】在 OneOneDAO 的 loadCompany 方法中出错：");
            ex.printStackTrace();
        }finally{
            HibernateSessionFactory.closeSession();
        }
        return company;
    }
```

```java
//添加客户信息
public void addClient(Client client) {
    Session session=HibernateSessionFactory.currentSession();
    Transaction ts=null;
    try{
        ts=session.beginTransaction();
        session.save(client);
        ts.commit();
    }catch(Exception ex){
        ts.rollback();
        System.out.println("【系统错误】在 OneOneDAO 的 addClient 方法中出错: ");
        ex.printStackTrace();
    }finally{
        HibernateSessionFactory.closeSession();
    }
}

//获取客户信息
public Client loadClient(Integer id) {
    Session session=HibernateSessionFactory.currentSession();
    Transaction ts=null;
    Client client=null;
    try{
        ts=session.beginTransaction();
        client=(Client)session.get(Client.class,id);
        ts.commit();
    }catch(Exception ex){
        ts.rollback();
        System.out.println("【系统错误】在 OneOneDAO 的 loadClient 方法中出错: ");
        ex.printStackTrace();
    }finally{
        HibernateSessionFactory.closeSession();
    }
    return client;
}
}
```

6. 关联关系测试类

为了对上述关联关系进行测试编写了一个测试类 TestBean。

【例 7-12】 TestBean 类(TestBean.java)。

```java
package test;
import PO.*;
import DAO.*;

public class TestBean {
    OneOneDAO dao=new OneOneDAO();
```

```java
//添加会员信息
public void addCompany(){
    Company company=new Company();
    Login login=new Login();
    login.setLoginname("QQ");
    login.setLoginpwd("123");
    company.setCompanyname("清华大学出版社");
    company.setLinkman("白立军");
    company.setTelephone("010-60772015");
    company.setEmail("bailj@163.com");
    //PO对象之间相互设置关联关系
    login.setCompany(company);
    company.setLogin(login);
    dao.addCompany(company);
}
//获取会员信息
public Company loadCompany(Integer id){
    return dao.loadCompany(id);
}
//添加客户信息
public void addClient(){
    Client client=new Client();
    Address address=new Address();
    address.setProvince("北京市");
    address.setCity("北京市");
    address.setStreet("清华园");
    address.setZipcode("100084");
    client.setClientname("李想");
    client.setPhone("010-56565566");
    client.setEmail("lixiang@163.com");
    //PO对象之间相互设置关联关系
    address.setClient(client);
    client.setAddress(address);
    dao.addClient(client);
}
//获取客户信息
public Client loadClient(Integer id){
    return dao.loadClient(id);
}
}
```

7. 测试页面

为了显示测试数据,编写一个JSP页面(index.jsp)。

【例7-13】 JSP页面(index.jsp)。

```jsp
<%@page contentType="text/html" pageEncoding="UTF-8"%>
```

```jsp
<%@page import="test.TestBean"%>
<%@page import="PO.*"%>
<html>
  <head>
    <title>Hibernate 的一对一关联关系映射</title>
  </head>
  <body>
    <h2>Hibernate 的一对一关联关系映射</h2>
    <hr>
    <!--调用 TestBean 测试-->
    <jsp:useBean id="test" class="test.TestBean" />
    <%
        test.addCompany();
        out.println("添加一个公司");
        test.addClient();
        out.println("添加一个客户");
        Integer id=new Integer(1);
        Company company=test.loadCompany(id);
        out.println("加载 id 为 1 的公司");
        Login login=company.getLogin();
        out.println("获取公司的登录账号");
        Client client=test.loadClient(id);
        out.println("获取 id 为 1 的客户");
        Address address=client.getAddress();
        out.println("获取该客户地址");
        out.println("<br>company.getCompanyname()="+
                    company.getCompanyname());
        out.println("<br>company.getLinkman()="+company.getLinkman());
        out.println("<br>company.getTelephone()="+company.getTelephone());
        out.println("<br>login.getLoginname()="+login.getLoginname());
        out.println("<br>login.getLoginpwd()="+login.getLoginpwd());
        out.println("<br>");
        out.println("<br>client.getClientname()="+client.getClientname());
        out.println("<br>client.getPhone()="+client.getPhone());
        out.println("<br>client.getEmail()="+client.getEmail());
        out.println("<br>address.getProvince()="+address.getProvince());
        out.println("<br>address.getCity()="+address.getCity());
        out.println("<br>address.getStreet()="+address.getStreet());
        out.println("<br>address.getZipcode()="+address.getZipcode());
    %>
  </body>
</html>
```

8. 项目部署和运行

项目部署后运行 index.jsp，运行效果如图 7-6 所示。

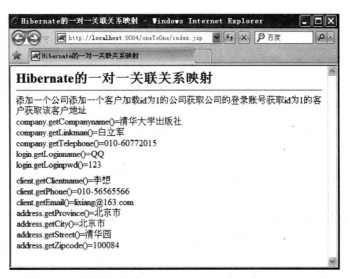

图 7-6 运行效果

7.1.3 一对多关联关系

一对多的关系很常见,例如,父亲和孩子、班级(Group)与学生(Student)、客户与订单的关系就是很典型的一对多的关系。在实际编写程序时,一对多关系有两种实现方式:单向关联和双向关联。单向的一对多关系只需在一方进行映射配置,而双向的一对多关系需要在关联的双方进行映射配置。下面以客户(Customer)与订单(Orders)为例讲解如何配置一对多的关系。

1. 单向关联

单向的一对多关系只需在一方进行映射配置,所以只需配置客户(Customer)的映射文件 Customer.hbm.xml,代码如例 7-14 所示。

【例 7-14】 Customer 类的映射文件(Customer.hbm.xml)。

```
<?xml version="1.0"?>
<!DOCTYPE hibernate-mapping PUBLIC
    "-//Hibernate/Hibernate Mapping DTD 3.0//EN"
    "http://hibernate.sourceforge.net/hibernate-mapping-3.0.dtd">
<hibernate-mapping package="PO">
    <class name="Customer" table="customer">
        <id column="ID" name="id" type="integer">
            <generator class="identity"/>
        </id>
        <property name="cname" column="CNAME" type="string"/>
        <property name="bank" column="BANK" type="string"/>
        <property name="phone" column="PHONE" type="string"/>
        <!--一对多双向关联映射 customer 到 orders,单的一方配置-->
        <set name="orders" table="orders " cascade="all" inverse="true" lazy=
        "false"
```

```xml
        cascade="all">
           <key column="CUSTOMER_ID"/>
           <one-to-many class="PO.Orders"/>
        </set>
     </class>
</hibernate-mapping>
```

<set>元素描述的字段(本例中为 orders 表)对应的类型为 java.util.Set,它的各个属性的含义如下。

(1) name：字段名；本例中的字段名为 orders，属于 java.util.Set 类型。

(2) table：关联表名；本例中，orders 的关联数据表名是 orders。

(3) lazy：是否延迟加载；lazy=false 表示立即加载。

(4) inverse：用于表示双向关联中的被动的一端。inverse 的值为 false 的一方负责维护关联关系。默认值为 false。

(5) cascade：级联关系；cascade=all 表示所有情况下均进行级联操作，即包括 save-update 和 delete 操作。

(6) sort：排序关系，其可选值为 unsorted(不排序)、natural(自然排序)和 comparatorClass (由某个实现了 java.util.comparator 接口的类型指定排序算法)。

(7) <key>子元素的 column 属性指定关联表(本例中为 orders 表)的外键。

(8) <one-to-many>子元素的 class 属性指定了关联类的名字。

2. 双向关联

如果要设置一对多双向关联关系，那么还需要在"多"方的映射文件中使用<many-to-one>标记。例如，在 Customer 与 Orders 一对多的双向关联中，除了修改 Customer 的映射文件 Customer.hbm.xml 外，还需要在 Orders 的映射文件 Orders.hbm.xm 中添加如下代码，如例 7-15 所示。

【例 7-15】 Orders 类的映射文件(Orders.hbm.xml)。

```xml
<?xml version="1.0" encoding="UTF-8"?>
<!DOCTYPE hibernate-mapping PUBLIC
    "-//Hibernate/Hibernate Mapping DTD 3.0//EN"
    "http://hibernate.sourceforge.net/hibernate-mapping-3.0.dtd">
<hibernate-mapping package="PO">
    <class name="Orders" table="orders">
        <id column="ID" name="id" type="integer">
            <generator class="identity"/>
        </id>
        <property name="orderno" column="ORDERNO" type="string"/>
        <property name="money" column="MONEY" type="double"/>
        <!--一对多双向关联映射中的多的一方配置-->
        <many-to-one name="customer"class="PO.Customer"
            column="CUSTOMER_ID" lazy="false" not-null="true"/>
    </class>
</hibernate-mapping>
```

7.1.4 一对多关联关系的应用实例

本实例练习一对多关联关系的双向关联。

1. 项目介绍

该项目中客户(Customer)和订单(Orders)是一对多的关系,其中 PO 对象 Customer 和 Orders 对应的表 Customer 与 orders 使用一对多的双向关联,Customer 类的代码如例 7-16 所示,该 PO 对象对应的映射文件为 Customer.hbm.xml,代码如例 7-14 所示,Orders 类的代码如例 7-17 所示,该 PO 对象对应的映射文件 Orders.hbm.xml 代码如例 7-15 所示,表 customer 的字段以及数据类型如图 7-7 所示,orders 表的字段以及数据类型如图 7-8 所示;该项目连接 MySQL 数据库,数据库名为 onetomany,需要的配置文件为 hibernate.cfg.xml,将例 7-9 的代码简单修改即可;加载该配置文件的类为 HibernateSessionFactory,代码如例 7-10 所示;为了实现对数据库的操作封装一个类 OneManyDAO,代码如例 7-18 所示;为了对上述关联关系进行测试编写了一个测试类 TestBean,代码如例 7-19 所示,可以编译 JSP 页面对以上关联关系进行测试;项目提供一个 JSP 页面(index.jsp)把测试数据显示出来,代码如例 7-20 所示,在该页面中调用 TestBean 类。项目的文件结构如图 7-9 所示。

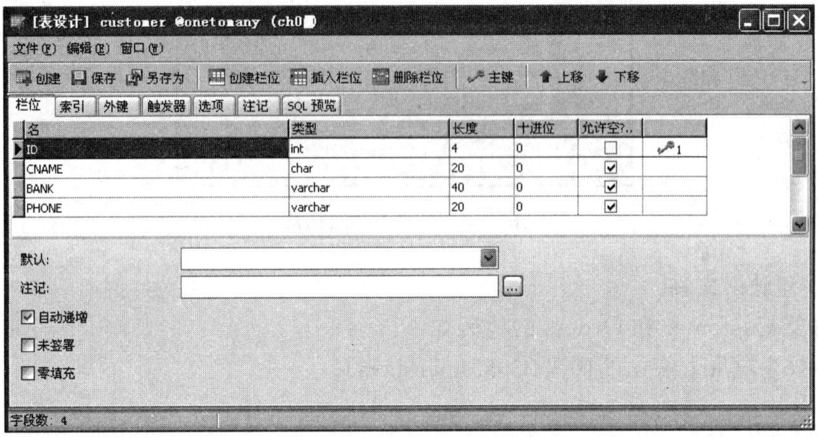

图 7-7 customer 表的字段以及数据类型

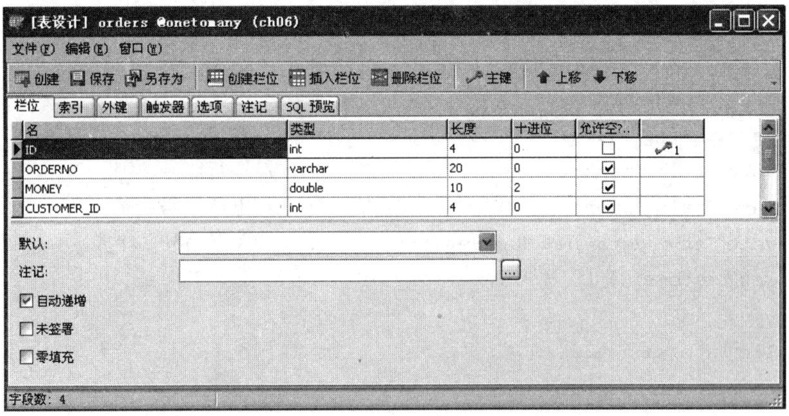

图 7-8 orders 表的字段以及数据类型

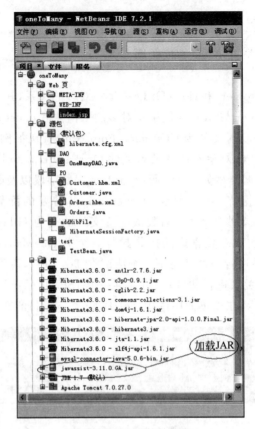

图 7-9 项目文件结构图

2. PO 与映射文件

PO 对象 Customer 和 Orders 的代码如下。

【例 7-16】 Customer 的代码(Customer.java)。

```
package PO;
import java.io.Serializable;
import java.util.*;

public class Customer implements Serializable {
    private Integer id;
    private String cname;                        //客户名称
    private String bank;                         //银行账号
    private String phone;                        //电话号码
    private Set orders=new HashSet();            //关联另外一个类
    public Customer() {
    }
    public Integer getId() {
        return id;
    }
    public void setId(Integer id) {
```

```java
        this.id=id;
    }
    public String getCname() {
        return cname;
    }
    public void setCname(String cname) {
        this.cname=cname;
    }
    public String getBank() {
        return bank;
    }
    public void setBank(String bank) {
        this.bank=bank;
    }
    public String getPhone() {
        return phone;
    }
    public void setPhone(String phone) {
        this.phone=phone;
    }
    public Set getOrders() {
        return orders;
    }
    public void setOrders(Set orders) {
        this.orders=orders;
    }
}
```

PO 对象 Customer 对应的映射文件 Customer.hbm.xml 代码如例 7-14 所示。

【例 7-17】 Orders 的代码(Orders.java)。

```java
package PO;
import java.io.Serializable;

public class Orders implements Serializable {
    private Integer id;
    private String orderno;                    //订单号
    private Double money;                      //所需资金
    private Customer customer;                 //关联另外一个 PO
    public Integer getId() {
        return id;
    }
    public void setId(Integer id) {
        this.id=id;
    }
    public String getOrderno() {
```

```
        return orderno;
    }
    public void setOrderno(String orderno) {
        this.orderno=orderno;
    }
    public Double getMoney() {
        return money;
    }
    public void setMoney(Double money) {
        this.money=money;
    }
    public Customer getCustomer() {
        return customer;
    }
    public void setCustomer(Customer customer) {
        this.customer=customer;
    }
}
```

PO 对象 Orders 对应的映射文件 Orders.hbm.xml 代码如例 7-15 所示。

3. Hibernate 的配置文件

配置文件参考例 7-9,只需修改数据库名为 onetomany。

4. 加载 Hibernate 配置文件的类

加载配置文件的类为 HibernateSessionFactory。代码与例 7-10 相同。

5. 封装对数据操作的类

为了实现对数据库的操作定义一个类 OneManyDAO 的 JavaBean,该类中封装了前述需要的功能。

【例 7-18】 OneManyDAO 类(OneManyDAO.java)。

```
package DAO;
import addHibFile.HibernateSessionFactory;
import org.hibernate.*;
import PO.*;

public class OneManyDAO {
    public void addCustomer(Customer customer) {
        Session session=HibernateSessionFactory.currentSession();
        Transaction ts=null;
        try{
            ts=session.beginTransaction();
            session.save(customer);
            ts.commit();
        }catch(Exception ex){
            ts.rollback();
            System.out.println("添加客户失败!");
```

```java
                ex.printStackTrace();
            }finally{
                HibernateSessionFactory.closeSession();
            }
    }
    public Customer loadCustomer(Integer id) {
        Session session=HibernateSessionFactory.currentSession();
        Transaction ts=null;
        Customer customer=null;
        try{
            ts=session.beginTransaction();
            customer= (Customer)session.get(Customer.class,id);
            ts.commit();
        }catch(Exception ex){
            ts.rollback();
            System.out.println("获取客户失败!");
            ex.printStackTrace();
        }finally{
            HibernateSessionFactory.closeSession();
        }
        return customer;
    }
    public void addOrders(Orders order) {
        Session session=HibernateSessionFactory.currentSession();
        Transaction ts=null;
        try{
            ts=session.beginTransaction();
            session.save(order);
            ts.commit();
        }catch(Exception ex){
            ts.rollback();
            System.out.println("添加订单失败!");
            ex.printStackTrace();
        }finally{
            HibernateSessionFactory.closeSession();
        }
    }
    public Orders loadOrders(Integer id) {
        Session session=HibernateSessionFactory.currentSession();
        Transaction ts=null;
        Orders order=null;
        try{
            ts=session.beginTransaction();
            order= (Orders)session.get(Orders.class,id);
            ts.commit();
```

```java
        }catch(Exception ex){
            ts.rollback();
            System.out.println("获取订单失败!");
            ex.printStackTrace();
        }finally{
            HibernateSessionFactory.closeSession();
        }
        return order;
    }
}
```

6. 关联关系测试类

为了对前述关联关系进行测试编写了一个测试类 TestBean。

【例 7-19】 TestBean 类(TestBean.java)。

```java
package test;
import java.util.Random;
import PO.*;
import DAO.*;

public class TestBean {
    OneManyDAO dao=new OneManyDAO();
    Random rnd=new Random();                          //用于产生订单号
    public void addCustomer(){
        Customer customer=new Customer();
        customer.setCname("清华大学出版社");
        customer.setBank("9559501012356789");
        customer.setPhone("010-62772015");
        dao.addCustomer(customer);
    }
    public Customer loadCustomer(Integer id){
        return dao.loadCustomer(id);
    }
    
    public void addOrders(Customer customer){
        Orders order=new Orders();
        order.setOrderno(new Long(System.currentTimeMillis()).toString());
        order.setMoney(new Double(rnd.nextDouble() * 10000));
        order.setCustomer(customer);
        customer.getOrders().add(order);
        dao.addOrders(order);
    }
    public Orders loadOrders(Integer id){
        return dao.loadOrders(id);
    }
}
```

7. 测试页面

为了显示测试数据,编写一个 JSP 页面(index.jsp)。

【例 7-20】 JSP 页面(index.jsp)。

```jsp
<%@page language="java" pageEncoding="gb2312"%>
<%@page import="test.TestBean"%>
<%@page import="PO.*"%>
<%@page import="java.util.*"%>
<%@page import="java.text.NumberFormat"%>
<html>
  <head>
    <title>Hibernate 的一对多双向关联关系映射</title>
  </head>
  <body>
    <h2>Hibernate 的一对多双向关联关系映射</h2>
    <hr>
    <jsp:useBean id="test" class="test.TestBean" />
    <%
        test.addCustomer();
        Integer id=new Integer(1);
        Customer customer=test.loadCustomer(id);
        test.addOrders(customer);
        test.addOrders(customer);
        test.addOrders(customer);
        //根据指定的客户,得到该客户的所有订单
        NumberFormat  nf=NumberFormat.getCurrencyInstance();
        out.println("<br>客户"+customer.getCname()+"的所有订单:");
          Iterator it=customer.getOrders().iterator();
        Orders order=null;
        while (it.hasNext()){
            order=(Orders)it.next();
            out.println("<br>订单号:"+order.getOrderno());
            out.println("<br>订单金额:"+nf.format(order.getMoney()));
        }
        //根据指定的订单,得到其所属的客户
        order=test.loadOrders(new Integer(1));
        customer=order.getCustomer();
        out.println("<br>");
        out.println("<br>订单号为"+order.getOrderno().trim()+"的所属客户为:"
                   +customer.getCname());
    %>
  </body>
</html>
```

8. 项目部署和运行

项目部署后运行 index.jsp,运行效果如图 7-10 所示。

图 7-10　运行效果

7.1.5　多对多关联关系

学生(Student)和课程(Course)、商品(Items)和订单(Orders)是典型的多对多关系。例如,某种商品可以存在于很多订单中,一个订单中也可以存在很多种商品。在映射多对多关系时,需要另外使用一个连接表(例如,selecteditems)。selecteditems 表包含两个字段:ORDERID 和 ITEMID。此外,在它们的映射文件中要使用<many-to-many>元素。

7.1.6　多对多关联关系的应用实例

本实例练习多对多关联关系。

1. 项目介绍

该项目中商品(Items)和订单(Orders)是多对多的关系,其中 PO 对象 Items 和 Orders 对应的表 items 和 orders 使用多对多的关联,Items 类的代码如例 7-21 所示,该 PO 对象对应的映射文件 Items.hbm.xml 代码如例 7-22 所示,Orders 类的代码如例 7-23 所示,该 PO 对象对应的映射文件 Orders.hbm.xml 代码如例 7-24 所示,表 items 的字段以及数据类型如图 7-11 所示,orders 表的字段以及数据类型如图 7-12 所示,关联关系表 selecteditems 的字段以及数据类型如图 7-13 所示;该项目连接 MySQL 数据库,数据库名为 manytomany,

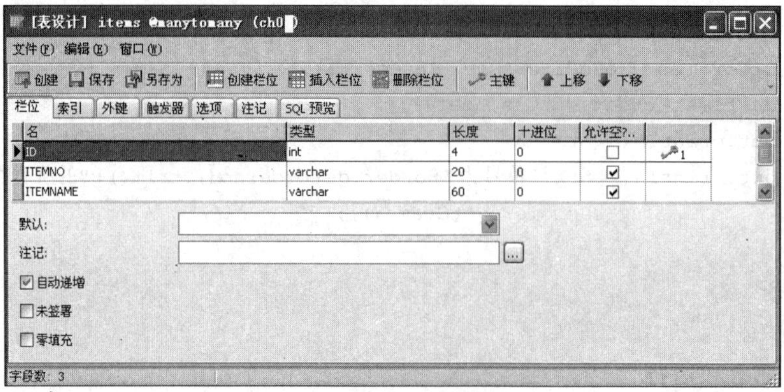

图 7-11　items 表的字段以及数据类型

需要的配置文件为 hibernate.cfg.xml,将例 7-9 代码简单修改即可;加载该配置文件的类为 HibernateSessionFactory,代码如例 7-10 所示;为了实现对数据的操作封装一个类 ManyManyDAO,代码如例 7-25 所示;为了对上述关联关系进行测试编写一个测试类 TestBean,代码如例 7-26 所示,可以通过 JSP 页面对以上关联关系进行测试;项目提供一个 JSP 页面(index.jsp)把测试数据显示出来,代码如例 7-27 所示,在该页面中调用 TestBean 类。项目的文件结构如图 7-14 所示。

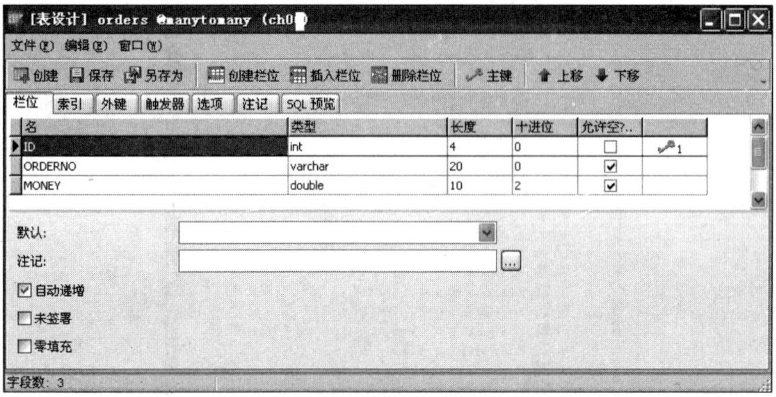

图 7-12　orders 表的字段以及数据类型

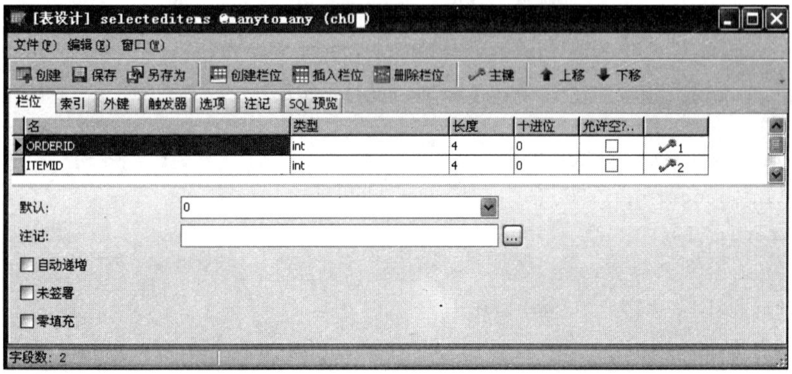

图 7-13　selecteditems 表的字段以及数据类型

2. PO 与映射文件

PO 对象 Items 和 Orders 的代码如下。

【例 7-21】　Items 的代码(Items.java)。

```
package PO;
import java.io.Serializable;
import java.util.*;

public class Items implements Serializable{
    private Integer id;
    private String itemno;                    //商品号
    private String itemname;                  //商品名称
```

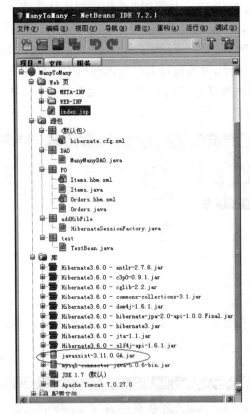

图 7-14 项目文件结构图

```
    private Set orders=new HashSet();
    public Integer getId() {
        return id;
    }
    public void setId(Integer id) {
        this.id=id;
    }
    public String getItemno() {
        return itemno;
    }
    public void setItemno(String itemno) {
        this.itemno=itemno;
    }
    public String getItemname() {
        return itemname;
    }
    public void setItemname(String itemname) {
        this.itemname=itemname;
    }
    public Set getOrders() {
```

```
        return orders;
    }
    public void setOrders(Set orders) {
        this.orders=orders;
    }
}
```

PO 对象 Items 对应的映射文件 Items.hbm.xml 代码如例 7-22 所示。

【例 7-22】 Items 对应的映射文件(Items.hbm.xml)。

```xml
<?xml version="1.0" encoding="UTF-8"?>
<!DOCTYPE hibernate-mapping PUBLIC
    "-//Hibernate/Hibernate Mapping DTD 3.0//EN"
    "http://hibernate.sourceforge.net/hibernate-mapping-3.0.dtd">
<hibernate-mapping package="PO">
    <class name="Items" table="items">
        <id column="ID" name="id" type="integer">
            <generator class="identity"/>
        </id>
        <property name="itemno" column="ITEMNO" type="string"/>
        <property name="itemname" column="ITEMNAME" type="string"/>
        <!--映射 Items 到 Orders 的多对多关联-->
        <set name="orders" table="selecteditems" cascade="save-update"
            inverse="true" lazy="true">
            <key column="ITEMID"/>
            <many-to-many class="PO.Orders" column="ORDERID"/>
        </set>
    </class>
</hibernate-mapping>
```

【例 7-23】 Orders 的代码(Orders.java)。

```java
package PO;
import java.io.Serializable;
import java.util.*;

public class Orders implements Serializable{
    private Integer id;
    private String orderno;
    private Double money;
    private Set items=new HashSet();
    public Integer getId() {
        return id;
    }
    public void setId(Integer id) {
        this.id=id;
    }
```

```java
    public String getOrderno() {
        return orderno;
    }
    public void setOrderno(String orderno) {
        this.orderno=orderno;
    }
    public Double getMoney() {
        return money;
    }
    public void setMoney(Double money) {
        this.money=money;
    }
    public Set getItems() {
        return items;
    }
    public void setItems(Set items) {
        this.items=items;
    }
}
```

PO 对象 Orders 对应的映射文件 Orders.hbm.xml 代码如例 7-24 所示。

【例 7-24】 Orders 对应的映射文件(Orders.hbm.xml)。

```xml
<?xml version="1.0" encoding="UTF-8"?>
<!DOCTYPE hibernate-mapping PUBLIC
    "-//Hibernate/Hibernate Mapping DTD 3.0//EN"
    "http://hibernate.sourceforge.net/hibernate-mapping-3.0.dtd">
<hibernate-mapping package="PO">
    <class name="Orders" table="orders">
        <id column="ID" name="id" type="integer">
            <generator class="identity"/>
        </id>
        <property name="orderno" column="ORDERNO" type="string"/>
        <property name="money" column="MONEY" type="double"/>
        <!--映射 Orders 到 Items 的多对多关联-->
        <set name="items" table="selecteditems" cascade="save-update" lazy=
          "true">
            <key column="ORDERID"/>
            <many-to-many class="PO.Items" column="ITEMID"/>
        </set>
    </class>
</hibernate-mapping>
```

3. Hibernate 的配置文件

配置文件参考例 7-9,只需修改数据库名为 manytomany。

4. 加载 Hibernate 配置文件的类

加载配置文件的类为 HibernateSessionFactory。代码和例 7-10 相同。

5. 封装对数据操作的类

为了实现对数据的操作封装一个类 ManyManyDAO,该类中封装了前述需要的功能。

【例 7-25】 ManyManyDAO 类(ManyManyDAO.java)。

```
package DAO;
import addHibFile.HibernateSessionFactory;
import PO.*;
import org.hibernate.*;

public class ManyManyDAO {
    public void addOrders(Orders order) {
        Session session=HibernateSessionFactory.currentSession();
        Transaction ts=null;
        try{
            ts=session.beginTransaction();
            session.save(order);
            ts.commit();
        }catch(Exception ex){
            ts.rollback();
            System.out.println("addOrders方法异常!");
            ex.printStackTrace();
        }finally{
            HibernateSessionFactory.closeSession();
        }
    }
    public Orders loadOrders(Integer id) {
        Session session=HibernateSessionFactory.currentSession();
        Transaction ts=null;
        Orders order=null;
        try{
            ts=session.beginTransaction();
            order= (Orders)session.get(Orders.class,id);
            Hibernate.initialize(order.getItems());
            ts.commit();
        }catch(Exception ex){
            ts.rollback();
            System.out.println("loadOrders方法异常!");
            ex.printStackTrace();
        }finally{
            HibernateSessionFactory.closeSession();
        }
        return order;
    }
    public void addItems(Items item) {
        Session session=HibernateSessionFactory.currentSession();
```

```
        Transaction ts=null;
        try{
            ts=session.beginTransaction();
            session.save(item);
            ts.commit();
        }catch(Exception ex){
            ts.rollback();
            System.out.println("addItems方法异常!");
            ex.printStackTrace();
        }finally{
            HibernateSessionFactory.closeSession();
        }
    }
    public Items loadItems(Integer id) {
        Session session=HibernateSessionFactory.currentSession();
        Transaction ts=null;
        Items item=null;
        try{
            ts=session.beginTransaction();
            item=(Items)session.get(Items.class,id);
            Hibernate.initialize(item.getOrders());
            ts.commit();
        }catch(Exception ex){
            ts.rollback();
            System.out.println("loadItems方法异常!");
        }finally{
            HibernateSessionFactory.closeSession();
        }
        return item;
    }
}
```

6. 关联关系测试类

为了对上述关联关系进行测试编写了一个测试类 TestBean。

【例 7-26】 TestBean 类(TestBean.java)。

```
package test;
import java.util.*;
import PO.*;
import DAO.*;

public class TestBean {
    ManyManyDAO dao=new ManyManyDAO();
    public void addItem(String itemno,String itemname){
        Items item=new Items();
        item.setItemno(itemno);
```

```
            item.setItemname(itemname);
            dao.addItems(item);
        }
        public void addOrder(String orderno,Double money,Set items){
            Orders order=new Orders();
            order.setOrderno(orderno);
            order.setMoney(money);
            order.setItems(items);
            dao.addOrders(order);
        }
        public Items loadItems(Integer id){
            return dao.loadItems(id);
        }
        public Orders loadOrders(Integer id){
            return dao.loadOrders(id);
        }
}
```

7. 测试页面

为了显示测试数据,编写一个 JSP 页面(index.jsp)。

【例 7-27】 JSP 页面(index.jsp)。

```
<%@page language="java" pageEncoding="gb2312"%>
<%@page import="test.TestBean"%>
<%@page import="PO.*"%>
<%@page import="java.util.*"%>
<html>
  <head>
    <title>Hibernate 的多对多双向关联关系映射</title>
  </head>
  <body>
    <h2>Hibernate 的多对多双向关联关系映射</h2>
    <hr>
    <jsp:useBean id="test" class="test.TestBean" />
    <%
        //新增三个商品
        test.addItem("001","A 商品");
        test.addItem("002","B 商品");
        test.addItem("003","C 商品");
        //选购其中的两个商品
        Set items=new HashSet();
        items.add(test.loadItems(new Integer(1)));
        items.add(test.loadItems(new Integer(2)));
        //为选购的商品产生一张订单
        test.addOrder("A00001",new Double(2100.5),items);
        //选购其中的两个商品
```

```jsp
Set items1=new HashSet();
items1.add(test.loadItems(new Integer(2)));
items1.add(test.loadItems(new Integer(3)));
//为选购的商品产生另一张订单
test.addOrder("A00002",new Double(3680),items1);
//获取两张订单
Orders order1=test.loadOrders(new Integer(1));
Orders order2=test.loadOrders(new Integer(2));
out.println("<br>订单""+order1.getOrderno().trim()+""中的商品清单为:");
Iterator it=order1.getItems().iterator();
Items item=null;
while (it.hasNext()){
    item= (Items)it.next();
    out.println("<br>商品编号:"+item.getItemno().trim());
    out.println("<br>商品名称:"+item.getItemname().trim());
}
out.println("<br>订单""+order2.getOrderno().trim()+""中的商品清单为:");
it=order2.getItems().iterator();
item=null;
while (it.hasNext()){
    item= (Items)it.next();
    out.println("<br>商品编号:"+item.getItemno().trim());
    out.println("<br>商品名称:"+item.getItemname().trim());
}
//获取两个商品
Items item1=test.loadItems(new Integer(1));
Items item2=test.loadItems(new Integer(2));
out.println("<br>商品""+item1.getItemname().trim()+""所在的订单为:");
it=item1.getOrders().iterator();
order1=null;
while (it.hasNext()){
    order1= (Orders)it.next();
    out.println("<br>订单编号:"+order1.getOrderno().trim());
}
out.println("<br>商品""+item2.getItemname().trim()+""所在的订单为:");
it=item2.getOrders().iterator();
order2=null;
while (it.hasNext()){
    order2= (Orders)it.next();
    out.println("<br>订单编号:"+order2.getOrderno().trim());
}
%>
</body>
</html>
```

8. 项目部署和运行

项目部署后运行 index.jsp,运行效果如图 7-15 所示。

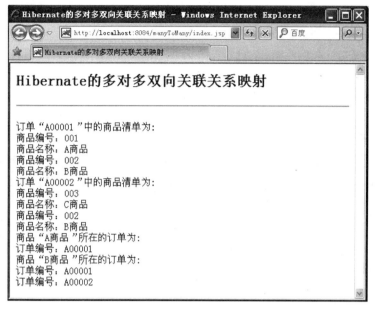

图 7-15　运行效果

7.2　Hibernate 数据查询

数据查询与检索是 Hibernate 的亮点之一。Hibernate 的数据查询方式主要有如下三种。

(1) Hibernate Query Language(HQL)。

(2) Criteria Query(CQ)。

(3) Native SQL。

下面对这三种查询方式分别进行介绍。

7.2.1　Hibernate Query Language

Hibernate Query Language(HQL)提供了十分强大的功能,使用 Hibernate 框架时一般推荐使用这种查询方式。HQL 具有与 SQL 语言类似的语法规范,只不过 SQL 是针对表中字段进行查询,而 HQL 是针对持久化对象(PO),用来取得对象,而不进行 update、delete 和 insert 等操作。而且 HQL 是完全面向对象的,具备继承、多态和关联等特性。

1. from 子句

from 子句是最简单的 HQL 语句,例如,from Stuinfo(参考 6.9 节),也可以写为 select s from Stuinfo s。结果返回 Stuinfo 类的所有实例。

除了 Java 类和属性的名称外,HQL 语句对大小写并不敏感,所以在上一句 HQL 语句中,from 与 FROM 是相同的,但是 Stuinfo 与 stuinfo 就不同了,所以上述语句写成 from stuinfo 就会报错。下列程序代码说明如何通过 from 子句取得所有的 Stuinfo 对象。

例如：

```
Query query=session.createQuery("from Stuinfo ");
List list=query.list();
for (int i=0;i<list.size(); i++){
    Stuinfo si=(Stuinfo)list.get(i);
    System.out.println(si.getName());
}
```

2. select 子句

有时并不需要得到对象的所有属性，这时可以使用 select 子句进行属性查询，例如，select s. name from Stuinfo s。下面程序代码说明如何使用这个子句：

```
Query query=session.createQuery("select s.name from Stuinfo s");
List list=query.list();
for (int i=0;i<list.size(); i++){
    String name=(String)list.get(i);
    System.out.println(name);
}
```

如果要查询两个以上的属性，查询结果会以数组的方式返回，如下面代码所示：

```
Query query=session.createQuery("select s.name, s.sex from Stuinfo as s");
List list=query.list();
for (int i=0;i<list.size(); i++) {
    Object obj[]=(Object[])list.get(i);
    System.out.println(ame(obj[0]+"的性别是："+obj[1]));
}
```

在使用属性查询时，由于使用对象数组，操作和理解都不太方便，如果将一个 object[] 中所有成员封装成一个对象就方便多了。下面的程序代码将查询结果进行了实例化：

```
Query query=session.createQuery("select new Stuinfo (s.name, s.sex) from Stuinfo s");
List list=query.list();
for (int i=0;i<list.size(); i++){
    Stuinfo si=(Stuinfo)list.get(i);
    System.out.println(si.getName());
}
```

要正确运行以上程序，还需要在 Stuinfo 类中加入一个如下的构造函数：

```
public Stuinfo(String name, String sex) {
    this.name=name;
    this.sex=sex;
}
```

3. 统计函数查询

可以在 HQL 中使用函数，经常使用的函数如下。

(1) count()：统计记录条数。

(2) min()：求最小值。

(3) max()：求最大值。

(4) sum()：求和。

(5) avg()：求平均值。

例如，要取得 Stuinfo 实例的数量，可以编写如下 HQL 语句：

```
select count(*) from Stuinfo
```

取得平均年龄的 HQL 语句如下：

```
select avg(s.age) from Stuinfo as s
```

可以使用 distinct 去除重复数据：

```
select distinct s.age from Stuinfo as s
```

4. where 子句

HQL 也支持子查询，它通过 where 子句实现这一机制。where 子句可以让用户缩小要返回的实例的列表范围，例如，下面语句会返回所有名字为 zzf 的 Stuinfo 实例：

```
Query query=session.createQuery("from Stuinfo as s where s.name='zzf'");
```

where 子句中允许出现的表达式包括了 SQL 中可以使用的大多数表达式。

(1) 数学操作：＋、－、＊、/。

(2) 关系比较操作：＝、>=、<=、<>、!=、like。

(3) 逻辑操作：and、or、not。

(4) 字符串连接：‖。

(5) SQL 标量函数：例如，upper()和 lower()。

如果子查询返回多条记录，可以用以下的关键字来量化。

① all：表示所有的记录。

② any：表示所有记录中的任意一条。

③ some：与 any 用法相同。

④ in：与 any 等价。

⑤ exists：表示子查询至少要返回一条记录。

5. order by 子句

查询返回的列表可以按照任何返回的类或者组件的属性排序：

```
from Stuinfo s order by s.name asc
```

asc 和 desc 是可选的，分别代表升序或者降序。

6. 连接查询

与 SQL 查询一样，HQL 也支持连接查询，如内连接、外连接和交叉连接。

(1) inner join：内连接。

(2) left outer join：左外连接。
(3) right outer join：右外连接。
(4) full join：全连接，不常用。

7.2.2 Criteria Query 方式

当查询数据时，人们往往需要设置查询条件。在 SQL 或 HQL 语句中，查询条件常常放在 where 子句中。此外，Hibernate 还支持 Criteria 查询（Criteria Query），这种查询方式把查询条件封装为一个 Criteria 对象。在实际应用中，可以使用 Session 的 createCriteria() 方法构建一个 Criteria 实例，然后把具体的查询条件通过 Criteria 的 add() 方法加入到 Criteria 实例中。这样，程序员可以在不使用 SQL 甚至 HQL 的情况下进行数据查询。

7.2.3 Native SQL 查询

本地 SQL 查询（Native SQL Query）指的是直接使用本地数据库（如 Oracle）的 SQL 语言进行查询。这对于把原来直接使用 SQL/JDBC 的程序迁移到基于 Hibernate 的应用很有帮助。Hibernate 使得用户可以使用手写的 SQL 来完成所有的操作（包括存储过程）。

7.3 Hibernate 的事务管理

事务（Transaction）是数据库操作的基本逻辑单位，可以用于确保数据库能够被正确修改，避免数据只修改了一部分而导致数据不完整，或者在修改时受到用户干扰。作为一名软件设计师，必须了解事务并合理利用，以确保数据库保存正确、完整的数据。数据库向用户提供保存当前程序状态的方法称为事务提交；在事务执行过程中，使数据库忽略当前的状态并回到前面保存的状态的方法称为事务回滚。

7.3.1 事务的特性

事务具备原子性（Atomicity）、一致性（Consistency）、隔离性（Isolation）和持久性（Durability）4 个属性，简称 ACID。下面对这 4 个特性分别进行说明。

(1) 原子性：将事务中所做的操作捆绑成一个原子单元，即对于事务所进行的数据修改等操作，要么全部执行，要么全部不执行。

(2) 一致性：事务在完成时，必须使所有的数据都保持一致状态，而且在相关数据中，所有规则都必须应用于事务的修改，以保持所有数据的完整性。事务结束时，所有的内部数据结构都应该是正确的。

(3) 隔离性：由并发事务所做的修改必须与任何其他事务所做的修改相隔离。事务查看数据时数据所处的状态，要么是被另一并发事务修改之前的状态，要么是被另一并发事务修改之后的状态，即事务不会查看由另一个并发事务正在修改的数据。这种隔离方式也称为可串行性。

(4) 持久性：事务完成之后，它对系统的影响是永久的，即使出现系统故障也是如此。

7.3.2 事务隔离

事务隔离意味着对于某个正在运行的事务来说,好像系统中只有这一个事务,其他并发的事务都不存在。大部分情况下,很少使用完全隔离的事务。但不完全隔离的事务会带来如下一些问题。

(1) 更新丢失(Lost Update):两个事务都企图去更新一行数据,导致事务抛出异常退出,两个事务的更新都无效。

(2) 脏数据(Dirty Data):如果第二个应用程序使用了第一个应用程序修改过的数据,而这个数据处于未提交状态,这时就会发生脏读。第一个应用程序随后可能会请求回滚被修改的数据,从而导致第二个事务使用的数据被损坏,即所谓的"变脏"。

(3) 不可重读(Unrepeatable Read):一个事务两次读同一行数据,可是这两次读到的数据不一样,就叫不可重读。如果一个事务在提交数据之前,另一个事务可以修改和删除这些数据,就会发生不可重读。

(4) 幻读(Phantom Read):一个事务执行了两次查询,发现第二次查询结果比第一次查询多出了一行,这可能是因为另一个事务在这两次查询之间插入了新行。

针对由事务的不完全隔离所引起的上述问题,人们提出了一些隔离级别,用来防范这些问题。

① 读操作未提交(Read Uncommitted):说明一个事务在提交前,其变化对于其他事务来说是可见的。这样脏读、不可重读和幻读都是允许的。当一个事务已经写入一行数据但未提交,其他事务都不能再写入此行数据;但是,任何事务都可以读任何数据。这个隔离级别使用排写锁实现。

② 读操作已提交(Read Committed):读取未提交的数据是不允许的,它使用临时的共读锁和排写锁实现。这种隔离级别不允许脏读,但不可重读和幻读是允许的。

③ 可重读(Repeatable Read):说明事务保证能够再次读取相同的数据而不会失败。此隔离级别不允许脏读和不可重读,但幻读会出现。

④ 可串行化(Serializable):提供最严格的事务隔离。这个隔离级别不允许事务并行执行,只允许串行执行。这样,脏读、不可重读或幻读都不可能发生。

在实际应用中,开发者经常不能确定使用什么样的隔离级别。太严格的级别将降低并发事务的性能,但是不足够的隔离级别又会产生一些小的Bug,不过Bug只会在系统重负荷(也就是并发情形较多时)的情况下才会出现。

一般来说,读操作未提交(Read Uncommitted)是很危险的。一个事务的回滚或失败会影响另一个并行的事务,或者说在内存中留下和数据库中不一致的数据。这些数据可能会被另一个事务读取并提交到数据库中。这是完全不允许的。

另外,大部分程序并不需要可串行化隔离(Serializable Isolation)。虽然它不允许幻读,但一般来说,幻读并不是一个大问题。可串行化隔离需要很大的系统开支,很少有人在实际开发中使用这种事务隔离模式。

实际可选的隔离级别是读操作已提交(Read Committed)和可重读(Repeatable Read)。Hibernate可以很好地支持可重读(Repeatable Read)隔离级别。

7.3.3 在 Hibernate 配置文件中设置隔离级别

JDBC 连接数据库使用的是默认隔离级别,即读操作已提交(Read Committed)和可重读(Repeatable Read)。在 Hibernate 的配置文件 hibernate.properties 中,可以修改隔离级别:

```
# hibernate.connection.isolation 4
```

在上面语句中,Hibernate 事务的隔离级别是 4,这是什么意思? 级别的数字意义如下。

1:读操作未提交(Read Uncommitted)。
2:读操作已提交(Read Committed)。
4:可重读(Repeatable Read)。
8:可串行化(Serializable)。

因此,数字 4 表示"可重读"隔离级别。如果要使以上语句有效,应把此语句行前的注释符 # 去掉:

```
hibernate.connection.isolation 4
```

也可以在配置文件 hibernate.cfg.xml 中加入以下代码:

```
<session-factory>.....                                   //把隔离级别设置为 4
    <property name=" hibernate.connection.isolation">4</property>
    ⋮
</session-factory>
```

在开始一个事务之前,Hibernate 从配置文件中获得隔离级别的值。

7.3.4 在 Hibernate 中使用 JDBC 事务

Hibernate 对 JDBC 进行了轻量级的封装,它本身在设计时并不具备事务处理功能。Hibernate 将底层的 JDBCTransaction 或 JTATransaction 进行了封装,在外面套上 Transaction 和 Session 的外壳,其实是通过委托底层的 JDBC 或 JTA 来实现事务处理功能的。

要在 Hibernate 中使用事务,可以在它的配置文件中指定使用 JDBCTransaction 或者 JTATransaction。在 hibernate.properties 中,查找 transaction.factory_class 关键字,可得到以下配置信息:

```
#hibernate.transaction.factory_class       org.hibernate.transaction.
JTATransactionFactory
#hibernate.transaction.factory_class       org.hibernate.transaction.
JDBCTransactionFactory
```

Hibernate 的事务工厂类可以设置成 JDBCTransactionFactory 或者 JTATransactionFactory。如果不进行配置,Hibernate 就会认为系统使用的是 JDBC 事务。

在 JDBC 的提交模式(Commit Mode)中,如果数据库连接是自动提交模式(Auto commit Mode),那么在每一条 SQL 语句执行后事务都将被提交,提交后如果还有任务,那么一个新的事务又开始了。

Hibernate 在 Session 控制下取得数据库连接后，就立刻取消自动提交模式，即 Hibernate 在执行 Session 的 beginTransaction()方法后，就自动调用 JDBC 层的 setAutoCommit(false)。如果想自己提供数据库连接并使用自己的 SQL 语句，为了实现事务，那么一开始就要把自动提交关掉(setAutoCommit(false))，并在事务结束时提交事务。

使用 JDBC 事务是进行事务管理最简单的实现方式，Hibernate 对于 JDBC 事务的封装也很简单。

7.3.5 在 Hibernate 中使用 JTA 事务

JTA(Java Transaction API)是事务服务的 J2EE 解决方案。本质上，它是描述事务接口的 Java EE 模型的一部分，开发人员直接使用该接口或者通过 Java EE 容器使用该接口来确保业务逻辑能够可靠地运行。

JTA 有 3 个接口，它们分别是 UserTransaction 接口、TransactionManager 接口和 Transaction 接口。这些接口共享公共的事务操作，例如，commit()和 rollback()，但也包含特殊的事务操作，例如，suspend()、resume()和 enlist()，它们只出现在特定的接口上，以便在实现中允许一定程度的访问控制。

在一个具有多个数据库的系统中，可能一个程序会调用几个数据库中的数据，需要一种分布式事务，或者准备用 JTA 来管理跨 Session 的长事务，那么就需要使用 JTA 事务。下面介绍如何在 Hibernate 的配置文件中配置 JTA 事务。JTA 设置如下面代码所示（把 JTATransactionFactory 所在配置行的注释符#取消）：

```
hibernate.transaction.factory_class org.hibernate.transaction.JTATransactionFactory
#hibernate.transaction.factory_class org.hibernate.transaction.JDBCTransactionFactory
```

或者在 hibernate.cfg.xml 文件中配置如下：

```
<session-factory>
    ⋮
    <property name=" hibernate.transaction.factory_class">
        org.hibernate.transaction.JTATransactionFactory
    </property>
    ⋮
</session-factory>
```

7.4　Hibernate 的 Cache 管理

Cache 就是缓存，它往往是提高系统性能的最重要手段，对数据起到一个蓄水池和缓冲的作用。Cache 对于大量依赖数据读取操作的系统而言尤其重要。在大并发量的情况下，如果每次程序都需要对数据库直接做查询操作，它们所带来的性能开销显而易见，频繁的网络传输、数据库磁盘的读写操作都会大大降低系统的整体性能。此时如果能让数据在本地内存中保留一个镜像,下次访问时只需从内存中直接获取,那么显然可以带来不小的性能提升。引入 Cache 机制的难点是如何保证内存中数据的有效性,否则脏数据的出现将会给系统带来难以预知的严重后果。虽然一个设计得很好的应用程序不用 Cache 也可以表现出让人接受的性能,但毫无疑问,一些对读操作要求很高的应用程序可以通过 Cache 取得更高的性能。对于应用程序,Cache 通过内存或磁盘保存了数据库中的当前有关数据状态,它是一个存储在本地的数据备份。Cache 位于数据库和应用程序之间,从数据库更新数据,并给程序提供数据。

Hibernate 实现了良好的 Cache 机制,可以借助 Hibernate 内部的 Cache 迅速提高系统的数据读取性能。Hibernate 中的 Cache 可分为两层:一级 Cache 和二级 Cache。

7.4.1 一级 Cache

Session 实现了第一级 Cache,它属于事务级数据缓冲。一旦事务结束,这个 Cache 也随之失效。一个 Session 的生命周期对应一个数据库事务或一个程序事务。

Session-cache 保证在一个 Session 中两次请求同一个对象时,取得的对象是同一个 Java 实例,有时它可以避免不必要的数据冲突。另外,它还能为另一些重要的性能提供保证。

(1) 在对一个对象进行循环引用时,不至于产生堆栈溢出。

(2) 当数据库事务结束时,对于同一数据表行,不会产生数据冲突,因为对于数据库中的一行,最多只有一个对象来表示它。

(3) 一个事务中可能会有很多个处理单元,在每一个处理单元中做的操作都会立即被另外的处理单元得知。

不用刻意去打开 Session-cache,它总是被打开并且不能被关闭。当使用 save()、update() 或 saveOrUpdate() 来保存数据更改,或通过 load()、find()、list() 等方法来得到对象时,对象就会被加入到 Session-cache。

如果要同步很大数量的对象,就需要有效地管理 Cache。可以用 Session 的 evict() 方法从一级 Cache 中移除对象。

7.4.2 二级 Cache

二级 Cache 是 SessionFactory 范围内的缓存,所有的 Session 共享同一个二级 Cache。在二级 Cache 中保存持久性实例的散装形式的数据。二级 Cache 的内部如何实现并不重要,重要的是采用哪种正确的缓存策略,以及采用哪种 Cache 提供器。持久化不同的数据需要不同的 Cache 策略,比如一些因素将影响 Cache 策略选择:数据的读/写比例、数据表是否能被其他的应用程序所访问等。对于一些读/写比例高的数据可以打开它的缓存,允许这些数据进入二级缓存容器有利于系统性能的优化;而对于能被其他应用程序访问的数据对象,最好将此对象的二级 Cache 选项关闭。

设置 Hibernate 的二级 Cache 需要分两步进行:首先确认使用什么数据并发策略,然后配置缓存过期时间并设置 Cache 提供器。

有 4 种内置的 Hibernate 数据并发冲突策略,代表了数据库隔离级别,如下所示。

(1) 事务(Transactional):仅在受管理的环境中可用。它保证可重读的事务隔离级别,可以对读/写比例高、很少更新的数据采用该策略。

（2）读写(Read-write)：使用时间戳机制维护已提交事务隔离级别。可以对读/写比例高、很少更新的数据采用该策略。

（3）非严格读写(Nonstrict-read-write)：不保证 Cache 和数据库之间的数据一致性。使用此策略时，应该设置足够短的缓存过期时间，否则可能从缓存中读出脏数据。当一些数据极少改变，并且当这些数据和数据库有一部分不一致但影响不大时，可以使用此策略。

（4）只读(Read-only)：当确保数据永不改变时，可以使用此策略。

确定了 Cache 策略之后，就要挑选一个高效的 Cache 提供器，它将作为插件被 Hibernate 调用。Hibernate 允许使用下述几种缓存插件。

① EhCache：可以在 JVM 中作为一个简单进程范围内的缓存，它可以把缓存的数据放入内存或磁盘，并支持 Hibernate 中可选用的查询缓存。

② OpenSymphony OSCache：和 EhCache 相似，并且提供了丰富的缓存过期策略。

③ SwarmCache：可作为集群范围的缓存，但不支持查询缓存。

④ JBossCache：可作为集群范围的缓存，但不支持查询缓存。

默认情况下，Hibernate 使用 EhCache 进行 JVM 级别的缓存。用户可以通过设置 Hibernate 配置文件中的 hibernate.cache.provider_class 的属性，指定其他的缓存策略，该缓存策略必须实现 org.hibernate.cache.CacheProvider 接口。

7.5 本章小结

本章介绍了 Hibernate 中的关联对象操作、数据查询、事务管理及 Cache 管理。通过本章的学习应了解和掌握以下内容。

（1）使用关联关系操作对象。
（2）Hibernate 框架中的数据查询方式。
（3）Hibernate 中的事务管理。
（4）Hibernate 中的 Cache 管理。

7.6 习　　题

7.6.1 选择题

1. 一对一关联关系在 Hibernate 中的实现有两种方式，它们是(　　)。
　　A. 单向和双向关联　　　　　　　　B. 主键和外键关联
　　C. 多向关联　　　　　　　　　　　D. 多对多

2. 一对多关联关系在 Hibernate 中的实现有(　　)两种方式。
　　A. 单向和双向关联　　　　　　　　B. 主键和外键关联
　　C. 多向关联　　　　　　　　　　　D. 多对多

3. Hibernate 框架中最常用的数据查询方式是(　　)。
　　A. CQ　　　　　B. NSQL　　　　　C. HQL　　　　　D. SQL

7.6.2 填空题

1. 数据对象之间的关联关系有_____、_____和_____。
2. Hibernate 的数据查询方式有_____、_____和_____。
3. Hibernate 中 Cache 管理分为_____和_____。

7.6.3 简答题

1. 简述一对一关联关系两种方式的区别。
2. 简述事务的特性。

7.6.4 实训题

1. 把 7.1.4 节中的一对多的关联关系改为单向关联并增加事务和缓存管理。
2. 利用关联关系开发一个学生管理系统,其中学生(Student)和学生证(Card)是一对一关系,学生和班级(Group)是一对多关系,学生和课程(Course)是多对多关系。

第8章 基于 Struts2＋Hibernate 的教务管理系统项目实训

本章围绕一个基于 Struts2＋Hibernate 的教务管理系统项目的设计与实现,介绍 Struts2 和 Hibernate 框架技术在实际项目开发中的综合应用。通过本项目的整合训练,培养熟练运用 Struts2 和 Hibernate 框架知识开发 Java Web 项目的实践能力。本项目既可以作为学生在课程学习结束后的实训项目,也可以作为学习过程中的大作业。

8.1 项目需求说明

在日常教学活动中,为了能够方便快捷地服务广大师生,许多高校都部署使用了教务管理系统。由于每个学校的管理理念和管理方式不同,所以各个高校的教务管理系统各不相同。本项目只是简单模拟教务管理系统的基本功能,通过熟悉的教务管理系统开发来综合训练 Struts2 和 Hibernate 框架技术的整合应用,并进一步提高项目实践能力。

项目实现的功能包括学生管理部分、教师管理部分和管理员管理部分。管理员管理部分实现对学生、教师以及课程的管理。

学生管理部分的功能主要包括学生学籍管理、必修课成绩查询、修改个人信息和密码、选课功能(选修课选课)、查询选修成绩,并提供 QQ 留言和校园论坛等功能。

教师管理部分的功能主要包括教师基本信息管理、修改个人信息和密码、查询必修课课程信息、成绩录入、查询选修课程以及 QQ 留言和校园论坛等。

管理员管理部分的功能主要包括学生管理、教师管理、课程管理和修改密码等。

另外,为了实现相对美观的系统页面,本项目开发过程中还使用了 CSS 和 JavaScript 技术。如果不熟悉 CSS 和 JavaScript 技术,也可以不使用这些技术。

8.2 项目系统分析

根据前述需求分析,教务管理系统功能描述如下。

1. 学生管理功能模块

学生功能模块实现的功能主要包括如下。

1) 学生学籍管理

学生学籍基本信息包括学号、姓名、籍贯、电话、电子邮件、学分、院系、性别以及照片等。

2) 必修课成绩查询

查询必修课中的考试课程及考试分数。

3) 修改个人信息

修改学生学籍信息。

4）修改秘密

修改个人密码。

5）进入选课

选择计划修习的选修课程。

6）已选课程

查询已选的选修课程。

7）选课成绩

查询选修课程名称及考试成绩。

8）其他服务

提供 QQ 留言和校园论坛（BBS）功能。可以通过 QQ 给管理员或者教师留言，也可以访问学校 BBS 论坛，在论坛中与同学、教师和管理员进行交流。

2. 教师管理功能模块

教师功能模块实现的功能主要包括如下。

1）教师基本信息管理

教师基本信息包括教师名、性别、年龄、职称、所在学院等。

2）修改个人信息

修改教师个人基本信息和密码。

3）查询必修课程信息

查询必修课程、上课时间、上课地点以及开课专业、选修学生信息等。

4）成绩录入

录入必修课程考试成绩。

5）查询选修课信息

查询选修课程信息以及录入成绩。

6）其他服务

QQ 留言，回复留言，访问学校 BBS 论坛，在论坛中与学生、教师和管理员进行交流。

3. 管理员管理功能模块

管理员功能模块实现的功能主要包括如下。

1）学生管理

主要包括查看所有学生信息、添加学生信息、导入学生信息（以 Excel 表格形式将批量学生信息导入数据库中）。

2）教师管理

主要包括查看所有教师信息、添加教师信息、导入教师信息（以 Excel 表格形式将批量教师信息导入数据库中）。

3）课程管理

主要包括查看所有课程信息、添加课程信息、导入课程信息（以 Excel 表格形式将批量课程信息导入数据库中）。

4）修改密码

修改管理员密码。

系统功能模块结构如图 8-1 所示。

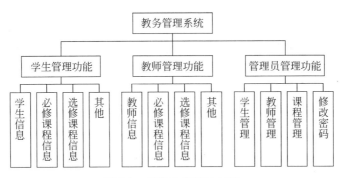

图 8-1 系统功能模块结构

8.3 项目数据库设计

如果已经掌握 DBMS 相关知识,读者可按照数据库优化的思想自行选择相应 DBMS 并设计项目的数据库及表结构。本章提供的数据库设计仅供参考,读者可根据需要进一步优化。

本项目使用的数据库是 MySQL。项目中用到的数据库(lqmsql)和表(admin、classes、score、student、student_classes、teacher)如图 8-2 所示。

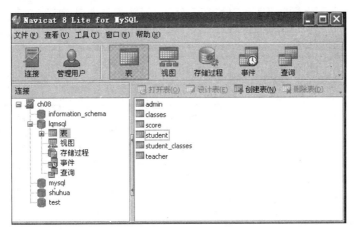

图 8-2 数据库和表

管理员表(admin)用于管理管理员账号和密码,如表 8-1 所示。

表 8-1 管理员表(admin)

字段名称	字段含义	数据类型	是否主键	是否外键	是否为空
id	标识	int(20)	是	否	否
username	用户名	varchar(50)	否	否	否
password	密码	varchar(50)	否	否	否

课程表(classes)用于管理课程的信息,如表 8-2 所示。

表 8-2 课程表(classes)

字段名称	字段含义	数据类型	是否主键	是否外键	是否为空
cs_id	课程号	int(30)	是	是	否
tea_id	教师号	int(50)	否	否	是
chooseMax	最大选课人数	int(11)	否	否	是
chooseCurNum	已选人数	int(11)	否	否	是
room_id	教师号	varchar(50)	否	否	是
cour_time	上课时间	varchar(50)	否	否	是
cmark	学分	varchar(50)	否	否	是
cname	课程名	varchar(60)	否	否	是

学生表(student)保存学生的有关信息,如表 8-3 所示。

表 8-3 学生表(student)

字段名称	字段含义	数据类型	是否主键	是否外键	是否为空
st_id	学生标识	int(50)	是	是	否
sno	学号	varchar(50)	否	否	否
username	用户名	varchar(50)	否	否	是
sex	性别	varchar(10)	否	否	是
password	密码	varchar(20)	否	否	是
department	院系	varchar(30)	否	否	是
jiguan	籍贯	varchar(60)	否	否	是
mark	学分	varchar(50)	否	否	是
email	电子邮件	varchar(50)	否	否	是
image	照片	varchar(100)	否	否	是
tel	电话	varchar(50)	否	否	是
maxClasses	最大选课数	int(11)	否	否	是

学生选课表(student_classes)保存学生成绩、学生标识和课程号,是把学生成绩与学生表和课程表关联起来的一个中间表,如表 8-4 所示。

表 8-4 学生选课表(student_classes)

字段名称	字段含义	数据类型	是否主键	是否外键	是否为空
cscore	成绩	int(11)	否	否	是
st_id	学生标识	int(50)	是	是	否
cs_id	课程号	int(50)	是	是	否

教师表(teacher)保存教师的相关信息,如表 8-5 所示。

表 8-5 教师表(teacher)

字段名称	字段含义	数据类型	是否主键	是否外键	是否为空
tid	教师号	int(50)	是	是	否
tname	教师名	varchar(50)	否	否	否
age	年龄	int(50)	否	否	是
email	电子邮件	varchar(50)	否	否	是
tel	电话	varchar(50)	否	否	是
tpassword	教师密码	varchar(50)	否	否	是
tea_id	教师类型	varchar(50)	否	否	是

8.4 项 目 实 现

8.4.1 项目文件结构

该项目命名为 jwzzuli,项目"Web 页"文件夹中有一个登录页面(index.jsp),如图 8-3 所示。项目的页面和包文件结构如图 8-4 所示。项目库文件结构如图 8-5 所示,其 lib 文件夹中包含了项目所需的 Struts2、Hibernate 3.6 以及用于导入 Excel 文件到数据库时用到的第三方 JAR 文件。

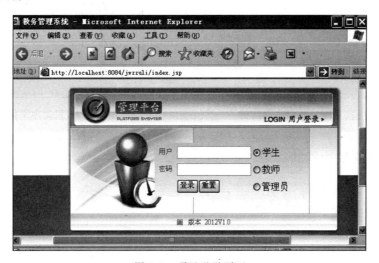

图 8-3 项目登录页面

登录页面提供了 3 种不同权限的角色,根据登录用户的不同权限跳转到不同的管理页面。学生管理功能模块的相关页面在图 8-4 中的 student 文件夹中,该文件夹中页面对应的业务控制器类在 Action 包中,该包中还保存有选课功能相关页面(ChooseCourse 文件夹中的页面)对应的业务控制器;该包中的业务控制器在 struts.xml 中配置。教师管理功能模块相关页面在图 8-4 中的 teacher 文件夹中,该文件夹中页面对应的业务控制器类在

Taction 包中,该包中业务控制器的配置在 Teacher.xml 中。管理员功能模块相关页面在图 8-4 中的 admin 文件夹中,该文件夹中页面对应的业务控制器类在 AdminAction 包中,该包中业务控制器的配置在 Admin.xml 中。文件夹 image 和 images 保存项目所需的图片;文件夹 upload 保存上传的照片;upload.jsp 和 uploadSuccess.jsp 是用于上传照片的页面;Tdao 包中的类用于提供教师业务控制器中的查询功能;config 包中的类在加载 hibernate 配置文件时使用;dao 包中的类用于学生、教师和管理员操作数据库;entity 包中的类是常用的 PO 对象及 PO 对象对应的映射文件;interceptor 包中的类是用于数据验证的拦截器;services 包中的类是选课功能常用方法的封装,在 Action 包中使用。本项目的完整源代码可以在清华大学出版社网站下载。

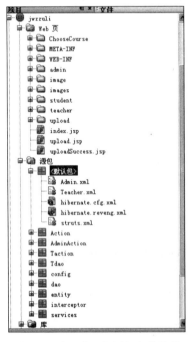

图 8-4　项目页面和源包文件结构

图 8-5　项目类库文件结构

8.4.2　用户登录功能的实现

本系统登录页面如图 8-3 所示。

登录页面(index.jsp)代码如下:

```
<%@page contentType="text/html" pageEncoding="UTF-8"%>
<%@taglib  prefix="s" uri="/struts-tags" %>
<!DOCTYPE HTML PUBLIC "-//W3C//DTD HTML 4.01 Transitional//EN"
    "http://www.w3.org/TR/html4/loose.dtd">
<html>
    <head>
        <meta http-equiv="Content-Type" content="text/html; charset=UTF-8">
        <title>教务管理系统</title>
        <style type="text/css">
```

```html
        <!--
        body {margin-left: 0px;margin-top: 0px;margin-right: 0px;
            margin-bottom: 0px;overflow:hidden;}
        .STYLE3 {color: #528311; font-size: 12px; }
        .STYLE4 {color: #42870a;font-size: 12px;}
        .STYLE5 {font-size: 24px;color: #66CC00;}
        -->
    </style>
</head>
<body>
    <s:form  action="login" method="post">
        <table width="100%" height="100%" border="0" cellpadding="0"
            cellspacing="0">
            <tr>
                <td bgcolor="#e5f6cf"> </td>
            </tr>
            <tr>
                <td height="608" background="images/login_03.gif">
                    <table width="862" border="0" align="center"
                        cellpadding="0" cellspacing="0">
                        <tr>
                            <td height="266" background=
                                "images/login_04.gif">
                                <div align="center" class="STYLE5">教务管
                                理系统</div>
                            </td>
                        </tr>
                        <tr>
                            <td height="95">
                                <table width="100%" border="0"
                                    cellspacing="0" cellpadding="0">
                                    <tr>
                                        <td width="342" height="95"
                                background="images/login_06.gif">
                                             </td>
                                        <td width="265"
                                background="images/login_07.gif">
                                            <table width="100%"
                                border="0" cellspacing="0" cellpadding="0">
                                                <tr>
                                                    <td width="21%"
                                                    height="30"><div align=
                                                    "center">
                                                    <span class="STYLE3">用户
                                                    </span></div></td>
```

```html
            <td width="39% " height="30"><input type="text" name="sno" style="height:18px; width:130px; border: solid 1px #cadcb2; font-size:12px; color:#81b432;"></td>
            <td width="40%"><label><input type="radio" name="radiobutton" value="1" checked="checked">学生</label></td>
          </tr>
          <tr>
            <td height="30"><div align="center"><span class="STYLE3">密码</span></div></td>
            <td height="30"><input type="password" name="password" style="height:18px; width:130px; border: solid 1px #cadcb2; font-size:12px; color:#81b432;"></td>
            <td height="30"><label><input type="radio" name="radiobutton" value="2">教师</label></td>
          </tr>
          <tr>
            <td height="30"> </td>
            <td height="30"><input type="submit" value="登录"><label><input type="reset" name="Submit" value="重置"></label></td>
            <td height="30"><input type="radio" name="radiobutton" value="3">管理员</td>
```

```html
                            </tr>
                        </table>
                    </td>
                    <td width="255"
background="images/login_08.gif"> </td>
                </tr>
            </table>
        </td>
    </tr>
    <tr>
        <td height="247" valign="top"
            background="images/login_09.gif">
            <table width="100%" border="0"
            cellspacing="0" cellpadding="0">
                <tr>
                    <td width="22%"
                        height="30"> </td>
                    <td width="56%">
                         </td>
                    <td width="22%">
                         </td>
                </tr>
                <tr>
                    <td> </td>
                    <td height="30">
                        <table width="100%"
border="0" cellspacing="0" cellpadding="0">
                            <tr>
                                <td
width="44%" height="20"> </td>
                                <td
width="56%" class="STYLE4">版本 2012V1.0 </td>
                            </tr>
                        </table>
                    </td>
                    <td> </td>
                </tr>
            </table>
        </td>
    </tr>
</table>
</s:form>
```

```
        </body>
</html>
```

登录页面(index.jsp)对应的业务控制器类 Login 在 Action 包中。Login.java 的代码如下：

```
package Action;
import com.opensymphony.xwork2.ActionSupport;
import dao.AdminDao;
import dao.TeacherDao;
import entity.Student;
import javax.servlet.http.HttpServletRequest;
import javax.servlet.http.HttpSession;
import dao.Usermanager;
import entity.Admin;
import entity.Teacher;
import org.apache.struts2.ServletActionContext;

public class Login   extends ActionSupport{
    private String username;
    private String sno;
    private Integer id;
    private String password;
    private Integer radiobutton;
    private HttpServletRequest request;
    //该类是 dao 包中的类,主要封装学生管理功能的相关操作
    Usermanager sm=new Usermanager();
    public String getUsername() {
        return username;
    }
    public void setUsername(String username) {
        this.username=username;
    }
    public String getSno() {
        return sno;
    }
    public void setSno(String sno) {
        this.sno=sno;
    }
    public String getPassword() {
        return password;
    }
     public Integer getId() {
        return id;
    }
    public void setId(Integer id) {
```

```java
        this.id=id;
    }
    public void setPassword(String password) {
        this.password=password;
    }
    public Integer getRadiobutton() {
        return radiobutton;
    }
    public void setRadiobutton(Integer radiobutton) {
        this.radiobutton=radiobutton;
    }
    //学生登录
    public String execute() {
        if(radiobutton==1){
            //该类是 entity 包中的类,PO 对象
            Student s=new Student();
            s.setSno(getSno());
            s.setPassword(getPassword());
            if(sm.stuLogin(s)) {
                request=ServletActionContext.getRequest();
                HttpSession session=request.getSession();
                Student sn=sm.getstudent1(sno);
                session.setAttribute("username", sn.getUsername());
                session.setAttribute("sno", this.sno);
                session.setAttribute("id", sn.getStId());
                return SUCCESS;
            }
            return INPUT;
        //教师登录
        }else if(radiobutton==2){
            //该类是 dao 包中的类,主要封装教师管理功能的相关操作
            TeacherDao td=new TeacherDao();
            //该类是 entity 包中的类,PO 对象
            Teacher tc=new Teacher();
            tc.setTeaId(sno);
            tc.setTpassword(password);
            if(td.tcLogin(tc))
            {
                request=ServletActionContext.getRequest();
                HttpSession session=request.getSession();
                Teacher tec=td.getTeacher1(sno);
                session.setAttribute("tname", tec.getTname());
                session.setAttribute("tid", tec.getTid());
                return "tsuccess";
            }
```

```java
            return INPUT;
        }
        //管理员登录
        else if(radiobutton==3){
            //该类是dao包中的类,主要封装管理员管理功能的相关操作
            AdminDao ad=new AdminDao();
            //该类是entity包中的类,PO对象
            Admin am=new Admin();
            am.setUsername(sno);
            am.setPassword(password);
            if(ad.adminLogin(am)){
                request=ServletActionContext.getRequest();
                HttpSession session=request.getSession();
                Admin adm=ad.getAdmin(sno);
                session.setAttribute("username", adm.getUsername());
                session.setAttribute("id", adm.getId());
                return "asuccess";
            }
            return INPUT;
        }
        else{
            return INPUT;
        }
    }
}
```

业务控制器类Login的配置文件(struts.xml)代码如下：

```xml
<!DOCTYPE struts PUBLIC
"-//Apache Software Foundation//DTD Struts Configuration 2.0//EN"
"http://struts.apache.org/dtds/struts-2.0.dtd">
<struts>
    <include file="Admin.xml"/>
    <include file="Teacher.xml"/>
    <!--Configuration for the default package.-->
    <package name="default" extends="struts-default">
        <!--登录-->
        <action name="login" class="Action.Login">
            <result name="success">/student/index.jsp</result>
            <result name="tsuccess" type="redirect">/teacher/index.jsp</result>
            <result name="asuccess" type="redirect">/admin/index.jsp</result>
            <result name="input" type="redirect">/index.jsp</result>
        </action>
        <action name="editStudent" class="Action.GetStudent"
            method="editStudent">
            <result>/student/editStudent.jsp</result>
```

```xml
</action>
<!--增加学生-->
<action name="saveStudent" class="Action.SaveStudent">
    <result>/student/addStudent.jsp</result>
    <result name="input">/student/addStudent.jsp</result>
</action>
<action name="updateStudent" class="Action.UpdateStudent">
    <result type="redirect">/getStudent.Action</result>
</action>
<!--查看个人信息-->
<action name="showStudent" class="Action.GetStudent"
    method="showstudent1 " >
    <result name="success">/student/getStudent.jsp</result>
    <result name="input">/student/error.jsp</result>
</action>
<action name="skanStudent" class="Action.GetStudent"
    method="showstudent1 " >
    <result name="success">/student/skanstudent.jsp</result>
    <result name="input">/student/error.jsp</result>
</action>
<!--修改密码-->
<action name="chang" class="Action.GetStudent" method="showstudent1 " >
    <result name="success">/student/changpassword.jsp</result>
    <result name="input">/student/student.jsp</result>
</action>
<action name="updateStudent" class="Action.UpdateStudent">
    <result name="input" type="redirect">/chang</result>
    <result name="success" >/student/changpasssuccess.jsp</result>
</action>
<!--个人信息修改-->
<action name="upstudent" class="Action.UpStudent">
    <result name="success">/student/success.jsp</result>
    <result name="input">/student/student.jsp</result>
</action>
<!--上传照片-->
<action name="myUpload" class="Action.MyUpload">
    <param name="path">/upload</param>
    <result name="success" >/uploadSuccess.jsp</result>
    <result name="input">/student/student.jsp</result>
</action>
<!--选课-->
<action name="choose" class="Action.ChooseCourseAction">
    <result name="success" >/student/Mainchoose.jsp</result>
    <result name="myclass">/student/myClasses.jsp</result>
    <result name="cancelSuccess" type="redirectAction" >
```

```xml
                choose!myClasses.action</result>
            <result name="chooseSuccess" type="redirectAction" >choose</result>
            <result name="chooseOver">/student/chooseOver.jsp</result>
            <result name="viewDetail">/student/CourseDetail.jsp</result>
            <result name="error">/student/error.jsp</result>
            <result name="exist">/student/exist.jsp</result>
        </action>
        <!--选课成绩-->
        <action name="cscore" class="Action.Score">
            <result name="SUCCESS">/student/stscore.jsp</result>
        </action>
        <!--退出登录-->
        <action name="clear"  class="Action.ClearUser">
            <result name="success" type="redirect">index.jsp</result>
            <result name="input">/student/student.jsp</result>
        </action>
    </package>
</struts>
```

Usermanager 类的代码如下：

```java
package dao;
import config.HibernateSessionFactory;
import entity.Classes;
import java.util.*;
import org.hibernate.Query;
import org.hibernate.Session;
import org.hibernate.Transaction;
import entity.Student;

public class Usermanager {
    private Session session;
    private Transaction transaction;
    private Query query;
    public void saveStudent(Student st){
        session=HibernateSessionFactory.getSession();
        try{
            transaction=session.beginTransaction();
            session.save(st);
            transaction.commit();
        }
        catch (Exception e){
            e.printStackTrace();
         }
        HibernateSessionFactory.closeSession();
```

```java
    }
    public void updateStudent(Student st){
        session=HibernateSessionFactory.getSession();
        try{
            transaction=session.beginTransaction();
            session.update(st);
            transaction.commit();
        }
        catch (Exception e){
            e.printStackTrace();
        }
        HibernateSessionFactory.closeSession();
    }
    public void deleteStudent(String  sno){
        session=HibernateSessionFactory.getSession();
        try{
            transaction=session.beginTransaction();
            session.delete(sno);
            transaction.commit();
        }
        catch(Exception e){
            e.printStackTrace();
        }
        HibernateSessionFactory.closeSession();
    }
    public Student getStudent(int stId){
        session=HibernateSessionFactory.getSession();
        Student student= (Student)session.get(Student.class, stId);
        HibernateSessionFactory.closeSession();
        return student;
    }
    public List<Student>allStudent(int pageNumber){
        List<Student>allStudent=new ArrayList<Student>();
        String hql="from Student as st";
        session=HibernateSessionFactory.getSession();
        query=session.createQuery(hql);
        query.setFirstResult((pageNumber-1) * 10);
        query.setMaxResults(10);
        allStudent=query.list();
        HibernateSessionFactory.closeSession();
        return allStudent;
    }
    public List<Classes>allClasses(int pageNumber){
        List<Classes>allClasses=new ArrayList<Classes>();
        String hql="from Classes as cs";
```

```java
        session=HibernateSessionFactory.getSession();
        query=session.createQuery(hql);
        query.setFirstResult((pageNumber-1)*10);
        query.setMaxResults(10);
        allClasses=query.list();
        HibernateSessionFactory.closeSession();
        return allClasses;
    }
    public Student getstudent1(String m){
        session=HibernateSessionFactory.getSession();
        Query query=(Query) session.createQuery("from Student as s where sno=
                                                '"+m+"'");
        Student st=(Student) query.uniqueResult();
        HibernateSessionFactory.closeSession();
            return st;
    }
    public boolean stuLogin(Student stu){
        if(stu.getSno()!=null&&stu.getPassword()!=null){
            session=HibernateSessionFactory.getSession();
            Query query=(Query) session.createQuery("from Student as s where
            sno='"+stu.getSno()+"' and password='"+stu.getPassword()+"'
                ");
            Student s=(Student) query.uniqueResult();
            if(s!=null){
                HibernateSessionFactory.closeSession();
                return true;
            }
        }
        return false;
    }
    public Student getstudent2(int stId){
    session=HibernateSessionFactory.getSession();
    Query query=(Query) session.createQuery ("from Student as s where stId='"+
                                            stId+"'");
    Student st=(Student) query.uniqueResult();
    HibernateSessionFactory.closeSession();
        return st;
    }
    public int getStudentAmount(){
        int studentAmount=0;
        session=HibernateSessionFactory.getSession();
        String hql="select count(*) from Student as st";
        query=session.createQuery(hql);
        long count=(Long) query.uniqueResult();
        studentAmount=(int)count;
```

```
            HibernateSessionFactory.closeSession();
            return studentAmount;
        }
}
```

TeacherDao 类的代码如下：

```
package dao;
import config.HibernateSessionFactory;
import java.util.*;
import org.hibernate.Query;
import org.hibernate.Session;
import org.hibernate.Transaction;

import entity.Teacher;
public class TeacherDao {
    private Session session;
    private Transaction transaction;
    private Query query;
    public void saveTeacher(Teacher tc){
        session =HibernateSessionFactory.getSession();
        try{
            transaction =session.beginTransaction();
            session.save(tc);
            transaction.commit();
        }
        catch (Exception e){
            e.printStackTrace();
        }
        HibernateSessionFactory.closeSession();
    }
    public void updateTeacher(Teacher tc)
    {
        session=HibernateSessionFactory.getSession();
        try{
            transaction=session.beginTransaction();
            session.update(tc);
            transaction.commit();
        }
        catch (Exception e){
            e.printStackTrace();
        }
        HibernateSessionFactory.closeSession();
    }
    public void deleteTeacher(int tid)
    {
```

```java
        session=HibernateSessionFactory.getSession();
        try{
            transaction=session.beginTransaction();
            session.delete(session.get(Teacher.class,tid));
            transaction.commit();
        }
        catch(Exception e){
            e.printStackTrace();
        }
        HibernateSessionFactory.closeSession();
    }
    public Teacher getTeacher(int tid)
    {
        session=HibernateSessionFactory.getSession();
        Teacher teacher=(Teacher)session.get(Teacher.class,tid);
        HibernateSessionFactory.closeSession();
        return teacher;
    }
    public List<Teacher>allTeacher(int pageNumber){
        List<Teacher>allTeacher=new ArrayList<Teacher>();
        String hql="from Teacher as tc";
        session=HibernateSessionFactory.getSession();
        query=session.createQuery(hql);
        query.setFirstResult((pageNumber-1)*10);
        query.setMaxResults(10);
        allTeacher=query.list();
        HibernateSessionFactory.closeSession();
        return allTeacher;
    }
    public boolean tcLogin(Teacher tc)
    {
        if(tc.getTeaId()!=null&&tc.getTpassword()!=null){
            session = HibernateSessionFactory.getSession();
            try{
                Query query = (Query) session.createQuery("from Teacher as t
                        where t.teaId = '" + tc.getTeaId () +"' and t.
                        tpassword='"+tc.getTpassword()+"' ");
                Teacher t=(Teacher)query.uniqueResult();
                if (t!=null) {
                    HibernateSessionFactory.closeSession();
                    return true;
                }
            }
            catch(Exception e){
                e.printStackTrace();
```

```
                }
                HibernateSessionFactory.closeSession();
        }
        return false;
    }
    public Teacher getTeacher1(String m)
    {
        session = HibernateSessionFactory.getSession();
        Query query= (Query) session.createQuery("from Teacher as t where teaId=
                                                '"+m+"'");
        Teacher tc=(Teacher) query.uniqueResult();
        HibernateSessionFactory.closeSession();
            return tc;
    }
    public int getTeacherAmount(){
        int teacherAmount=0;
        session=HibernateSessionFactory.getSession();
        String hql="select count(*) from Teacher as tc";
        query=session.createQuery(hql);
        long count = (Long) query.uniqueResult();
        teacherAmount=(int)count;
        HibernateSessionFactory.closeSession();
        return teacherAmount;
    }
}
```

AdminDao 类的代码如下：

```
package dao;
import config.HibernateSessionFactory;
import entity.Admin;
import org.hibernate.Query;
import org.hibernate.Session;
import org.hibernate.Transaction;

public class AdminDao {
    private Session session;
    private Transaction transaction;
    private Query query;
    public boolean adminLogin(Admin ad){
        if(ad.getUsername()!=null&&ad.getPassword()!=null){
            session = HibernateSessionFactory.getSession();
            try{
                Query query= (Query) session.createQuery("from Admin as a where
                    a.username='"+ad.getUsername()+"' and a.password
                        ='"+ad.getPassword()+"' ");
```

```java
                    Admin s= (Admin) query.uniqueResult();
                    if (s!=null) {
                        HibernateSessionFactory.closeSession();
                        return true;
                    }
                }catch(Exception e){
                    e.printStackTrace();
                }
                HibernateSessionFactory.closeSession();
            }
        return false;
    }
    public Admin getAdmin(String m){
        session = HibernateSessionFactory.getSession();
        Query query= (Query) session.createQuery("from Admin as a where
                                            a.username='"+m+"'");
        Admin ad= (Admin) query.uniqueResult();
        HibernateSessionFactory.closeSession();
            return ad;
    }
    public void saveAdmin(Admin a){
        session =HibernateSessionFactory.getSession();
        try{
            transaction =session.beginTransaction();
            session.save(a);
            transaction.commit();
        }
        catch (Exception e){
            e.printStackTrace();

        }
        HibernateSessionFactory.closeSession();
    }
}
```

Classes 类(PO 对象)的代码如下:

```java
package entity;
import java.util.HashSet;
import java.util.Set;

public class Classes   implements java.io.Serializable {
    private Integer csId;
    private String cname;
    private String cmark;
    private String courTime;
```

```java
    private String roomId;
    private String teaId;
    private Integer chooseMax;
    private Integer chooseCurNum;
    private Set students=new HashSet();
    public Classes() {
    }
    public Classes(String cname,String cmark,String courTime,String roomId,String teaId,Integer chooseMax,Integer chooseCurNum) {
        this.cname=cname;
        this.cmark=cmark;
        this.courTime=courTime;
        this.roomId=roomId;
        this.teaId=teaId;
        this.chooseMax=chooseMax;
        this.chooseCurNum=chooseCurNum;
    }
    public Integer getCsId() {
        return this.csId;
    }
    public void setCsId(Integer csId) {
        this.csId=csId;
    }
    public String getCname() {
        return this.cname;
    }
    public void setCname(String cname) {
        this.cname=cname;
    }
    public String getCmark() {
        return this.cmark;
    }
    public void setCmark(String cmark) {
        this.cmark=cmark;
    }
    public String getCourTime() {
        return this.courTime;
    }
    public void setCourTime(String courTime) {
        this.courTime=courTime;
    }
    public String getRoomId() {
        return this.roomId;
    }
    public void setRoomId(String roomId) {
```

```java
            this.roomId=roomId;
        }
        public String getTeaId() {
            return this.teaId;
        }
        public void setTeaId(String teaId) {
            this.teaId=teaId;
        }
        public Integer getChooseMax() {
            return this.chooseMax;
        }

        public void setChooseMax(Integer chooseMax) {
            this.chooseMax=chooseMax;
        }
        public Integer getChooseCurNum() {
            return this.chooseCurNum;
        }
        public void setChooseCurNum(Integer chooseCurNum) {
            this.chooseCurNum=chooseCurNum;
        }
        public Set getStudents() {
            return students;
        }
        public void setStudents(Set students) {
            this.students=students;
        }
}
```

Classes 类（PO 对象）对应映射文件（Classes.hbm.xml）的代码如下：

```xml
<?xml version="1.0"?>
<!DOCTYPE hibernate-mapping PUBLIC
    "-//Hibernate/Hibernate Mapping DTD 3.0//EN"
    "http://hibernate.sourceforge.net/hibernate-mapping-3.0.dtd">
<hibernate-mapping>
    <class name="entity.Classes" table="classes" catalog="lqmsql">
        <id name="csId" type="java.lang.Integer">
            <column name="cs_id" />
            <generator class="identity" />
        </id>
        <property name="cname" type="string">
            <column name="cName" length="60" />
        </property>
        <property name="cmark" type="string">
            <column name="cmark" length="50" />
```

```xml
        </property>
        <property name="courTime" type="string">
            <column name="cour_time" length="50" />
        </property>
        <property name="roomId" type="string">
            <column name="room_id" length="50" />
        </property>
        <property name="teaId" type="string">
            <column name="tea_id" length="50" />
        </property>
        <property name="chooseMax" type="java.lang.Integer">
            <column name="chooseMax" />
        </property>
        <property name="chooseCurNum" type="java.lang.Integer">
            <column name="chooseCurNum" />
        </property>
        <set name="students" table="student_classes"  lazy="false"
              cascade="save-update">
            <key column="cs_id" />
            <many-to-many class="entity.Student" column="st_id" />
        </set>
    </class>
</hibernate-mapping>
```

Student 类的代码如下：

```java
package entity;
import java.util.HashSet;
import java.util.Set;

public class Student   implements java.io.Serializable {
    private int stId;
    private String username;
    private String sno;
    private String email;
    private String tel;
    private String mark;
    private String sex;
    private String department;
    private String jiguan;
    private String password;
    private String image;
    private Integer maxClasses;
    private Set classes=new HashSet();
    private Set Score=new HashSet();
    public Student() {
```

```java
    }
    public Student(String username, String sno) {
        this.username=username;
        this.sno=sno;
    }
    public Student(int stId,String username, String sno, String email, String tel,
String mark, String sex, String department, String jiguan, String password,
String image, Integer maxClasses) {
        this.stId=stId;
        this.username=username;
        this.sno=sno;
        this.email=email;
        this.tel=tel;
        this.mark=mark;
        this.sex=sex;
        this.department=department;
        this.jiguan=jiguan;
        this.password=password;
        this.image=image;
        this.maxClasses=maxClasses;
    }
    public int getStId() {
        return this.stId;
    }
    public void setStId(int stId) {
        this.stId=stId;
    }
    public String getUsername() {
        return this.username;
    }
    public void setUsername(String username) {
        this.username=username;
    }
    public String getSno() {
        return this.sno;
    }
    public void setSno(String sno) {
        this.sno=sno;
    }
    public String getEmail() {
        return this.email;
    }
    public void setEmail(String email) {
        this.email=email;
    }
```

```java
public String getTel() {
    return this.tel;
}
public void setTel(String tel) {
    this.tel=tel;
}
public String getMark() {
    return this.mark;
}
public void setMark(String mark) {
    this.mark=mark;
}
public String getSex() {
    return this.sex;
}
public void setSex(String sex) {
    this.sex=sex;
}
public String getDepartment() {
    return this.department;
}
public void setDepartment(String department) {
    this.department=department;
}
public String getJiguan() {
    return this.jiguan;
}
public void setJiguan(String jiguan) {
    this.jiguan=jiguan;
}
public String getPassword() {
    return this.password;
}
public void setPassword(String password) {
    this.password=password;
}
public String getImage() {
    return this.image;
}
public void setImage(String image) {
    this.image=image;
}
public Integer getMaxClasses() {
    return this.maxClasses;
}
```

```
        public void setMaxClasses(Integer maxClasses) {
            this.maxClasses=maxClasses;
        }
        public Set getClasses() {
            return classes;
        }
        public void setClasses(Set classes) {
            this.classes=classes;
        }
        public Set getScore() {
            return Score;
        }
        public void setScore(Set Score) {
            this.Score=Score;
        }
    }
```

Student 类(PO 对象)对应映射文件(Student.hbm.xml)的代码如下：

```xml
<?xml version="1.0"?>
<!DOCTYPE hibernate-mapping PUBLIC
    "-//Hibernate/Hibernate Mapping DTD 3.0//EN"
    "http://hibernate.sourceforge.net/hibernate-mapping-3.0.dtd">
<hibernate-mapping>
    <class name="entity.Student" table="student" catalog="lqmsql">
        <id name="stId" type="java.lang.Integer">
            <column name="st_id" />
            <generator class="identity" />
        </id>
        <property name="username" type="string">
            <column name="username" length="100" not-null="true" />
        </property>
        <property name="sno" type="string">
            <column name="sno" length="50" not-null="true" />
        </property>
        <property name="email" type="string">
            <column name="email" length="50" />
        </property>
        <property name="tel" type="string">
            <column name="tel" length="50" />
        </property>
        <property name="mark" type="string">
            <column name="mark" length="50" />
        </property>
        <property name="sex" type="string">
            <column name="sex" length="10" />
```

```xml
        </property>
        <property name="department" type="string">
            <column name="department" length="30" />
        </property>
        <property name="jiguan" type="string">
            <column name="jiguan" length="60" />
        </property>
        <property name="password" type="string">
            <column name="password" length="20" />
        </property>
        <property name="image" type="string">
            <column name="image" length="100" />
        </property>
        <property name="maxClasses" type="java.lang.Integer">
            <column name="maxClasses" />
        </property>
        <set name="classes" table="student_classes" cascade="save-update" >
            <key column="st_id" />
            <many-to-many class="entity.Classes" column="cs_id" />
        </set>
    </class>
</hibernate-mapping>
```

Teacher 类(PO 对象)的代码如下：

```java
package entity;

public class Teacher implements java.io.Serializable {
    private Integer tid;
    private Integer age;
    private String email;
    private String tel;
    private String teaId;
    private String tpassword;
    private String tname;
    public Teacher() {
    }
    public Teacher(String tel, String teaId, String tpassword, String tname) {
        this.tel=tel;
        this.teaId=teaId;
        this.tpassword=tpassword;
        this.tname=tname;
    }
    public Teacher (Integer age, String email, String tel, String teaId, String tpassword, String tname) {
        this.age=age;
```

```java
            this.email=email;
            this.tel=tel;
            this.teaId=teaId;
            this.tpassword=tpassword;
            this.tname=tname;
        }
        public Integer getTid() {
            return this.tid;
        }
        public void setTid(Integer tid) {
            this.tid=tid;
        }
        public Integer getAge() {
            return this.age;
        }
        public void setAge(Integer age) {
            this.age=age;
        }
        public String getEmail() {
            return this.email;
        }
        public void setEmail(String email) {
            this.email=email;
        }
        public String getTel() {
            return this.tel;
        }
        public void setTel(String tel) {
            this.tel=tel;
        }
        public String getTeaId() {
            return this.teaId;
        }
        public void setTeaId(String teaId) {
            this.teaId=teaId;
        }
        public String getTpassword() {
            return this.tpassword;
        }
        public void setTpassword(String tpassword) {
            this.tpassword=tpassword;
        }
        public String getTname() {
            return this.tname;
        }
```

```java
    public void setTname(String tname) {
        this.tname=tname;
    }
}
```

Teacher 类(PO 对象)对应映射文件(Teacher.hbm.xml)的代码如下：

```xml
<?xml version="1.0"?>
<!DOCTYPE hibernate-mapping PUBLIC
    "-//Hibernate/Hibernate Mapping DTD 3.0//EN"
    "http://hibernate.sourceforge.net/hibernate-mapping-3.0.dtd">
<hibernate-mapping>
    <class name="entity.Teacher" table="teacher" catalog="lqmsql">
        <id name="tid" type="java.lang.Integer">
            <column name="tid" />
            <generator class="identity" />
        </id>
        <property name="age" type="java.lang.Integer">
            <column name="age" />
        </property>
        <property name="email" type="string">
            <column name="email" length="50" />
        </property>
        <property name="tel" type="string">
            <column name="tel" length="50" not-null="true" />
        </property>
        <property name="teaId" type="string">
            <column name="tea_id" length="50" not-null="true" />
        </property>
        <property name="tpassword" type="string">
            <column name="tpassword" length="50" not-null="true" />
        </property>
        <property name="tname" type="string">
            <column name="tname" length="50" not-null="true" />
        </property>
    </class>
</hibernate-mapping>
```

Admin 类(PO 对象)的代码如下：

```java
package entity;

public class Admin   implements java.io.Serializable {
    private Integer id;
    private String username;
    private String password;
    public Admin() {
```

```java
    }
    public Admin(String username, String password) {
        this.username=username;
        this.password=password;
    }
    public Integer getId() {
        return this.id;
    }
    public void setId(Integer id) {
        this.id=id;
    }
    public String getUsername() {
        return this.username;
    }
    public void setUsername(String username) {
        this.username=username;
    }
    public String getPassword() {
        return this.password;
    }
    public void setPassword(String password) {
        this.password=password;
    }
}
```

Admin 类(PO 对象)对应映射文件(Admin.hbm.xml)的代码如下：

```xml
<?xml version="1.0"?>
<!DOCTYPE hibernate-mapping PUBLIC
    "-//Hibernate/Hibernate Mapping DTD 3.0//EN"
    "http://hibernate.sourceforge.net/hibernate-mapping-3.0.dtd">
<hibernate-mapping>
    <class name="entity.Admin" table="admin" catalog="lqmsql">
        <id name="id" type="java.lang.Integer">
            <column name="id" />
            <generator class="identity" />
        </id>
        <property name="username" type="string">
            <column name="username" length="50" not-null="true" />
        </property>
        <property name="password" type="string">
            <column name="password" length="50" not-null="true" />
        </property>
    </class>
</hibernate-mapping>
```

此外，该项目使用 Hibernate 框架，连接数据库所需的配置文件为 Hibernate.cfg.xml。

加载配置文件的类为 HibernateSessionFactory(该类在 config 包中)。

Hibernate 配置文件(Hibernate.cfg.xml)的代码如下：

```xml
<?xml version="1.0" encoding="UTF-8"?>
<!DOCTYPE hibernate-configuration PUBLIC "-//Hibernate/Hibernate Configuration DTD 3.0//EN" "http://hibernate.sourceforge.net/hibernate-configuration-3.0.dtd">
<hibernate-configuration>
    <session-factory>
        <property name="hibernate.dialect">
            org.hibernate.dialect.MySQLDialect
        </property>
        <property name="hibernate.connection.driver_class">
            com.mysql.jdbc.Driver
        </property>
        <property name="hibernate.connection.url">
            jdbc:mysql://localhost:3306/lqmsql
        </property>
        <property name="hibernate.connection.username">root</property>
        <property name="hibernate.connection.password">root</property>
    </session-factory>
</hibernate-configuration>
```

HibernateSessionFactory 类的代码如下：

```java
package config;
import org.hibernate.HibernateException;
import org.hibernate.Session;
import org.hibernate.cfg.Configuration;

public class HibernateSessionFactory {
    private static String CONFIG_FILE_LOCATION="/config/hibernate.cfg.xml";
    private static final ThreadLocal<Session>threadLocal=new
                                       ThreadLocal<Session>();
    private static Configuration configuration=new Configuration();
    private static org.hibernate.SessionFactory sessionFactory;
    static{
        try{
            configuration.configure(CONFIG_FILE_LOCATION);
            sessionFactory=configuration.buildSessionFactory();

        }
        catch(Exception e)
        {
            e.printStackTrace();
        }
    }
```

```java
        private HibernateSessionFactory(){
        }
        public static Session getSession() throws HibernateException{
            Session session=(Session)threadLocal.get();
            if(session==null||!session.isOpen()){
                if(sessionFactory==null){
                    rebuildSessionFactory();
                }
                session=(sessionFactory!=null)?sessionFactory.openSession():null;
                threadLocal.set(session);
            }
            return session;
        }
        public static void rebuildSessionFactory()
        {
            try{
                configuration.configure(CONFIG_FILE_LOCATION);
                sessionFactory=configuration.buildSessionFactory();
            }
            catch(Exception e)
            {
                    e.printStackTrace();
            }
        }
        public static void closeSession() throws HibernateException{
            Session session=(Session)threadLocal.get();
            threadLocal.set(null);
            if(session!=null)
            {
                session.close();
            }
        }
        public static org.hibernate.SessionFactory getSessionFactory(){
            return sessionFactory;
        }
        public static Configuration getConfiguration(){
            return configuration;
        }
    }
```

8.4.3 学生管理功能的实现

在图 8-3 所示登录页面中输入正确的学生用户名和密码并单击"登录"按钮后，页面跳转到学生管理主页面，如图 8-6 所示。该页面由 student 文件夹中的 index.jsp 实现。该学生管理主页面使用框架，把整个页面分为 3 个子窗口，对应的 3 个页面文件分别是 top.jsp、 menu.jsp 和 main.jsp。

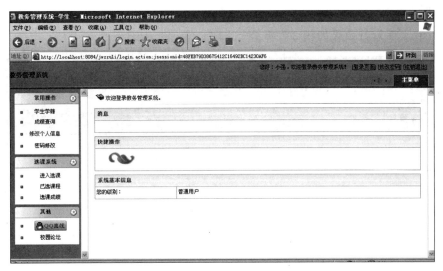

图 8-6 学生管理主页面

学生管理主页面(index.jsp)的代码如下:

```
<%@page contentType="text/html" pageEncoding="UTF-8"%>
<!DOCTYPE html PUBLIC "-//W3C//DTD HTML 4.0//EN"
    "http://www.w3.org/TR/REC-html40/strict.dtd">
<html>
    <head>
        <meta http-equiv="Content-Type" content="text/html; charset=UTF-8">
        <title>教务管理系统-学生</title>
        <style>
            body
            {
                scrollbar-base-color:#C0D586;
                scrollbar-arrow-color:#FFFFFF;
                scrollbar-shadow-color:DEEFC6;
            }
        </style>
    </head>
    <frameset rows="60,*" cols="*" frameborder="no" border="0" framespacing="0">
        <frame src="student/top.jsp" name="topFrame" scrolling="no">
        <frameset cols="180,*" name="btFrame" frameborder="NO" border="0"
            framespacing="0">
            <frame src="student/menu.jsp" noresize name="menu" scrolling="yes">
            <frame src="student/main.jsp" noresize name="main" scrolling="yes">
        </frameset>
    </frameset>
    <noframes>
        <body>您的浏览器不支持框架!</body>
```

```
        </noframes>
</html>
```

top.jsp 页面运行效果如图 8-7 所示。

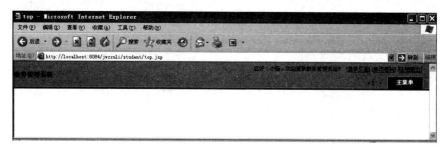

图 8-7 top.jsp 页面运行效果

top.jsp 页面的代码如下：

```
<%@page contentType="text/html" pageEncoding="UTF-8"%>
<%@taglib prefix="s" uri="/struts-tags" %>
<!DOCTYPE html PUBLIC "-//W3C//DTD HTML 4.0//EN"
    "http://www.w3.org/TR/REC-html40/strict.dtd">
<html>
    <head>
        <meta http-equiv="Content-Type" content="text/html; charset=UTF-8">
        <title>top</title>
        <link href="skin/css/base.css" rel="stylesheet" type="text/css">
        <script language='javascript'>
            var preFrameW='206,*';
            var FrameHide=0;
            var curStyle=1;
            var totalItem=9;
            function ChangeMenu(way){
                var addwidth=10;
                var fcol=top.document.all.btFrame.cols;
                if(way==1) addwidth=10;
                else if(way==-1) addwidth=-10;
                else if(way==0){
                    if(FrameHide ==0){
                        preFrameW=top.document.all.btFrame.cols;
                        top.document.all.btFrame.cols='0,*';
                        FrameHide=1;
                        return;
                    }else{
                        top.document.all.btFrame.cols=preFrameW;
                        FrameHide=0;
                        return;
                    }
```

```
                }
                fcols=fcol.split(',');
                fcols[0]=parseInt(fcols[0])+addwidth;
                top.document.all.btFrame.cols=fcols[0]+', * ';
            }
            function mv(selobj,moveout,itemnum)
            {
                if(itemnum==curStyle) return false;
                if(moveout=='m') selobj.className='itemsel';
                if(moveout=='o') selobj.className='item';
                return true;
            }
            function changeSel(itemnum)
            {
                curStyle=itemnum;
                for(i=1;i<=totalItem;i++)
                {
                    if(document.getElementById('item'+i))
                        document.getElementById('item'+i).className='item';
                }
                document.getElementById('item'+itemnum).className='itemsel';
            }
    </script>
<style>
        body { padding:0px; margin:0px; }
        #tpa {
            color: #009933;
            margin:0px;
            padding:0px;
            float:right;
            padding-right:10px;
        }
        #tpa dd {
            margin:0px;
            padding:0px;
            float:left;
            margin-right:2px;
        }
        #tpa dd.ditem {
                margin-right:8px;
        }
        #tpa dd.img {
            padding-top:6px;
        }
        div.item
```

```css
{
    text-align:center;
    background:url(skin/images/frame/topitembg.gif) 0px 3px no-repeat;
    width:82px;
    height:26px;
    line-height:28px;
}
.itemsel {
    width:80px;
    text-align:center;
    background:#226411;
    border-left:1px solid #c5f097;
    border-right:1px solid #c5f097;
    border-top:1px solid #c5f097;
    height:26px;
    line-height:28px;
}
.itemsel {
    height:26px;
    line-height:26px;
}
a:link,a:visited {
    text-decoration: underline;
}
.item a:link, .item a:visited {
    font-size: 12px;
    color: #ffffff;
    text-decoration: none;
    font-weight: bold;
}
.itemsel a:hover {
    color: #ffffff;
    font-weight: bold;
    border-bottom:2px solid #E9FC65;
}
.itemsel a:link, .itemsel a:visited {
    font-size: 12px;
    color: #ffffff;
    text-decoration: none;
    font-weight: bold;
}
.itemsel a:hover {
    color: #ffffff;
    border-bottom:2px solid #E9FC65;
}
```

```
        .rmain {
            padding-left:10px;
        }
    </style>
</head>
<body bgColor='#ffffff'>
    <table width="100%" border="0" cellpadding="0" cellspacing="0"
        background="skin/images/frame/topbg.gif">
        <tr>
            <td width='20%' height="60"><h2>教务管理系统</h2></td>
            <td width='80%' align="right" valign="bottom">
                <table width="750" border="0" cellspacing="0"
                    cellpadding="0">
                    <tr>
                        <td align="right" height="26"
                            style="padding-right:10px;line-height:26px;">
                            您好:<span class="username">
                            <s:property value="#session.username"/>
                            </span>,欢迎登录教务管理系统!
                              [<a href="<%=request.getContextPath()%>
                              /index.jsp" target="_blank">登录页面</a>]
                              [<a href="chang" target="main">修改密码</a>]
                              [<a href="clear" target="_top">注销退出</a>]
                               
                        </td>
                    </tr>
                    <tr>
                        <td align="right" height="34" class="rmain">
                            <dl id="tpa">
                                <dd class='img'>
                                <a href="javascript:ChangeMenu(-1);"><img
                                vspace="5" src="skin/images/frame/arrl.gif"
                                border="0" width="5" height="8" alt="缩小左框
                                架" title="缩小左框架" /></a></dd>
                                <dd class='img'><a href="javascript:
                                ChangeMenu(0);"><img vspace="3" src="skin/
                                images/frame/arrfc.gif" border="0" width="12"
                                height="12" alt="显示/隐藏左框架" title="显示/
                                隐藏左框架" /></a></dd>
                                <dd class='img' style="margin-right:10px;"><a
                                href="javascript:ChangeMenu(1);"><img vspace
                                ="5" src="skin/images/frame/arrr.gif" border=
                                "0" width="5" height="8" alt="增大左框架" title
                                ="增大左框架" /></a></dd>
                                <dd><div class='itemsel' id='item1'
```

```
                        onMouseMove="mv(this,'move',1);" onMouseOut="
                        mv(this,'o',1);"><a href="menu.asp" onclick=
                        "changeSel(1)" target="menu">主菜单</a></div>
                      </dd>
                    </dl>
                  </td>
                </tr>
              </table>
            </td>
          </tr>
        </table>
      </body>
</html>
```

备注：如果不熟悉 CSS 和 JavaScript 技术，也可以不使用 CSS 和 JavaScript 实现该页面。

menu.jsp 页面运行效果如图 8-8 所示。

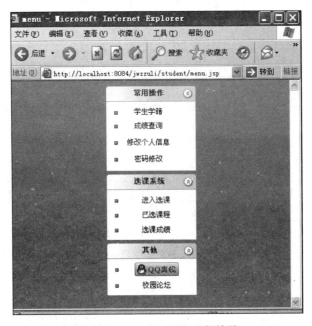

图 8-8　menu.jsp 页面运行效果

menu.jsp 页面的代码如下：

```
<%@page contentType="text/html" pageEncoding="UTF-8"%>
<!DOCTYPE html PUBLIC "-//W3C//DTD HTML 4.0//EN"
    "http://www.w3.org/TR/REC-html40/strict.dtd">
<html>
    <head>
        <meta http-equiv="Content-Type" content="text/html; charset=UTF-8">
        <title>menu</title>
```

```html
<link rel="stylesheet" href="skin/css/base.css" type="text/css" />
<link rel="stylesheet" href="skin/css/menu.css" type="text/css" />
<meta http-equiv="Content-Type" content="text/html; charset=gb2312" />
<script language='javascript'>var curopenItem='1';</script>
<script language="javascript" type="text/javascript"
    src="skin/js/frame/menu.js"></script>
<base target="main" />
</head>
<body target="main">
    <table width='99%' height="100%"border='0'cellspacing='0'cellpadding='0'>
        <tr>
            <td style='padding-left:3px;padding-top:8px' valign="top">
                <!--常用操作开始-->
                <dl class='bitem'>
                    <dt onClick='showHide("items1_1")'><b>
                        常用操作</b></dt>
                    <dd style='display:block' class='sitem' id='items1_1'>
                        <ul class='sitemu'>
                            <li>
                                <div class='items'>
                                    <div class='fllct'><a href='skanStudent'
                                        target='main'>学生学籍</a></div>
                                </div>
                            </li>
                            <li><div class='fllct'><a href='cscore'
                                target='main'>成绩查询</a></div></li>
                            <li>
                                <div class='fllct'><a href='showStudent'
                                    target='main'>修改个人信息</a></div>
                            </li>
                            <li><div class='fllct'><a href='chang'
                                target='main'>密码修改</a></div></li>
                        </ul>
                    </dd>
                </dl>
                <!--常用操作结束-->
                <!--选课系统开始-->
                <dl class='bitem'>
                    <dt onClick='showHide("items2_1")'><b>
                        选课系统</b></dt>
                    <dd style='display:block' class='sitem' id='items2_1'>
                        <ul class='sitemu'>
                            <li><a href='choose.action' target='main'>
                                进入选课</a></li>
                            <li><a href='choose!myClasses.action'
```

```
                        target='main'>已选课程</a></li>
                    <li><a href='cscore' target='main'>
                        选课成绩</a></li>
                </ul>
            </dd>
        </dl>
        <!--选课系统结束-->
        <!--其他开始-->
        <dl class='bitem'>
            <dt onClick='showHide("items2_1")'><b>其他</b></dt>
            <dd style='display:block' class='sitem' id='items2_1'>
                <ul class='sitemu'>
                    <li><a target="_blank"
                        href="http://wpa.qq.com/msgrd?v=3&uin=
                        330262363&site=qq&menu=yes"><img border="0"
                        src="http://wpa.qq.com/pa?p=2:330262363:41
                        &r=0.7815125500474443" alt="请留言" title="
                        请留言"></a></li>
                    <li><a href='' target='main'>校园论坛</a></li>
                </ul>
            </dd>
        </dl>
        <!--其他结束-->
        </td>
      </tr>
    </table>
  </body>
</html>
```

main.jsp 页面运行效果如图 8-9 所示。

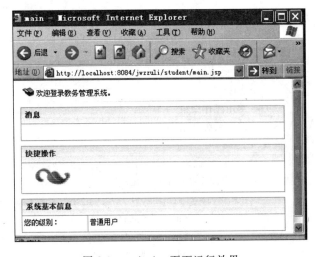

图 8-9 main.jsp 页面运行效果

main.jsp 页面的代码如下：

```jsp
<%@page contentType="text/html" pageEncoding="UTF-8"%>
<!DOCTYPE html PUBLIC "-//W3C//DTD HTML 4.0//EN"
    "http://www.w3.org/TR/REC-html40/strict.dtd">
<html>
    <head>
        <meta http-equiv="Content-Type" content="text/html; charset=UTF-8">
        <title>main</title>
        <base target="_self">
        <link rel="stylesheet" type="text/css" href="skin/css/base.css" />
        <link rel="stylesheet" type="text/css" href="skin/css/main.css" />
    </head>
    <body leftmargin="8" topmargin='8'>
        <table width="98%" border="0" align="center" cellpadding="0"
            cellspacing="0">
            <tr>
                <td><div style='float:left'><img height="14"
                        src="skin/images/frame/book1.gif" width="20" />
                         欢迎登录教务管理系统。</div>
                    <div style='float:right;padding-right:8px;'>
                        <!--//保留接口  -->
                    </div>
                </td>
            </tr>
            <tr>
                <td height="1" background="skin/images/frame/sp_bg.gif"
                    style='padding:0px'></td>
            </tr>
        </table>
        <table width="98%" align="center" border="0" cellpadding="3"
cellspacing="1" bgcolor="#CBD8AC" style="margin-bottom:8px;margin-top:8px;">
            <tr>
                <td background="skin/images/frame/wbg.gif" bgcolor="#EEF4EA"
                    class='title'><span>消息</span></td>
            </tr>
            <tr bgcolor="#FFFFFF">
                <td> </td>
            </tr>
        </table>
        <table width="98%" align="center" border="0" cellpadding="4"
          cellspacing="1" bgcolor="#CBD8AC" style="margin-bottom:8px">
            <tr>
                <td colspan="2" background="skin/images/frame/wbg.gif"
                  bgcolor="#EEF4EA" class='title'>
                    <div style='float:left'><span>快捷操作</span></div>
```

```html
                <div style='float:right;padding-right:10px;'></div>
            </td>
        </tr>
        <tr bgcolor="#FFFFFF">
            <td height="30" colspan="2" align="center" valign="bottom">
                <table width="100%" border="0" cellspacing="1"
                    cellpadding="1">
                    <tr>
                        <td width="15%" height="31" align="center">
<img src="skin/images/frame/qc.gif" width="90" height="30" /></td>
                        <td width="85%" valign="bottom">
                            <!--   //保留接口   -->
                            <div class='icoitem'>
                                <div class='ico'></div>
                                <div class='txt'>
                                    <a href=''><u></u></a></div>
                            </div>
                            <div class='icoitem'>
                                <div class='ico'></div>
                                <div class='txt'>
                                    <a href=''><u></u></a></div>
                            </div>
                            <div class='icoitem'>
                                <div class='ico'></div>
                                <div class='txt'>
                                    <a href=''><u></u></a></div>
                            </div>
                            <div class='icoitem'>
                                <div class='ico'></div>
                                <div class='txt'>
                                    <a href=''><u></u></a></div>
                            </div>
                            <div class='icoitem'>
                                <div class='ico'></div>
                                <div class='txt'>
                                    <a href=''><u></u></a></div>
                            </div>
                            <div class='icoitem'>
                                <div class='ico'></div>
                                <div class='txt'>
                                    <a href=''><u></u></a></div>
                            </div>
                        </td>
                    </tr>
                </table>
```

```html
                </td>
            </tr>
        </table>
        <table width="98%" align="center" border="0" cellpadding="4"
            cellspacing="1" bgcolor="#CBD8AC" style="margin-bottom:8px">
            <tr bgcolor="#EEF4EA">
                <td colspan="2" background="skin/images/frame/wbg.gif"
                    class='title'><span>系统基本信息</span></td>
            </tr>
            <tr bgcolor="#FFFFFF">
                <td width="25%" bgcolor="#FFFFFF">您的级别:</td>
                <td width="75%" bgcolor="#FFFFFF">普通用户</td>
            </tr>
        </table>
        <table width="98%" align="center" border="0" cellpadding="4"
            cellspacing="1" bgcolor="#CBD8AC">
        </table>
    </body>
</html>
```

备注：该页面的设计实现基于项目整体美观以及功能扩展考虑，使用了 CSS 和 JavaScript 技术。也可以使用简单的 JSP 页面实现。

单击图 8-6 所示页面中"常用操作"中的"学生学籍"，出现如图 8-10 所示的页面。可以查看学生个人基本信息，并可以修改个人基本信息、上传照片。该部分还实现了必修课程的"成绩查询"和"密码修改"功能。

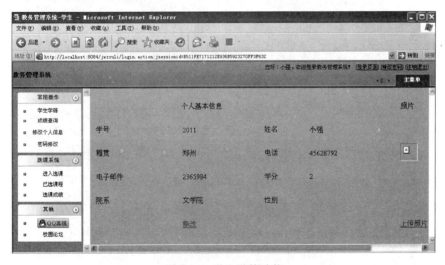

图 8-10 学生学籍功能

单击图 8-10 所示页面中"选课系统"中的"进入选课"，出现如图 8-11 所示的页面，可以选择要选修的课程。如单击"计算机组成原理"，出现如图 8-12 所示的页面。单击图 8-12 所示页面中的"选择"，因当前登录用户已选修该课程，会出现如图 8-13 所示的页面；若还没有选修该课程，将确定选修该课程。

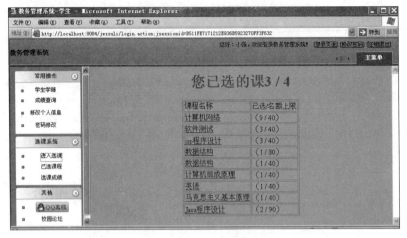

图 8-11　进入选课页面

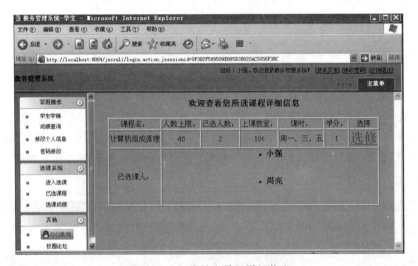

图 8-12　查看所选课程详细信息

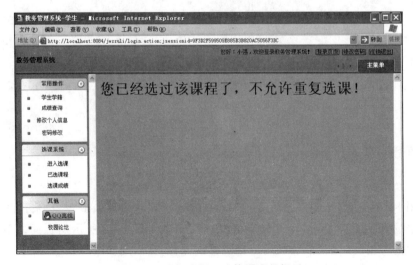

图 8-13　重复选择已选修课程的提示

可以查看已选修的选修课程信息。在图 8-13 所示页面中单击"已选课程",出现如图 8-14 所示的页面,也可以在该页面取消已选课程。要查看选修课的成绩,可以在图 8-14 所示页面中单击"选课成绩",出现如图 8-15 所示的页面。

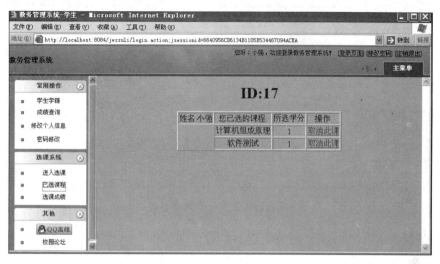

图 8-14　查询选修课信息

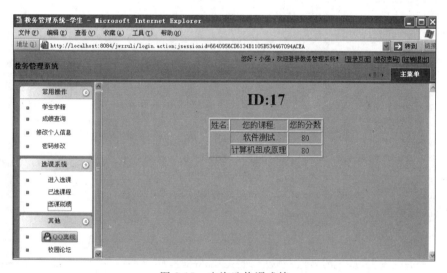

图 8-15　查询选修课成绩

图 8-15 所示页面中显示的"QQ 离线"表示学生要留言的 QQ 不在线。该功能可以实现给老师留言或者在线聊天功能;"校园论坛"链接一个 BBS 校园论坛网站。

8.4.4　管理员管理功能的实现

在图 8-3 所示页面中输入正确的管理员用户名和密码并单击"登录"按钮后,页面跳转到管理员管理主页面,如图 8-16 所示。该页面由 admin 文件夹中的 index.jsp 实现。该主页面使用框架,把页面分为 3 个子窗口,对应的 3 个页面文件分别是 top.jsp、menu.jsp 和 main.jsp。

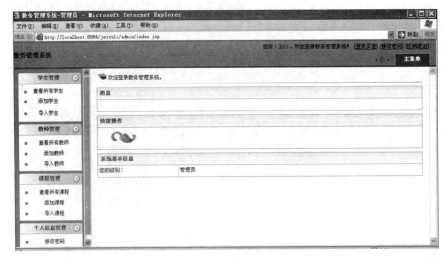

图 8-16 管理员管理主页面

管理员管理主页面(index.jsp)的代码如下：

```jsp
<%@page contentType="text/html" pageEncoding="UTF-8"%>
<!DOCTYPE html PUBLIC "-//W3C//DTD HTML 4.0//EN"
    "http://www.w3.org/TR/REC-html40/strict.dtd">
<html>
    <head>
        <meta http-equiv="Content-Type" content="text/html; charset=UTF-8">
        <title>教务管理系统-管理员</title>
        <style>
            body
            {
                scrollbar-base-color:#C0D586;
                scrollbar-arrow-color:#FFFFFF;
                scrollbar-shadow-color:DEEFC6;
            }
        </style>
    </head>
    <frameset rows="60,*" cols="*" frameborder="no" border="0" framespacing="0">
        <frame src="top.jsp" name="topFrame" scrolling="no">
        <frameset cols="180,*" name="btFrame" frameborder="NO" border="0"
            framespacing="0">
            <frame src="menu.jsp" noresize name="menu" scrolling="yes">
            <frame src="main.jsp" noresize name="main" scrolling="yes">
        </frameset>
    </frameset>
    <noframes>
        <body>您的浏览器不支持框架！</body>
    </noframes>
</html>
```

页面 top.jsp 和 main.jsp 的运行效果和代码与 student 文件夹中的 top.jsp 和 main.jsp 几乎一样,这里不再介绍,请参考前面代码。

menu.jsp 页面运行效果如图 8-17 所示。

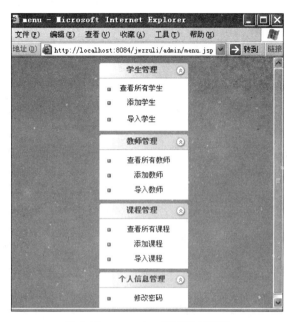

图 8-17 menu.jsp 页面运行效果

menu.jsp 页面的代码如下:

```
<%@page contentType="text/html" pageEncoding="UTF-8"%>
<!DOCTYPE html PUBLIC "-//W3C//DTD HTML 4.0//EN"
    "http://www.w3.org/TR/REC-html40/strict.dtd">
<html>
    <head>
        <meta http-equiv="Content-Type" content="text/html; charset=UTF-8">
        <title>menu</title>
        <link rel="stylesheet" href="skin/css/base.css" type="text/css" />
        <link rel="stylesheet" href="skin/css/menu.css" type="text/css" />
        <meta http-equiv="Content-Type" content="text/html; charset=gb2312" />
        <script language='javascript'>var curopenItem='1';</script>
        <script language="javascript" type="text/javascript"
            src="skin/js/frame/menu.js"></script>
        <base target="main" />
    </head>
    <body target="main">
        <table width='99%'height="100%"border='0'cellspacing='0'cellpadding='0'>
            <tr>
                <td style='padding-left:3px;padding-top:8px' valign="top">
                    <dl class='bitem'>
                        <dt onClick='showHide("items1_1")'><b>
```

```html
            学生管理</b></dt>
        <dd style='display:block' class='sitem' id='items1_1'>
            <ul class='sitemu'>
                <li>
                    <div class='items'>
                        <div class='fllct'><a href='pageAction'
                            target='main'>查看所有学生</a></div>
                    </div>
                </li>
                <li><div class='fllct'><a href='addstudent.jsp'
                    target='main'>添加学生</a></div></li>
                <li>
                    <div class='fllct'>
<a href='importStudent.jsp' target='main'>导入学生</a></div>
                </li>
            </ul>
        </dd>
    </dl>
    <dl class='bitem'>
        <dt onClick='showHide("items2_1")'><b>
            教师管理</b></dt>
        <dd style='display:block' class='sitem' id='items2_1'>
            <ul class='sitemu'>
                <li><a href='tpageAction' target='main'>
                    查看所有教师</a></li>
                <li><a href='addteacher.jsp' target='main'>
                    添加教师</a></li>
                <li><a href='importTeacher.jsp'
                    target='main'>导入教师</a></li>
            </ul>
        </dd>
    </dl>
    <dl class='bitem'>
        <dt onClick='showHide("items2_1")'><b>
            课程管理</b></dt>
        <dd style='display:block' class='sitem' id='items2_1'>
            <ul class='sitemu'>
                <li><a href='cpageAction' target='main'>
                    查看所有课程</a></li>
                <li><a href='addclasses.jsp' target='main'>
                    添加课程</a></li>
                <li><a href='' target='main'>
                    导入课程</a></li>
            </ul>
        </dd>
```

```
                </dl>
                <dl class='bitem'>
                    <dt onClick='showHide("items2_1")'><b>
                        个人信息管理</b></dt>
                    <dd style='display:block' class='sitem' id='items2_1'>
                        <ul class='sitemu'>
                            <li><a href='changPass.jsp' target='main'>
                                修改密码</a></li>
                        </ul>
                    </dd>
                </dl>
            </td>
        </tr>
    </table>
</body>
</html>
```

单击图 8-16 所示页面中"学生管理"中的"查看所有学生",出现如图 8-18 所示的页面。可以查看所有学生信息并实现分页功能,单击"查看用户"可查看学生的个人基本信息,单击"删除"可以删除学生信息;单击"添加学生"可以添加学生信息;单击"导入学生"可以把存储在 Excel 表格中的数据导入到数据库表中,如图 8-19 所示。需要注意的是 Excel 文件的数据格式应参考给定的模板。从 Excel 中导入数据使用的是一个插件,该插件对应的 JAR 文件已经添加到"库"中,JAR 文件如图 8-20 所示。

图 8-18 查看所有学生信息

教师管理功能中"查看所有教师"的页面效果如图 8-21 所示。"添加教师"和"导入教师"的功能与学生管理中的"添加学生"和"导入学生"功能类似。

课程管理功能中"查看所有课程"的页面效果如图 8-22 所示。"添加课程"和"导入课程"的功能与学生管理中的"添加学生"和"导入学生"功能类似。

单击个人信息管理功能中的"修改密码"可以修改管理员密码。

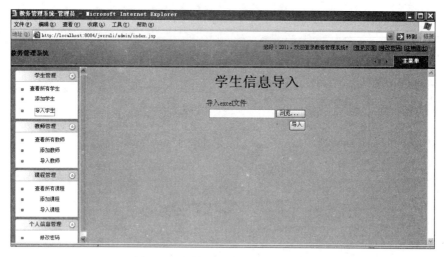

图 8-19 按模板格式导入学生信息

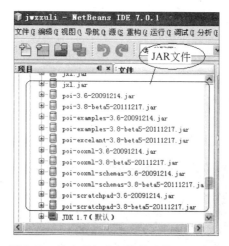

图 8-20 导入 Excel 文件所需的 JAR 文件

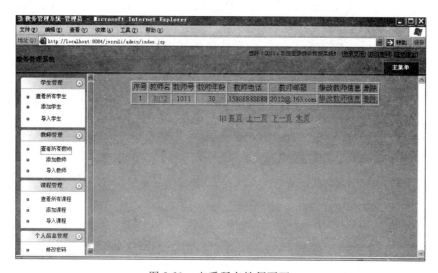

图 8-21 查看所有教师页面

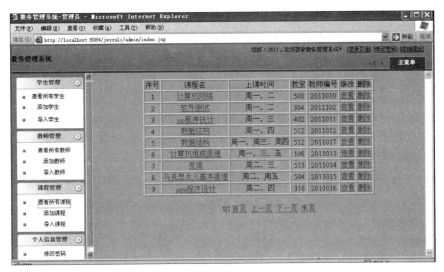

图 8-22 查看所有课程页面

8.4.5 教师管理功能的实现

教师管理功能的实现请参考前述需求分析自行编码实现。

8.5 本章小结

本章基于"以项目为驱动的教学模式",提供了一个具有一定实际应用价值的教务管理系统项目,详细介绍了该项目的设计和开发过程。在本项目实现过程中完成对 Struts2 和 Hibernate 框架技术应用的系统训练,达到整合知识体系,培养解决工程实践问题能力的教学目的。

8.6 习 题

实训题

1. 请根据业务需求编程实现教师管理模块的功能。
2. 使用 Struts2 中的验证功能对项目中的输入数据进行验证。

第9章 Spring3 框架技术入门

在软件设计中通常用耦合度和内聚度作为衡量模块独立程度的标准。划分模块的一个准则就是高内聚、低耦合。耦合度是指模块之间联系的紧密程度，模块间的耦合度简单地说是模块之间的依赖性；模块之间的耦合度越高，维护成本越高。降低模块间的耦合度能减少模块间的影响，防止对某一模块修改所引起的"牵一发而动全身"的水波效应，因此模块之间的耦合应最小。Spring3 的核心是 IoC 和 AOP，其中 IoC 的主要作用就是降低模块之间的耦合度。本章主要介绍 Spring3 的基础内容。

本章主要内容如下所示。
(1) Spring3 框架的体系结构。
(2) IoC 框架的原理。
(3) IoC 框架的主要组件。
(4) IoC 的应用和注入。
(5) 基于 Struts2＋Hibernate＋Spring3 的登录和注册系统。

9.1 Spring3 基础知识

Spring3 是 Java Web 三大经典框架(Struts、Spring、Hibernate，SSH)中主要用于降低模块之间耦合度的框架，实际上 Spring3 除了能够通过 IoC 降低模块之间的耦合度外，还提供了其他功能。

9.1.1 Spring3 的由来与发展

2002 年，Rod Johnson 编著出版了《Expert one to one J2EE design and development》，该书中对 Java EE 框架臃肿、低效、脱离现实的种种现状提出了质疑，并积极寻求探索革新之道。以此书为指导思想，他编写了 interface21 框架，这是一个力图冲破 J2EE 传统开发的困境，从实际需求出发，着眼于轻便、灵巧，易于开发、测试和部署的轻量级开发框架。Spring 框架即以 interface21 框架为基础，经过重新设计，并不断丰富其内涵，2003 年推出 Spring 1.0 测试版；2004 年 3 月 24 日，发布了 1.0 正式版；同年他又推出了一部堪称经典的力作《Expert one-to-one J2EE Development without EJB》，该书在 Java 世界掀起了轩然大波，不断改变着 Java 开发者程序设计和开发的思考方式。在该书中，作者根据自己多年丰富的实践经验，对 EJB 的各种笨重臃肿的结构进行了逐一的分析和否定，并分别以简洁实用的方式替换之。至此一战功成，Rod Johnson 成为一个改变 Java 世界的大师级人物。传统 J2EE 应用的开发效率低，应用服务器厂商对各种技术的支持并没有真正统一，导致 J2EE 的应用没有真正实现"一次编写，到处运行"的承诺。Spring 作为开源的中间件，独立于各种应用服务器，甚至无须应用服务器的支持，也能提供应用服务器的功能，如声明式事务、事务处理等。Spring 致力于 J2EE 应用的各层的解决方案，而不是仅仅专注于某一层的方案。可以说 Spring 是企业应用开发的"一站式"选择,并贯穿表现层、业务层及持久层。然而,Spring 并不想取代那些已有的框架,而是与它们无缝地整合。

2006 年,推出 Spring 2.0;2007 年推出 Spring 2.5;2008 年推出 Spring 3.0;2011 年推出 Spring 3.1。目前,Spring 最新版本是 Spring 3.1.3,2012 年 10 月发布。

9.1.2 Spring3 的下载与配置

1. 软件包下载

Spring 官方下载地址是 www.springsource.org,如图 9-1 所示,单击 Spring Framework,出现如图 9-2 所示的页面。

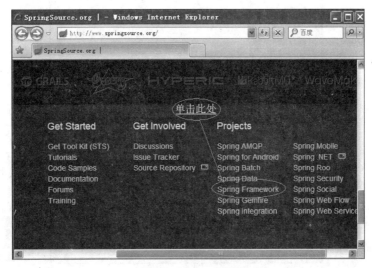

图 9-1　Spring3 下载页面(一)

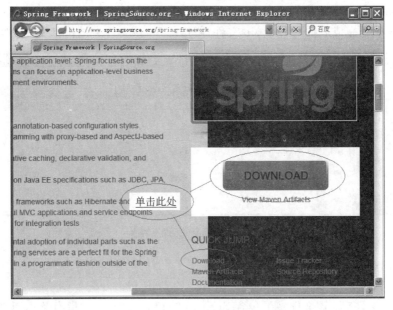

图 9-2　Spring3 下载页面(二)

单击图 9-2 中的 DOWNLOAD 按钮，出现如图 9-3 所示的选择 Spring 版本的页面，可在该页面中选择项目开发需要的 Spring 版本并下载。

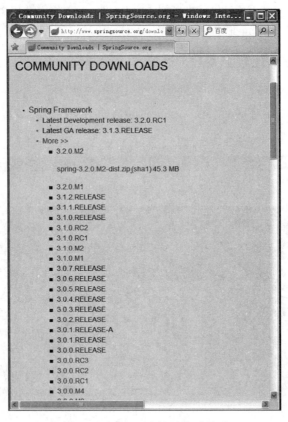

图 9-3　Spring3 选择下载项页面

2. Spring3 软件包中主要文件

Spring3 下载完成后得到一个 zip 文件，解压缩后，该文件夹结构如图 9-4 所示。

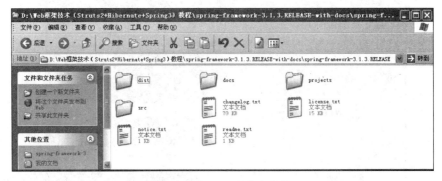

图 9-4　Spring3 文件夹结构

文件夹内容如下。

（1）dist 文件夹：该文件夹下包含 Spring3 的 JAR 包。

（2）docs 文件夹：该文件夹下包含 Spring3 的相关文档、开发指南及 API 参考文档。

（3）projects 文件夹：该文件夹包含 Spring3 提供的应用实例，这些实例是 Spring3 入门学习的案例。

（4）src 文件夹：该文件夹下包含 Spring3 的全部源文件，如果开发过程中有地方无法把握，可以参考该源文件，了解其底层实现。

3. Spring3 的配置

图 9-5　添加库

NetBeans 7.2、MyEclipse 10.6 和 Eclipse 4.2 中都集成了 Spring2 版本和 Spring3 版本，可以直接使用。例如，要在图 9-5 所示的 NetBeans 7.2 中添加自带的 Spring 类库，单击项目 ch09"库"→"添加库"，弹出如图 9-6 所示的"添加库"对话框。在该对话框中单击"导入"按钮，将弹出"导入库"对话框，选择需要的 Spring3 类库，如图 9-7 所示。

图 9-6　"添加库"对话框

图 9-7　"导入库"对话框

如果项目需要使用最新的 Spring3 类库或者其他版本的 Spring，需要把 Spring3 中 dist 文件夹下的 JAR 文件添加到库中，添加方法和 Struts2 的配置类似，这里不再介绍。

9.1.3　Spring3 框架的体系结构

Spring3 是一个轻量级的控制反转（IoC）和面向切面（AOP）的容器框架。轻量级是指从软件大小与开销两方面而言 Spring3 都是轻量的。完整的 Spring3 框架可以在一个大小只有 1MB 多的 JAR 文件里发布。并且 Spring3 所需的处理开销也是微不足道的。此外，Spring3 是非侵入式的，Spring3 应用中的对象不依赖于 Spring3 的特定类。一般说 EJB 是重量级的容器，Spring3 是轻量级容器。Spring3 通过一种称为控制反转（IoC）的技术促进了耦合度降低。当应用了 IoC，一个对象依赖的其他对象会通过被动的方式传递进来，而不是这个对象自己创建或者查找依赖对象。AOP 是指 Spring3 提供了对面向切面编程的丰富支持，允许通过分离应用的业务逻辑与系统级服务（如事务管理）进行内聚性的开发。应用对象只实现它们应该做的即完成业务逻辑，仅此而已。它们并不负责其他的系统级关注点，例如，日志或事务支持。容器是指 Spring3 包含并管理应用对象的配置和生命周期，在这个意义上它是一种容器，可以配置每个 Bean 如何被创建以及它们是如何相互关联的。然而，Spring3 不应该被混同于传统的重量级的 EJB 容器，EJB 经常是庞大与笨重的，难以使用。框架是指 Spring3 可以将简单的组件配置、组合成为复杂的应用。在 Spring3 中，应用对象被声明式地组合，典型地是在一个 XML 文件里。Spring3 也提供了很多基础功能（事务管理、持久化框架集成等），将应用逻辑的开发留给开发者。

Spring3 框架的主要优势之一就是其分层架构，分层架构允许选择使用任何一个组件，同时也可以集成其他框架。Spring3 框架由 7 个定义良好的模块组成分层架构。组成 Spring3 框架的每个组件（模块）都可以单独存在，也可以与其他一个或多个组件联合实现。Spring3 组件构建在核心容器之上，核心容器定义了创建、配置和管理 Bean 的方式，如图 9-8 所示。

图 9-8 Spring3 框架的体系结构

Spring3 框架各个组件的功能如下。

1．核心容器

核心容器提供 Spring3 框架的基本功能。核心容器的主要组件是 BeanFactory 和 ApplicationContext。容器使用控制反转（IoC）模式将应用程序的配置和依赖性规范与实际的应用程序代码分开。

2．Spring 上下文

Spring 上下文是一个配置文件，向 Spring3 框架提供上下文信息。Spring 上下文包括企业服务，例如，校验、JNDI、EJB、电子邮件、国际化和任务调度等。

3．Spring AOP

通过配置管理特性，Spring AOP 模块直接将面向切面的编程功能集成到了 Spring3 框架中。所以，可以很容易地使 Spring3 框架管理的任何对象支持 AOP。Spring AOP 模块为基于 Spring3 的应用程序中的对象提供了事务管理服务。通过使用 Spring AOP，不必依赖 EJB 组件，就可以将声明性事务管理集成到应用程序中。

4．Spring DAO

JDBC DAO 抽象层提供了有意义的异常层次结构，可用该结构来管理异常处理和不同数据库供应商抛出的错误消息。异常层次结构简化了错误处理，并且极大地降低了需要编写的异常代码数量（如打开和关闭连接）。

5．Spring ORM

Spring3 框架插入了若干个 ORM 框架，从而提供了 ORM 的工具，其中包括 JDO、

Hibernate 和 iBatisSQL Map。所有这些都遵从 Spring 的通用事务和 DAO 异常层次结构。

6. Spring Web

Web 上下文模块建立在应用程序上下文模块之上,为基于 Web 的应用程序提供了上下文。所以,Spring3 框架支持与 Struts 的集成。Web 模块还简化了处理多部分请求以及将请求参数绑定到域对象的工作。

7. Spring MVC Web 框架

MVC 框架是一个全功能的构建 Web 应用程序的 MVC 实现。通过策略接口,MVC 框架成为高度可配置的。MVC 容纳了大量视图技术,其中包括 JSP、Velocity、Tiles、iText 等。

Spring3 框架的功能可以用在任何 Java EE 服务器中,大多数功能也适用于不受管理的环境。Spring3 的核心要点是:支持不绑定到特定 Java EE 服务的可重用业务和数据访问对象。毫无疑问,这样的对象可以在不同 Java EE 环境(Web 或 EJB)、独立应用程序、测试环境之间重用。

Spring3 框架提供了"一站式服务"。Spring3 框架既可用在 Java 程序设计中,也可以用在 Java Web 程序设计中,甚至可以用在.NET 程序设计中。另外,Spring3 框架中提供了多种框架技术,也可以集成其他框架技术,例如,Struts2、Hibernate 等。Spring3 框架具有以下特点。

(1) 方便解耦,简化开发。通过 Spring3 提供的 IoC 容器,可以将对象之间的依赖关系交由 Spring3 进行控制,避免硬编码所造成的程序过度耦合。

(2) AOP 编程的支持。通过 Spring3 提供的 AOP 功能,方便进行面向切面的编程,许多不容易使用面向对象程序设计(Object Oriented Programming,OOP)实现的功能可以通过 AOP 轻松实现。

(3) 声明式事务的支持。在 Spring3 中,可以从单调烦闷的事务管理代码中解脱出来,通过声明方式灵活地进行事务的管理,提高开发效率和质量。

(4) 方便程序的测试。可以用非容器依赖的编程方式进行几乎所有的测试工作,在 Spring3 中,测试不再是昂贵的操作,而是随手可做的事情。

(5) 方便集成各种优秀框架。Spring3 不排斥各种优秀的开源框架,相反,Spring3 可以降低各种框架的使用难度,Spring3 提供了对各种优秀框架(如 Struts2、Hibernate、Quartz 等)的直接支持。

(6) 降低 Java EE API 的使用难度。Spring3 对很多难用的 Java EE API(如 JavaMail、远程调用等)提供了一个薄薄的封装层,通过 Spring3 的简单封装,这些 Java EE API 的使用难度大大降低。

(7) 框架源码是经典学习范例。Spring3 的源码设计精妙、结构清晰、匠心独具,处处体现着大师对 Java 设计模式的灵活运用以及在 Java 技术上的高深造诣。Spring3 框架源码无疑是 Java 技术的最佳实践范例。如果想在短时间内迅速提高自己的 Java 技术水平和应用开发水平,学习和研究 Spring3 源码将会使你收到意想不到的效果。

9.2 Spring3 IoC 的原理和主要组件

Spring3 框架的核心是实现了 IoC 模式的轻量级容器，IoC 是 Spring3 框架的核心。

9.2.1 IoC 的基础知识以及原理

1. IoC 理论的背景

在采用面向对象方法设计的软件系统中，底层实现都是由 N 个对象组成的，所有的对象通过彼此的合作，最终实现系统的业务逻辑。

如果打开机械式手表的后盖，就会看到与上面类似的情形，各个齿轮分别带动时针、分针和秒针顺时针旋转，从而在表盘上产生正确的时间。图 9-9 中描述的就是这样的一个齿轮组，它拥有多个独立的齿轮，这些齿轮相互啮合在一起，协同工作，共同完成某项任务。可以看到，在这样的齿轮组中，如果有一个齿轮出了问题，就可能会影响整个齿轮组的正常运转。

齿轮组中齿轮之间的啮合关系，与软件系统中对象之间的耦合关系非常相似。对象之间的耦合关系是无法避免的，也是必要的，这是协同工作的基础。现在，伴随着工业级应用的规模越来越庞大，对象之间的依赖关系也越来越复杂，经常会出现对象之间的多重依赖性关系，因此，架构师和设计师对于系统的分析和设计，将面临更大的挑战。对象之间耦合度过高的系统，必然会出现"牵一发而动全身"的情形。

耦合关系不仅会出现在对象与对象之间，也会出现在软件系统的各模块（组件）之间，以及软件系统和硬件系统之间。对象之间可能存在复杂的依赖关系，如图 9-10 所示。如何降低系统之间、模块之间及对象之间的耦合度，是软件工程永远追求的目标之一。为了解决对象之间耦合度过高的问题，软件专家 Michael Mattson 提出了 IoC 理论，用来实现对象之间的"解耦"，目前这个理论已经被成功地应用到实践当中，很多 Java EE 项目均采用了 IoC 框架产品 Spring3。

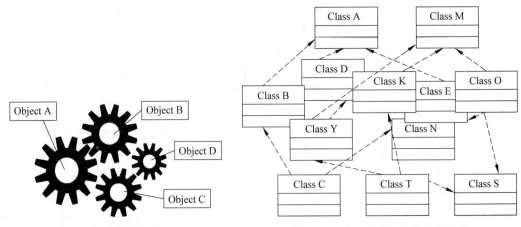

图 9-9 软件系统中耦合的对象　　　　图 9-10 对象之间复杂的依赖关系

2. 控制反转

IoC 是 Inversion of Control 的缩写，有的翻译成"控制反转"，还有翻译为"控制反向"或者"控制倒置"。

1996 年,Michael Mattson 在一篇探讨有关面向对象框架的文章中,首先提出了 IoC 这个概念。简单来说就是把复杂系统分解成相互合作的对象,这些对象类通过封装以后,内部实现对外部是透明的,从而降低了解决问题的复杂度,而且可以灵活地被重用和扩展。IoC 理论提出的观点大体是这样的:借助于"第三方"实现具有依赖关系的对象之间的解耦,如图 9-11 所示。

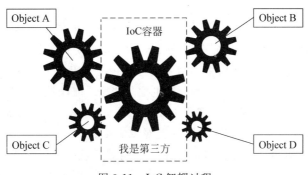

图 9-11　IoC 解耦过程

从图 9-11 中可以看出,由于引进了中间位置的"第三方",也就是 IoC 容器,使得 A、B、C、D 这 4 个对象没有了耦合关系,齿轮之间的传动全部依靠"第三方"了,所有对象的控制权全部上缴给"第三方"IoC 容器,所以,IoC 容器成了整个系统的关键核心,它起到了一种类似"黏合剂"的作用,把系统中的所有对象黏合在一起发挥作用,如果没有这个"黏合剂",对象与对象之间会彼此失去联系,这就是有人把 IoC 容器比喻成"黏合剂"的由来。

如果把上图 9-11 中间的 IoC 容器拿掉,如图 9-12 所示。

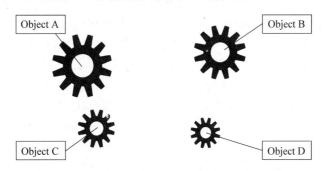

图 9-12　去掉 IoC 容器后的系统

图 9-12 所示就是要实现整个系统所需要完成的全部内容。这时候,A、B、C、D 这 4 个对象之间已经没有了耦合关系,彼此毫无联系。这样的话,在实现 A 的时候,根本无须再去考虑 B、C 和 D,对象之间的依赖关系已经降低到了最低程度。所以,如果真能实现 IoC 容器,对于系统开发而言,这将是一件多么美好的事情,参与开发的每一成员只要实现自己的类就可以了,跟别人没有任何关系。

再来看看,控制反转(IoC)到底为什么要起这么个名字?

来对比一下:软件系统在没有引入 IoC 容器之前,如图 9-9 所示,对象 A 依赖于对象 B,那么对象 A 在初始化或者运行到某一点的时候,自己必须主动去创建对象 B 或者使用已经创建的对象 B。无论是创建还是使用对象 B,控制权都在自己手上。

软件系统在引入 IoC 容器之后,这种情形就完全改变了,如图 9-12 所示,由于 IoC 容器的加入,对象 A 与对象 B 之间失去了直接联系,所以,当对象 A 运行到需要对象 B 的时候,IoC 容器会主动创建一个对象 B 注入到对象 A 需要的地方。

通过前后的对比,不难看出,对象 A 获得依赖对象 B 的过程,由主动行为变为了被动行为,控制权颠倒过来了,这就是"控制反转"这个名称的由来。

3. IoC 的别名——依赖注入

2004 年,Martin Fowler 探讨了同一个问题,既然 IoC 是控制反转,那么到底是"哪些方面的控制被反转了呢?",经过详细的分析和论证后,他得出了答案:"获得依赖对象的过程被反转了"。控制被反转之后,获得依赖对象的过程由自身管理对象变为由 IoC 容器主动注入。于是,他给"控制反转"取了一个更合适的名字叫做"依赖注入(Dependency Injection,DI)"。他的这个答案,实际上给出了实现 IoC 的方法:注入。所谓依赖注入,就是由 IoC 容器在运行期间,动态地将某种依赖关系注入到对象之中。

所以,依赖注入(DI)和控制反转(IoC)是从不同的角度描述了同一件事情,就是指通过引入 IoC 容器,利用依赖关系注入的方式,实现对象之间的解耦。

图 9-13 USB 接口和 USB 设备

下面举一个生活中的例子,来帮助理解依赖注入的过程。大家对 USB 接口和 USB 设备都很熟悉,USB 为人们使用计算机提供了很大的方便,现在有很多的外部设备都支持 USB 接口,如图 9-13 所示。

现在,利用计算机主机和 USB 接口来实现一个任务:从外部 USB 设备读取一个文件。计算机主机读取文件的时候,它一点也不会关心 USB 接口上连接的是什么外部设备,而且它确实也无须知道。它的任务就是读取 USB 接口,挂接的外部设备只要符合 USB 接口标准即可。所以,如果给计算机主机连接上一个 U 盘,那么主机就从 U 盘上读取文件;如果给计算机主机连接上一个移动硬盘,那么计算机主机就从移动硬盘上读取文件。挂接外部设备的权力由我们做主,即控制权归我们,至于 USB 接口挂接的是什么设备,计算机主机是决定不了,它只能被动地接受。计算机主机需要外部设备的时候,根本不用它告诉我们,我们就会主动帮它挂上它想要的外部设备。这就是我们生活中常见的一个依赖注入的例子。在这个过程中,我们就起到了 IoC 容器的作用。

通过这个例子,依赖注入的思路已经非常清楚:当计算机主机读取文件的时候,人们就把它所要依赖的外部设备,帮它挂接上。整个外部设备注入的过程和一个被依赖的对象在系统运行时被注入另外一个对象内部的过程完全一样。

把依赖注入应用到软件系统中,再来描述一下这个过程:对象 A 依赖于对象 B,当对象 A 需要用到对象 B 的时候,IoC 容器就会立即创建一个对象 B 送给对象 A。IoC 容器就是一个对象制造工厂,你需要什么,它会给你送去,你直接使用就行了,再也不用去关心你所用的东西是如何制成的,也不用关心最后是怎么被销毁的,这一切全部由 IoC 容器完成。

在传统的实现中,由程序内部代码来控制组件之间的关系。经常使用 new 关键字来实现两个组件之间关系的组合,这种实现方式会造成组件之间的耦合。IoC 很好地解决了该问题,它实现了将组件间关系从程序内部提到外部容器,也就是说由容器在运行期将组件间的某种依赖关系动态注入组件中。

4．使用 IoC 的好处

还是以 USB 的例子为例,使用 USB 外部设备相比使用内置硬盘,可带来以下好处。

(1) USB 设备作为计算机主机的外部设备,在插入主机之前,与计算机主机没有任何的关系,只有被人们连接在一起之后,两者才发生联系,具有相关性。所以,无论两者中的任何一方出现什么的问题,都不会影响另一方的运行。这种特性体现在软件工程中,就是可维护性比较好,非常便于进行单元测试,便于调试程序和诊断故障。代码中的每一个 Class 都可以单独测试,彼此之间互不影响,只要保证自身的功能无误即可,这就是组件之间低耦合或者无耦合带来的好处。

(2) USB 设备和计算机主机之间的无关性,还带来了另外一个好处,生产 USB 设备的厂商和生产计算机主机的厂商完全可以是互不相干的人,各干各事,他们之间唯一需要遵守的就是 USB 接口标准。这种特性体现在软件开发过程中,好处可是太大了。每个开发团队的成员都只需要关注自己要实现的业务逻辑,完全不用去关心其他人的工作进展,因为你的任务跟别人没有任何关系,你的任务可以单独测试,你的任务也不用依赖于别人的组件,再也不用扯不清责任了。所以,在一个大中型项目中,团队成员分工明确、责任明晰,很容易将一个大的任务划分为细小的任务,开发效率和产品质量必将得到大幅度的提高。

(3) 同一个 USB 外部设备可以插接到任何支持 USB 的设备,可以插接到计算机主机,也可以插接到 DV 机,USB 外部设备可以被反复利用。在软件工程中,这种特性就是可复用性好,人们可以把具有普遍性的常用组件独立出来,反复应用到项目中的其他部分,或者是其他项目,当然这也是面向对象的基本特征。显然,IoC 更好地贯彻了这个原则,提高了模块的可复用性。符合接口标准的实现,都可以插接到支持此标准的模块中。

(4) 同 USB 外部设备一样,模块具有热插拔特性。IoC 生成对象的方式转为外置方式,也就是把对象生成放在配置文件里进行定义,这样,当人们更换一个实现子类将会变得很简单,只要修改配置文件就可以了,完全具有热插拔的特性。

5．IoC 的原理

控制反转是 Spring 框架的核心。其原理是基于 OO 设计原则的 The Hollywood Principle：Don't call us, we'll call you(别找我,我会来找你的)。也就是说,所有的组件都是被动的,所有的组件初始化和调用都由容器负责。组件处在一个容器当中,由容器负责管理。简单来讲,就是由容器控制程序之间的关系,而非传统实现中,由程序代码直接操控,即在一个类中调用另外一个类。这也就是所谓"控制反转"的概念所在：控制权由应用代码中转到了外部容器,控制权的转移,即所谓反转。

6．工厂模式

在 Spring3 IoC 中经常用到一个设计模式,即工厂模式。工厂模式提供创建对象的接口。

工厂模式是指当应用程序中甲组件需要乙组件协助时,并不是在甲组件中直接实例化乙组件对象,而是通过乙组件的工厂获取,即该工厂可以生成某一类型组件的实例对象。在这种模式下,甲组件无须与乙组件以硬编码的方式耦合在一起,而只与乙组件的工厂耦合。下面举例说明工厂模式的应用。

1) 项目介绍

项目有一个接口 Person,代码如例 9-1 所示,工厂模式需要一个接口；有两个实现接口的类分别为 Chinese 类和 American 类,代码如例 9-2 和例 9-3 所示;Chinese 类和 American 类在工厂类 Factory 中进行关联,Chinese 类和 American 类之间并不直接关联,Factory 类的代码如例 9-4 所示;测试该项目的类为 Test,代码如例 9-5 所示。项目的文件结构以及运行结果如图 9-14 所示。

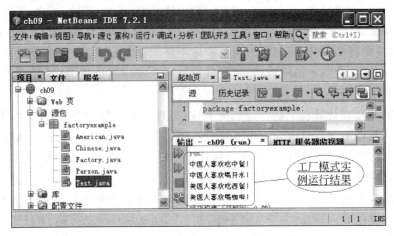

图 9-14 项目的文件结构以及运行结果

2) 接口 Person 的代码

【例 9-1】 Person 接口(Person.java)。

```
package factoryexample;

public interface Person {
    void eat();                    //定义抽象的吃方法
    void drink();                  //定义抽象的喝方法
}
```

3) 实现接口类的代码

【例 9-2】 Chinese 类(Chinese.java)。

```
package factoryexample;

public class Chinese implements Person{
    public void eat()
    {
        System.out.println("中国人喜欢吃中餐!");
    }
    public void drink()
    {
        System.out.println("中国人喜欢喝开水!");
    }
}
```

【例 9-3】 American 类(American.java)。

```java
package factoryexample;

public class American implements Person{
    public void eat()
    {
        System.out.println("美国人喜欢吃西餐!");
    }
    public void drink()
    {
        System.out.println("美国人喜欢喝咖啡!");
    }
}
```

4) 工厂类

【例 9-4】 Factory 类(Factory.java)。

```java
package factoryexample;

public class Factory {
    public Person getPerson(String name)
    {
        if(name.equals("中国人"))
            return new Chinese();
        else if(name.equals("美国人"))
            return new American();
        else
            throw new IllegalArgumentException("参数不正确!");
    }
}
```

5) 测试类

【例 9-5】 Test 类(Test.java)。

```java
package factoryexample;

public class Test {
    public static void main(String[] args) {
        Person person=null;
        person=new Factory().getPerson("中国人");
        person.eat();
        person.drink();
        person=new Factory().getPerson("美国人");
        person.eat();
        person.drink();
    }
}
```

6）测试结果

运行 Test 类,结果如图 9-14 所示。

9.2.2 IoC 的主要组件

Spring3 框架的两个最基本和最重要的包是 org.springframework.beans.factory(该包中的主要接口是 BeanFactory)和 org.springframework.context 包(该包中的主要接口是 ApplicationFactory)。这两个包中的代码提供了 Spring IoC 特性的基础。Spring IoC 框架主要组件如下所示。

(1) Beans。
(2) 配置文件(beans.xml 或 applicationContext.xml)。
(3) BeanFactory 接口及其相关类。
(4) ApplicationContext 接口及其相关类。

下面分别介绍 IoC 的主要组件技术。

1. Beans

Beans 是指项目中提供业务功能的 Bean,即容器要管理的 Bean。例 9-6 中的代码就是一个 Bean。参考 9.2.4 节项目。

【例 9-6】 Chinese 类(Chinese.java)。

```
package iocexample2;

public class Chinese implements Person{
    private Language language;
    public void speak()
    {
        System.out.println(language.kind());
    }
    public void setLanguage(Language language)
    {
        this.language=language;
    }
}
```

Bean 可以包含一些属性以及属性对应的 getter 和 setter 方法,也可以包含其他方法。

2. 配置文件

在 Spring3 中对 Bean 的管理是在配置文件中进行的。在 Spring3 容器内编辑配置文件管理 Bean 又称为 Bean 的装配;实际上装配就是告诉容器需要哪些 Bean,以及容器是如何使用 IoC 将它们配置起来。

Bean 的配置文件是一个 XML 文件,它可命名为 beans.xml、applicationContext.xml 或其他。一般习惯使用 applicationContext.xml。

配置文件包含 Bean 的 id、类、属性及其值;包含一个<beans>元素和数个<bean>子元素。Spring IoC 框架可根据 Bean 的 id 从 Bean 配置文件中取得该 Bean 的类,并生成该类的一个对象,继而从配置文件中获得该对象的属性和值。常见applicationContext.xml配置文件格式如例9-7所示。参考9.2.4节项目。

【例9-7】 配置文件(applicationContext.xml)。

```
<?xml version="1.0" encoding="UTF-8"?>
<!--配置文件的DTD信息,Spring的版本不一样,DTD信息会有一定的差异,请使用与版本对应的DTD文件-->
<!DOCTYPE beans PUBLIC "-//SPRING//DTD BEAN 2.0//EN"
"http://www.springframework.org/dtd/spring-beans-2.0.dtd"
[<!ENTITY contextInclude SYSTEM
"org/springframework/web/context/WEB-INF/contextInclude.xml">]>
<!--配置文件的根元素-->
<beans>
    <!--配置一个Bean,可以有多个bean子元素。向Spring3容器添加一个Bean只需在配置文件中添加一对<bean>元素。每个<bean>元素配置一个Bean,其中,id指定Bean的名字,在作用域内必须唯一,"中国人"为Chinese类命名的id名,由程序员自己命名。class指定Bean的类,iocexample2是包名,Chinese是类名-->
    <bean id="中国人" class="iocexample2.Chinese">
        <!--<property>元素用来指定需要容器注入的属性;name指定属性值为language;ref指定需要向language属性注入的id,即注入的对象"英语",该对象由English类生成-->
        <property name="language" ref="英语"></property>
    </bean>
    <!--配置另外一个Bean-->
    <bean id="英语" class="iocexample2.English"></bean>
</beans>
```

当通过Spring3容器创建一个Bean实例时,不仅可以完成Bean的实例化,也可以为Bean指定作用域。

Bean作用域可以设置如下。

(1) 原型模式和单实例模式:Spring3中的Bean默认情况下是单实例模式,容器分配Bean时,总返回同一个实例;但是,如果每次向ApplicationContext请求一个Bean时需要得到不同的实例,需将Bean定义为原型模式。可以使用<bean>元素的singleton属性设置,默认值为true,如果设置为false,就是把Bean定义为原型Bean。

(2) request和session:作用域是request和session。每次请求都会产生一个不同的Bean实例。

(3) global session:共享一个Bean实例。

3. BeanFactory接口及其相关类

BeanFactory采用了工厂设计模式,即Bean容器模式,负责读取Bean的配置文件,管理对象的生成、加载,维护Bean对象之间的依赖关系,负责Bean对象的生命周期,对于简单的应用程序来说,使用BeanFactory就已经足够管理Bean,在对象的管理上就可以获得许多便利性。

org.springframework.beans.factory.BeanFactory是一个顶级接口,它包含管理Bean的各种方法。Spring3框架也提供了一些实现该接口的类。

org.springframework.beans.factory.xml.XmlBeanFactory 是 BeanFactory 常用的实现类。其根据配置文件中的定义装载 Bean。要创建 XmlBeanFactory，需要传递一个 FileInputStream 对象，该对象把 XML 文件提供给工厂。

例如：

```
BeanFactory factory= new
XmlBeanFactory(new FileInputStream("applicationContext.xml"));
```

BeanFactory 的常用方法如下。

（1）getBean(String name)：该方法可根据 Bean 的 id 生成该 Bean 的对象。

（2）getBean(String name, Class requiredType)：该方法可根据 Bean 的 id 和相应类生成该 Bean 的对象。

4. ApplicationContext 接口及其相关类

作为一个应用程序框架，只提供 Bean 容器管理的功能是不够的。若要利用 Spring3 所提供的一些高级容器功能，则可以使用 ApplicationContext 接口，该接口是提供高级功能的容器。ApplicationContext 的基本功能与 BeanFactory 很相似，但它还有以下功能。

（1）提供访问资源文件的更方便的方法。

（2）支持国际化消息。

（3）提供文字消息解析的方法。

（4）可以发布事件，对事件感兴趣的 Bean 可以接收到这些事件。

ApplicationContext 接口的常用实现类有 3 个。

① FileSystemXmlApplicationContext：从文件系统中的 XML 文件加载上下文中定义的信息。

② ClassPathXmlApplicationContext：从类路径中的 XML 文件加载上下文中定义的信息，把上下文定义的文件当成类路径资源。

③ XmlWebApplicationContext：从 Web 系统中的 XML 文件加载上下文中定义的信息。

例如：

```
ApplicationContext context = new FileSystemXmlApplicationContext ( "d:/appcon.xml");
ApplicationContext context=new ClassPathXmlApplicationContext("appcon.xml");
```

FileSystemXmlApplicationContext 和 ClassPathXmlApplicationContext 的区别是：FileSystemXmlApplicationContext 只能在指定的路径中查找 appcon.xml 文件，而 ClassPathXmlApplicationContext 可以在整个类路径中查找 appcon.xml。

9.2.3 IoC 的应用实例

在 9.2.1 节中介绍了工厂模式，就是甲组件需要乙组件对象的时候，无须直接创建其实例，而是通过工厂获得，只要创建一个工厂即可。而 Spring3 容器则提供了更好的办法，开发人员不用创建工厂，可以直接应用 Spring3 提供的 IoC 方式。下面两个例子修改自 9.2.1 节中的实例，分别使用 NetBeans 7.2 中集成的 Spring 3.1.1 和另行下载的新版本的 Spring 3.1.3

来实现。

首先使用 NetBeans 7.2 中集成的 Spring 3.1.1 重新实现 9.2.1 中的实例。

1．项目介绍

项目有一个接口 Person,代码如例 9-8 所示;有两个实现接口的类分别为 Chinese 类和 American 类,代码如例 9-9 和例 9-10 所示;配置 Bean 的配置文件为 applicationContext.xml,代码如例 9-11 所示;测试该项目的类为 Test,代码如例 9-12 所示。

NetBeans 7.2 集成的 Spring 3.1.1 的加载过程可参考图 9-5～图 9-7。

可以通过 NetBeas 7.2 生成一个 applicationContext.xml,具体步骤如下。

(1) 选择"其他"命令,如图 9-15 所示。

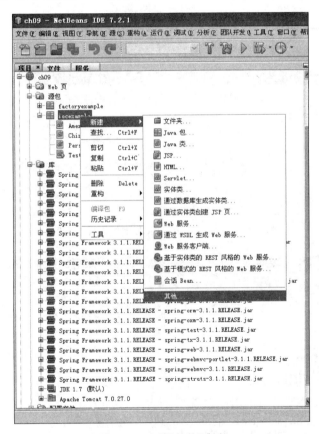

图 9-15　选择"其他"命令

(2) 单击图 9-15 中的"其他"命令,弹出如图 9-16 所示的对话框。

(3) 单击图 9-16 中"类别"框中的"Spring 框架"后,在"文件类型"框中选择"Spring XML 配置文件";单击"下一步"按钮,弹出图 9-17 所示的对话框。在该对话框"文件名"框中输入 applicationContext,即新建 applicationContext.xml 文件,该文件所在的文件夹为"src\java\iocexample",在程序中要使用该文件夹路径。单击图 9-17 中的"下一步"按钮,出现另外一个对话框,可以直接单击该对话框中的"完成"按钮,就会在文件夹"src\java\iocexample"下生成一个 applicationContext.xml 文件。

项目的文件结构以及库中文件和运行结果如图 9-18 所示。

图 9-16 "新建文件"对话框

图 9-17 "New Spring XML 配置文件"对话框

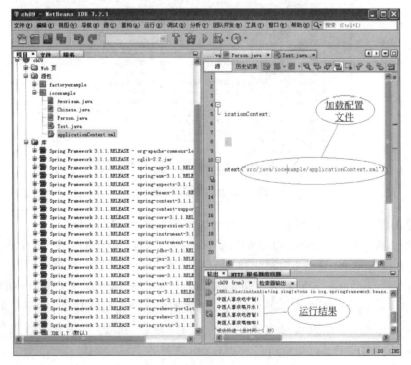

图 9-18 项目的文件结构以及库中文件和运行结果

2. 接口 Person 的代码

Person 接口的代码如例 9-8 所示。

【例 9-8】 Person 接口(Person.java)。

```java
package iocexample;

public interface Person {
    void eat();            //定义抽象的吃方法
    void drink();          //定义抽象的喝方法
}
```

3. 实现接口类的代码

Chinese 类和 American 类的代码如例 9-9 和例 9-10 所示。

【例 9-9】 Chinese 类(Chinese.java)。

```java
package iocexample;

public class Chinese implements Person{
    public void eat()
    {
        System.out.println("中国人喜欢吃中餐!");
    }
    public void drink()
    {
        System.out.println("中国人喜欢喝开水!");
    }
}
```

【例 9-10】 American 类(American.java)。

```java
package iocexample;

public class American implements Person{
    public void eat()
    {
        System.out.println("美国人喜欢吃西餐!");
    }
    public void drink()
    {
        System.out.println("美国人喜欢喝咖啡!");
    }
}
```

4. Bean 的配置文件

applicationContext.xml 配置文件的代码。

【例 9-11】 Bean 的配置文件(applicationContext.xml)。

```xml
<?xml version="1.0" encoding="UTF-8"?>
<beans xmlns="http://www.springframework.org/schema/beans"
    xmlns:xsi="http://www.w3.org/2001/XMLSchema-instance"
    xsi:schemaLocation="http://www.springframework.org/schema/beans http://www.springframework.org/schema/beans/spring-beans-2.5.xsd">
    <bean id="中国人" class="iocexample.Chinese"></bean>
    <bean id="美国人" class="iocexample.American"></bean>
</beans>
```

5. 测试类

【例 9-12】 Test 类（Test.java）。

```java
package iocexample;
import org.springframework.context.ApplicationContext;
import org.springframework.context.support.FileSystemXmlApplicationContext;

public class Test {
    public static void main(String[]args) {
        //从 applicationContext.xml 配置文件中实例化配置的 Bean 对象
        ApplicationContext ac=new
        FileSystemXmlApplicationContext("src/java/iocexample/applicationContext.xml");
        Person person=null;
        person=(Person)ac.getBean("中国人");
        person.eat();
        person.drink();
        person=(Person)ac.getBean("美国人");
        person.eat();
        person.drink();
    }
}
```

6. 测试结果

运行的结果如图 9-18 所示。

下面使用 Spring 3.1.3 重新实现 9.2.1 中的实例。

1）项目介绍

项目有 Person 接口，实现接口的类为 Chinese 和 American；测试该项目的类为 Test，与使用 NetBeans 7.2 中集成的 Spring 3.1.1 实现 9.2.1 中实例的代码几乎一样。不过本例中使用的是 Spring 3.1.3，Spring 3.1.3 中对 IoC 的配置文件有一点改动，代码如例 9-13 所示，该配置文件可在下载的 Spring 3.1.3 包中查询找到。

由于使用 Spring 3.1.3 中的类库，需要先加载 Spring3.1.3 的类库。

Spring 3.1.3 类库的加载过程如下：单击"库"→"添加 JAR/文件夹"，弹出"添加 JAR/文件夹"对话框，找到 spring-framework-3.1.0.RELEASE 文件夹下面的 dist 文件夹，选择 JAR 文件，如图 9-19 所示，在该对话框中单击"打开"按钮后，Spring3.1.3 类库即添加完成。

图 9-19 添加 Spring 3.1.3 类库

项目的文件结构以及库中文件如图 9-20 所示。

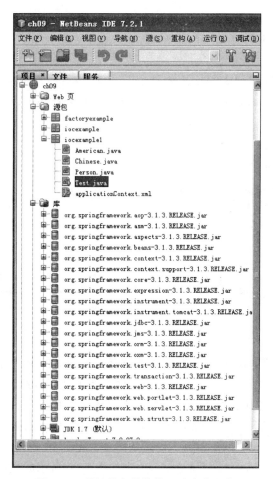

图 9-20 项目的文件结构以及库中文件

2)接口 Person 的代码

Person 接口的代码与例 9-8 几乎一样,只是包名改为"package iocexample1;"。

3)实现接口类的代码

Chinese 类和 American 类的代码与例 9-9 和例 9-10 几乎一样,只是包名改变而已。

4)Bean 的配置文件

applicationContext.xml 配置文件的代码。

【例 9-13】 Bean 的配置文件(applicationContext.xml)。

```
<?xml version="1.0" encoding="UTF-8"?>
<!DOCTYPE beans PUBLIC "-//SPRING//DTD BEAN 2.0//EN"
    "http://www.springframework.org/dtd/spring-beans-2.0.dtd"
    [<!ENTITY contextInclude SYSTEM
    "org/springframework/web/context/WEB-INF/contextInclude.xml">]>
<beans>
    <bean id="中国人" class="iocexample1.Chinese"></bean>
    <bean id="美国人" class="iocexample1.American"></bean>
</beans>
```

请比较例 9-11 和例 9-13 的异同。

5)测试类

Test 类的代码与例 9-12 几乎一样。只是修改了"ApplicationContext ac = new FileSystemXmlApplicationContext("src/java/iocexample1/applicationContext.xml");"和包名。

6)测试结果

运行时将出现如图 9-21 所示的异常。

图 9-21　出现异常

7)排除异常

出现上述异常是因为在 Spring 3.1.3 中没有提供所需的第三方 JAR 文件,即 Apache 的 LogFactory 类库,需要在 www.apache.org 网站上下载,下载页面如图 9-22 所示。

单击图 9-22 所示页面中的 Download,出现如图 9-23 所示的页面。

单击图 9-23 所示页面中的 Commons,出现如图 9-24 所示的选择 Logging。

单击图 9-24 所示页面中的 Logging,出现如图 9-25 所示的 Commons Logging 下载页面。

单击图 9-25 所示页面中的 Download,出现如图 9-26 所示的页面。

下载完成后对 zip 文件进行解压,然后把 commons-logging-1.1.1 文件夹下的 JAR 文件添加到项目"库"中。

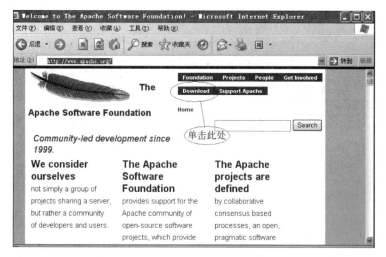

图 9-22 下载页面

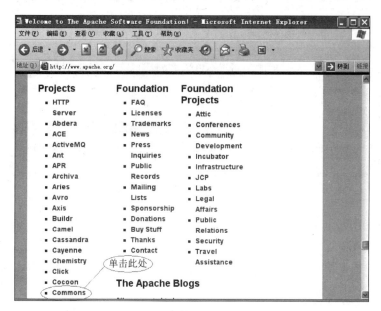

图 9-23 单击 Commons

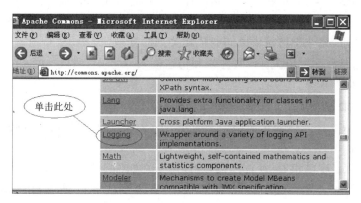

图 9-24 单击 Logging

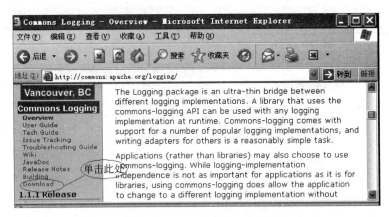

图 9-25　单击 Commons Logging 中的 Download

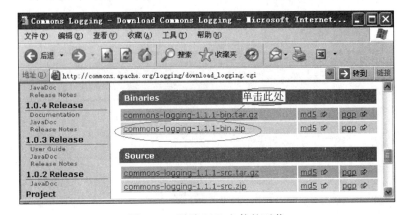

图 9-26　所需 JAR 文件的下载

8）重新测试

运行结果与图 9-18 所示一致。

备注：可以根据项目开发需要选择相应的 Spring 版本，但应注意它们之间的差异。

9.2.4　注入的两种方式

通过前面的学习可知，依赖注入是指程序在运行过程中，如果需要另外一个对象协助完成，无须在代码中创建被调用者，而是依赖外部的注入获取。Spring3 的依赖注入对调用者几乎没有任何要求，完全支持对象之间的依赖关系的管理。

依赖注入通常有两种方式：设置注入和构造注入。

1．设置注入

设置注入是通过 setter 方法注入被调用者的实例。该方式简单、直观，而且容易理解，所以 Spring3 的设置注入被大量使用。下面通过一个项目介绍设置注入的使用。

1）项目介绍

项目有 Person 接口和 Language 接口，代码分别如例 9-14 和例 9-15 所示；实现两个接口的类分别为 Chinese 和 English，代码分别如例 9-16 和例 9-17 所示，在 Chinese 类中要用到 Language 对象。因 Language 是接口，所以用其实现类来创建对象，在 Chinese 类中为其定义一个 setter 方法，即 setLanguage()；通过 Spring3 的配置文件来完成对象的注入，配置文件为 applicationContext.xml，代码如例 9-18 所示。测试该项目的类为 Test，代码如例 9-19 所示；项目的文件结构以及运行结果如图 9-27 所示。

图 9-27　项目的文件结构以及运行结果

2）接口 Person 和 Language 的代码

【例 9-14】　Person 接口（Person.java）。

```
package iocexample2;

public interface Person {
    void speak();                    //定义抽象的说方法
}
```

【例 9-15】　Language 接口（Language.java）。

```
package iocexample2;

ublic interface Language {
    String kind();                   //定义抽象的语言类型方法
}
```

3）实现接口类的代码

【例 9-16】　Chinese 类（Chinese.java）。

```
package iocexample2;

public class Chinese implements Person{
    private Language language;
    public void speak()
    {
        System.out.println(language.kind());
```

```
    }
    public void setLanguage(Language language)
    {
        this.language=language;
    }
}
```

【例9-17】 English 类(English.java)。

```
package iocexample2;

public class English implements Language{
    public String kind()
    {
        return "中国人喜欢说汉语,也会说英语!";
    }
}
```

4) Bean 的配置文件

【例9-18】 Bean 的配置文件(applicationContext.xml)。

```xml
<?xml version="1.0" encoding="UTF-8"?>
<!DOCTYPE beans PUBLIC "-//SPRING//DTD BEAN 2.0//EN"
"http://www.springframework.org/dtd/spring-beans-2.0.dtd"
[<!ENTITY contextInclude SYSTEM
"org/springframework/web/context/WEB-INF/contextInclude.xml">]>
<beans>
    <!--配置Bean,注入Chinese类对象-->
    <bean id="中国人" class="iocexample2.Chinese">
        <!--<property>元素用来指定需要容器注入的属性;name指定属性值为language;ref
        指定需要向language属性注入的id,即注入的对象"英语",该对象由English类生成-->
        <property name="language" ref="英语"></property>
    </bean>
    <!--注入English-->
    <bean id="英语" class="iocexample2.English"></bean>
</beans>
```

通过配置文件管理Bean的好处是,各Bean之间的依赖关系被放在配置文件中实现,而不是在代码中实现。通过配置文件,Spring3能很好地为每个Bean注入属性。

Spring3通过property元素为Bean注入属性值。

5) 测试类

【例9-19】 Test 类(Test.java)。

```
package iocexample2;
import org.springframework.context.ApplicationContext;
import org.springframework.context.support.FileSystemXmlApplicationContext;
```

```
public class Test {
    public static void main(String[] args) {
        ApplicationContext ac=new FileSystemXmlApplicationContext("src/java/iocexample2/applicationContext.xml");
        Person person=null;
        person= (Person)ac.getBean("中国人");
        person.speak();
    }
}
```

6）测试结果

运行 Test 类,结果如图 9-27 所示。

2. 构造注入

利用构造方法来设置依赖注入的方式称为构造注入。

下面实例介绍构造注入的使用。

接口和实现类的代码与前例类似,只需修改 Chinese 类,其他不变,修改后 Chinese 类的代码如例 9-20 所示。配置文件也需要简单修改,修改后代码如例 9-21 所示。

【例 9-20】 Chinese 类(Chinese.java)。

```
public class Chinese implements Person{
    private Language language;
    public Chinese(){}
    public Chinese(Language language)
    {
        this.language=language;
    }
    public void speak()
    {
        System.out.println(language.kind());
    }
}
```

【例 9-21】 Bean 的配置文件(applicationContext.xml)。

```
<?xml version="1.0" encoding="UTF-8"?>
<!DOCTYPE beans PUBLIC "-//SPRING//DTD BEAN 2.0//EN"
"http://www.springframework.org/dtd/spring-beans-2.0.dtd" [
<!ENTITY contextInclude SYSTEM
"org/springframework/web/context/WEB-INF/contextInclude.xml">]>
<beans>
    <bean id="中国人" class="iocexample2.Chinese">
        <constructor-arg ref="英语"></constructor-arg>
    </bean>
    <bean id="英语" class="iocexample2.English"></bean>
</beans>
```

测试类不变。运行结果与设置注入方式运行的结果一样,区别在于创建 Person 实例中 Language 属性的时间不同。设置注入是先创建一个默认的 Bean 实例,然后调用对应的 setter 方法注入依赖关系;构造注入是创建 Bean 实例时,已经完成了依赖关系的注入。

9.3 基于 Struts2＋Hibernate＋Spring3 的登录系统

9.3.1 项目介绍

本节使用 Struts 2.3.4、Hibernate 3.6.0 和 Spring 3.1.3 开发一个实现登录功能的项目,命名为 sshLogin。该项目的文件结构如图 9-28 所示。项目中用到的 Struts 2.3.4、Hibernate 3.6.0、Spring 3.1.3 和 MySQL 5.0 驱动的 JAR 文件已添加到项目 sshLogin 的"库"中,如图 9-29 所示。

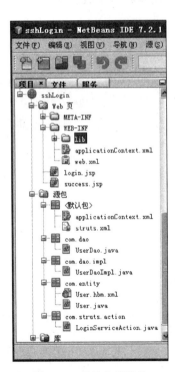

图 9-28 项目文件结构

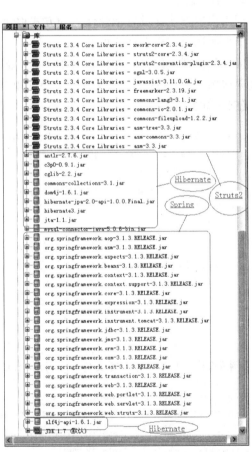

图 9-29 项目初始时加载的 SSH 的 JAR

在 WEB-INF 文件夹下新建一个 lib 文件夹,该文件夹中保存了 Hibernate 3.6.0 所需的 JAR 文件、Spring 3.1.3 所需的 JAR 文件、MySQL 的 JDBC 驱动。在库中加载 JAR 文件时可以直接从该文件夹中加载这些 JAR 文件到项目库中;另外,在 WEB-INF 文件夹下还有一个 applicationContext.xml 文件,这是因为在 web.xml 中配置的有加载该文件的参数,便于将来加载该配置文件。其中,web.xml 文件的代码如例 9-22 所示。有一个登录页面(login.jsp),代码如例 9-23 所示,登录页面对应的业务控制器类为 LoginServiceAction (在包 com.struts.action 中),代码如例 9-25 所示;如果登录成功跳转到登录成功页面 (success.jsp),代码如例 9-24 所示。在包 com.dao 中有一个接口(UserDao.java),代码如例 9-26 所示,该接口定义了登录功能的常用操作,该接口的实现类(UserDaoImpl.java)在包 com.dao.impl 中,代码如例 9-27 所示;PO 对象类(User.java)在包 com.entity 中,代码如例 9-28 所示,该 PO 对象对应的映射文件也在包 com.entity 中,名为 User.hbm.xml,代码如例 9-29 所示。Action 需要在 struts.xml 中配置,代码如例 9-30 所示。

另外还需要通过 hibernate.cfg.xml 和 applicationContext.xml 文件分别配置 Hibernate 3.6.0 和 Spring 3.1.3。在整合 SSH 技术的时候项目中共有三个配置文件,但是在实际项目开发中一般都会把对 Hibernate 的配置也放到 Spring 的配置文件中。本项目中对 Hibernate 和 Spring 的配置都在 applicationContext.xml 中完成,代码如例 9-31 所示。

本项目使用 MySQL 数据库,数据库名为 test,表名为 t_use_info,表的字段名称、类型以及长度定义如图 9-30 所示。

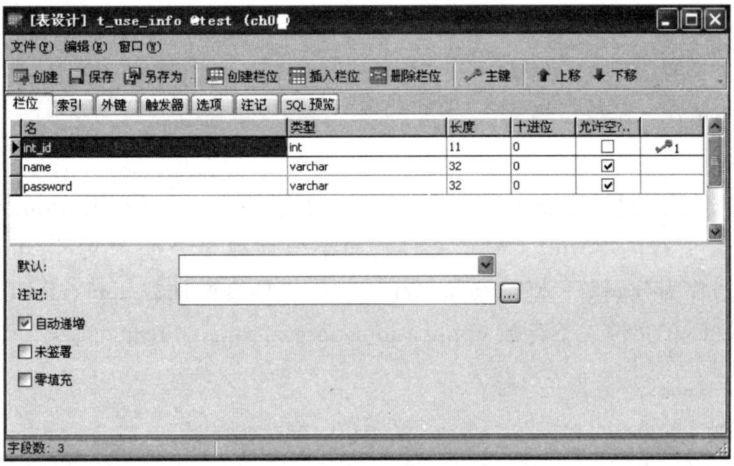

图 9-30 项目表的设计

9.3.2 在 web.xml 中配置 Struts2 和 Spring3

【例 9-22】 web.xml 的配置。

```
<?xml version="1.0" encoding="UTF-8"?>
<web-app version="3.0" xmlns="http://java.sun.com/xml/ns/javaee"
  xmlns:xsi="http://www.w3.org/2001/XMLSchema-instance" xsi:schemaLocation=
  "http://java.sun.com/xml/ns/javaee http://java.sun.com/xml/ns/javaee/web-app_
  3_0.xsd">
  <context-param>
      <param-name>contextConfigLocation</param-name>
      <param-value>/WEB-INF/applicationContext.xml</param-value>
  </context-param>
  <listener>
      <listener-class>
```

```xml
            org.springframework.web.context.ContextLoaderListener
        </listener-class>
    </listener>
    <filter>
        <filter-name>struts2</filter-name>
        <filter-class>org.apache.struts2.dispatcher.FilterDispatcher</filter-class>
    </filter>
    <filter-mapping>
        <filter-name>struts2</filter-name>
        <url-pattern>/*</url-pattern>
    </filter-mapping>
    <session-config>
        <session-timeout>
            30
        </session-timeout>
    </session-config>
    <welcome-file-list>
        <welcome-file>index.jsp</welcome-file>
    </welcome-file-list>
</web-app>
```

在 web.xml 中通过 applicationContext.xml 配置 Spring3，applicationContext.xml 中的元素定义了要加载的 Spring3 配置文件。如果想加载多个配置文件，可以在＜param-value＞中用逗号作为分隔符。默认会从/WEB-INF/文件夹中加载 applicationContext.xml，所以需要在 WEB-INF 文件夹下存放 applicationContext.xml，另外也可以如下配置：

```xml
<context-param>
    <param-name>contextConfigLocation</param-name>
    <param-value>classpath:applicationContext.xml</param-value>
</context-param>
```

这种形式的配置一般不需要在 WEB-INF 文件夹下保存 applicationContext.xml。

ContextLoaderListener 的作用就是启动 Web 容器时自动装配 ApplicationContext 的配置信息。因为它实现了 ServletContextListener 接口，在 web.xml 配置该监听器，启动容器时，就会默认执行它实现的方法。

9.3.3 编写视图组件

登录页面（login.jsp）代码如例 9-23 所示。

【例 9-23】 登录页面（login.jsp）。

```jsp
<%@page contentType="text/html" pageEncoding="UTF-8"%>
<%@taglib prefix="s" uri="/struts-tags"%>
<html>
    <head>
        <meta http-equiv="Content-Type" content="text/html; charset=UTF-8">
```

```
        <title>登录页面</title>
    </head>
    <body>
        <h1>欢迎登录:</h1>
        <div>
            <s:form action="login" method="post">
                <s:textfield name="username" label="账号"></s:textfield>
                <s:password name="password" label="密码"></s:password>
                <s:submit value="登录"></s:submit>
            </s:form>
        </div>
    </body>
</html>
```

登录成功页面代码如例 9-24 所示。

【例 9-24】 登录成功页面(success.jsp)。

```
<%@page contentType="text/html" pageEncoding="UTF-8"%>
<%@taglib prefix="s" uri="/struts-tags" %>
<html>
    <head>
        <meta http-equiv="Content-Type" content="text/html; charset=UTF-8">
        <title>JSP Page</title>
    </head>
    <body>
        <h1>欢迎您,<s:property value="username"></s:property>,登录成功!</h1>
    </body>
</html>
```

9.3.4 Action 和 JavaBean

【例 9-25】 登录页面对应的业务控制器类(LoginServiceAction.java)。

```
package com.struts.action;
import com.dao.UserDao;
import com.entity.User;
import com.opensymphony.xwork2.ActionSupport;
import java.util.List;
import org.springframework.context.ApplicationContext;
import org.springframework.context.support.ClassPathXmlApplicationContext;

public class LoginServiceAction extends  ActionSupport {
    private String username;
    private String password;
    ApplicationContext ctx=
    new ClassPathXmlApplicationContext("applicationContext.xml");
    UserDao userDao= (UserDao)ctx.getBean("userDao");
```

```java
    public String getUsername() {
        return username;
    }
    public void setUsername(String username) {
        this.username=username;
    }
    public String getPassword() {
        return password;
    }
    public void setPassword(String password) {
        this.password=password;
    }
    public String execute() throws Exception {
        //查找账号相符的用户
        List<User>userList=userDao.findByName(username);
        //使用简化的 for 语句对集合进行遍历并比较用户的密码
        for (User user : userList) {
            if (user.getPassword().equals(password)) {
                return  SUCCESS;
            }
            else
            {
                return ERROR;
            }
        }
        return  ERROR;
    }
}
```

【例 9-26】 登录功能的接口(UserDao.java)。

```java
package com.dao;
import com.entity.User;
import java.util.List;

public interface UserDao {
    /**
     * 加载 User 实例
     * @参数 id 指定需要加载的 User 实例的主键值
     * @return 返回加载的 User 实例
     */
    User get(Integer id);
    /**
     * 保存 User 实例
     * @参数 User 指定需要保存的 User 实例
     * @return 返回刚刚保存的 User 实例的标识属性值
```

```
     */
    Integer save(User user);
    /**
     * 根据用户名查找 User
     * @参数 name 指定查询的用户名
     * @return 返回用户名对应的全部 User
     */
    List<User> findByName(String name);
}
```

【例 9-27】 实现登录功能类(UserDaoImpl.java)。

```java
package com.dao.impl;
import com.dao.UserDao;
import com.entity.User;
import java.util.List;
import org.hibernate.SessionFactory;
import org.springframework.orm.hibernate3.HibernateTemplate;

public class UserDaoImpl implements UserDao {
    //实例化一个 HibernateTemplate 对象,用于执行持久化操作
    private  HibernateTemplate ht=null;
    //Hibernate 持久化操作所需 SessionFactory
    private SessionFactory sessionFactory=null;
    //用于依赖注入的 setter 方法
    public  void setSessionFactory(SessionFactory sessionFactory){
        this.sessionFactory=sessionFactory;
    }
    //初始化 HibernateTemplate 方法
    private HibernateTemplate getHibernateTemplate(){
        if (ht ==null) {
            ht=new HibernateTemplate(sessionFactory);
        }
        return ht;
    }
    public User get(Integer id) {
        //获取对应表中 id 为某个值的数据,id 为主键索引
        return getHibernateTemplate().get(User.class, id);
    }
    public Integer save(User user) {
        return  (Integer)getHibernateTemplate().save(user);
    }
    public List<User> findByName(String name) {
        //根据名称查找匹配的 User
        return (List<User>)getHibernateTemplate().find("from User u where u.name=?",
            name);
```

        }
    }

【例 9-28】 PO 对象(User.java)。

```java
package com.entity;
import java.io.Serializable;

public class User implements Serializable {
    private Integer intId;
    private String name;
    private  String password;
    public User(){}
    public User(Integer intId,String name,String password){
        this.intId=intId;
        this.name=name;
        this.password =password;
    }
    public Integer getIntId() {
        return intId;
    }
    public void setIntId(Integer intId) {
        this.intId=intId;
    }
    public String getName() {
        return name;
    }
    public void setName(String name) {
        this.name=name;
    }
    public String getPassword() {
        return password;
    }
    public void setPassword(String password) {
        this.password=password;
    }
}
```

【例 9-29】 PO 对象对应的映射文件(User.hbm.xml)。

```xml
<?xml version="1.0" encoding="UTF-8"?>
<!DOCTYPE hibernate-mapping PUBLIC
    "-//Hibernate/Hibernate Mapping DTD 3.0//EN"
    "http://hibernate.sourceforge.net/hibernate-mapping-3.0.dtd">
<hibernate-mapping>
  <class name="com.entity.User" table="t_use_info" catalog="test">
        <id name="intId" type="java.lang.Integer">
```

```xml
            <column name="int_id" />
            <generator class="increment"></generator>
        </id>
        <property name="name" type="java.lang.String">
            <column name="name" length="32" not-null="true" />
        </property>
        <property name="password" type="java.lang.String">
            <column name="password" length="32" not-null="true" />
        </property>
    </class>
</hibernate-mapping>
```

9.3.5 Struts2、Spring3 和 Hibernate 的配置文件

【例 9-30】 struts.xml 配置文件(struts.xml)。

```xml
<!DOCTYPE struts PUBLIC
"-//Apache Software Foundation//DTD Struts Configuration 2.0//EN"
"http://struts.apache.org/dtds/struts-2.0.dtd">
<struts>
    <package name="com.struts.action" extends="struts-default">
        <action name="login" class="com.struts.action.LoginServiceAction">
            <result name="success">/success.jsp</result>
            <result name="error">/login.jsp</result>
        </action>

    </package>
</struts>
```

【例 9-31】 Hibernate 和 Spring3 的配置(applicationContext.xml)。

```xml
<?xml version="1.0" encoding="UTF-8"?>
<beans xmlns="http://www.springframework.org/schema/beans"
       xmlns:xsi="http://www.w3.org/2001/XMLSchema-instance"
       xsi:schemaLocation="http://www.springframework.org/schema/beans
           http://www.springframework.org/schema/beans/spring-beans-3.0.xsd">
    <!--配置数据源-->
    <bean id="dataSource"
        class="com.mchange.v2.c3p0.ComboPooledDataSource">
        <property name="driverClass"
            value="com.mysql.jdbc.Driver"></property>
        <property name="jdbcUrl"
            value="jdbc:mysql://localhost:3306/test"></property>
        <property name="user" value="root"></property>
        <property name="password" value="root"></property>
        <!--指定数据库连接池的最大连接数-->
        <property name="maxPoolSize" value="40"/>
```

```xml
            <!--指定数据库连接池的最小连接数-->
            <property name="minPoolSize" value="1"/>
            <!--指定数据库连接池的初始化连接数-->
            <property name="initialPoolSize" value="1"/>
            <!--指定连接数据库连接池的连接的最大空闲时间-->
            <property name="maxIdleTime" value="20"/>
        </bean>
        <!--定义了Hibernate的SessionFactory-->
        <bean id="sessionFactory"
            class="org.springframework.orm.hibernate3.LocalSessionFactoryBean">
            <property name="dataSource">
                <ref bean="dataSource" />
            </property>
            <property name="mappingResources">
                <list>
                    <value>com/entity/User.hbm.xml</value>
                </list>
            </property>
            <property name="hibernateProperties">
                <props>
                    <prop key="hibernate.dialect">
                        org.hibernate.dialect.MySQLDialect
                    </prop>
                </props>
            </property>
        </bean>
        <!--HibernateTemplate类是简化Hibernate数据访问代码的辅助类,可以获取一个Session对象-->
        <bean id="hibernateTemplate"
            class="org.springframework.orm.hibernate3.HibernateTemplate">
            <property name="sessionFactory">
                <ref bean="sessionFactory"/>
            </property>
            <property name="allowCreate">
                <value>true</value>
            </property>
        </bean>
        <!--依赖注入-->
        <bean id="userDao" class="com.dao.impl.UserDaoImpl" >
            <!--注入持久化操作所需的SessionFactory-->
            <property name="sessionFactory">
                <ref bean="sessionFactory"/>
            </property>
        </bean>
</beans>
```

9.3.6　Struts2、Spring3 和 Hibernate 整合中常见问题

1. "java.lang.NoClassDefFoundError：org/apache/commons/logging/LogFactory"异常

java.lang.NoClassDefFoundError：org/apache/commons/logging/LogFactory 异常如图 9-31 所示,该异常与 9.2.3 节中图 9-21 所示异常类似,可通过加载 LogFactory 类解决,请参考 9.2.3 节。可以把需要的 JAR 文件先放到 WEB-INF 文件夹下,然后加载。

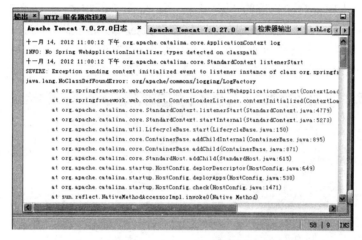

图 9-31　LogFactory 类异常

2. "java.lang.NoClassDefFoundError：org/objectweb/asm/ClassVisitor"异常

java.lang.NoClassDefFoundError：org/objectweb/asm/ClassVisitor 异常产生的主要原因是 Struts2、Spring3 和 Hibernate 三个框架中有重复的 JAR 文件。异常信息如图 9-32 所示。

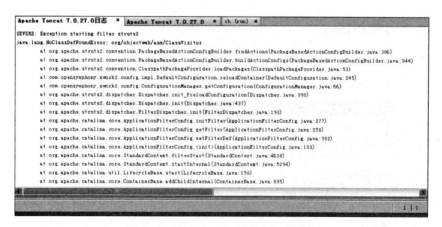

图 9-32　JAR 文件重复或者缺少异常

处理方案:出现该异常一般是因为 asm 和 cglib 在这三个框架中的 JAR 文件有重复。例如,asm 文件在 Struts 2.3.4 框架中对应的 JAR 文件为 asm-3.3.jar,在 Spring 3.1.3 中则为 org.springframework.asm-3.1.3.RELEASE.jar,这时需要删除 Struts 2.3.4 中的 asm-3.3.jar,但是有时在某些平台上没有出现冲突就不需要删除该文件。有时缺少该 JAR

文件也会出现图 9-31 中所示的异常。新版本的 Struts2 和 Spring3 通常都会在开发时已考虑解决冲突问题,所以类似的异常一般会出现在低版本的 Struts2 和 Spring3 中;cglib 文件在 Hibernate 3.6.0 框架中对应的 JAR 文件为 cglib-2.2.jar,在 Spring 3.1.1 中则为 org.springframework.cglib-2.2.RELEASE.jar,可删除 Hibernate 3.6.0 中的 cglib-2.2.jar 或者在加载 Hibernate 3.6.0 的 JAR 文件时不加载该文件。在 Spring 3.1.3 中已经弃用 cglib-2.2.jar 文件。

如果在开发平台上删除相应的 JAR 文件后还报图 9-32 中所示的异常,通常是因为与在开发平台上删除的 JAR 文件对应的、在系统生成的 lib 文件夹中的 JAR 文件并没有删除,还需到项目所在工作区找到如图 9-33 所示的 lib 文件夹,在 lib 文件夹中找到保存在文件夹中的重复 JAR 文件进行删除。删除后要重新编译或者发布项目。

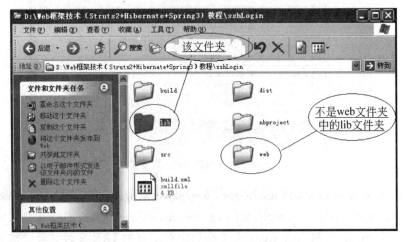

图 9-33　lib 文件夹

9.3.7　项目部署和运行

项目部署运行后,效果如图 9-34 所示,在其中输入账号和密码,如图 9-35 所示,单击"登录"按钮后,如果输入的数据在数据库中存在将成功登录,如图 9-36 所示。

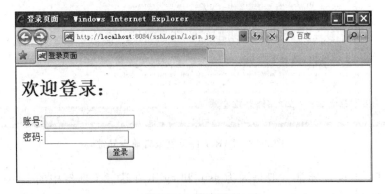

图 9-34　登录页面

图 9-35 输入数据

图 9-36 登录成功页面

9.4 本章小结

本章介绍了 Spring3 框架的基础知识并详细讲解了 Spring3 的核心技术 IoC,通过本章的学习,应了解和掌握以下内容。

(1) Spring3 框架的体系结构。
(2) IoC 框架的原理。
(3) IoC 框架的主要组件。
(4) IoC 的应用和注入。
(5) 基于 Struts2+Hibernate+Spring3 的登录系统。

9.5 习 题

9.5.1 选择题

1. Java Web 三大经典框架中用于降低模块之间耦合度的框架是(   )。
   A. Struts2   B. Spring3   C. Hibernate   D. JSF
2. Spring3 的核心部分是(   )。
   A. IoC   B. AOP   C. MVC 框架   D. ORM 框架
3. 依赖注入是(   )。
   A. AOP   B. ORM   C. DI   D. DAO

9.5.2 填空题

1. Spring 的上下文是一个_____。
2. IoC 主要组件有 Beans、配置文件、_____和_____。
3. 依赖注入的两种方式是_____和_____。

9.5.3 简答题

1. 简述 Spring3 框架各个组件的功能。
2. 简述 Spring3 框架的特点。

9.5.4 实训题

1. 把 9.2.4 节中设置注入实例改用构造注入方式完成。
2. 开发一个基于 SSH 框架的学生信息管理系统。

第 10 章 Spring3 的 AOP 框架

面向切面的编程(Aspect Oriented Programming,AOP)的主要目的是在业务处理过程中对切面进行提取,所关注的是处理过程中的某个阶段或步骤,降低各业务逻辑之间的耦合度。可以通过预编译方式和运行时动态代理,实现在不修改源代码的情况下统一给程序动态添加功能。

本章主要内容如下所示。

(1) AOP 基础知识以及 Spring AOP 的主要术语。
(2) 通知(Advice)及应用实例。
(3) 切点(Pointcut)及应用实例。
(4) 引入(Introduction)。

10.1 AOP 框架基础知识

AOP 是目前软件开发中的一个热点,也是 Spring3 框架中的一个重要内容。利用 AOP 可以对业务逻辑的各个部分进行隔离,从而使得业务逻辑各部分之间的耦合度降低,提高程序的可重用性,同时提高开发的效率。

10.1.1 AOP 框架简介

面向过程编程离人们已经有些"遥远",面向对象编程正"主宰着"软件世界。当每个新的软件设计师都被要求掌握如何将功能需求转换成一个个类,并且定义它们的数据变量、方法,以及它们之间的复杂关系的时候,面向切面编程(AOP)为人们带来了新的想法、新的思想、新的模式。

如果说面向对象编程是关注将功能需求划分为不同的并且相对独立、封装良好的类,并让它们有着属于自己的行为,依靠继承和多态等来定义彼此关系的话;那么面向切面编程则是希望能够将通用功能需求从不相关的类当中分离出来,能够使得很多类共享一个行为,一旦发生变化,不必修改很多类,而只需修改这个行为即可。

面向切面编程是一个令人兴奋不已的新模式。就开发软件系统而言,它的影响力必将会和有着十数年应用历史的面向对象编程一样巨大。面向切面编程和面向对象编程不但不是互相竞争的技术而且彼此还是很好的互补。面向对象编程主要用于为同一对象层次的公用行为建模。它的弱点是将公共行为应用于多个无关对象模型之间。而这恰恰是面向切面编程适合的地方。有了 AOP,人们可以定义交叉的关系,并将这些关系应用于跨模块的、彼此不同的对象模型。AOP 同时还可以让人们层次化功能性而不是嵌入功能性,从而使得代码有更好的可读性和易于维护。它会和面向对象编程合作得很好。

传统的程序通常表现出一些不能自然地适合单一的程序模块或者是几个紧密相关的程序模块的行为,AOP 将这种行为称为切面,它们跨越了给定编程模型中的典型职责界限。

切面行为的实现都是分散的,软件设计师会发现这种行为难以用正常的逻辑来思考、实现和更改。

AOP 主要的功能是日志记录、性能统计、安全控制、事务处理、异常处理等。主要的意图是将日志记录、性能统计、安全控制、事务处理、异常处理等代码从业务逻辑代码中划分出来,通过对这些行为的分离,希望可以将它们独立到非指导业务逻辑的方法中,进而改变这些行为的时候不影响业务逻辑的代码。

在初次接触 AOP 的时候可能会考虑,AOP 能做到的,一个定义良好的 OOP 的接口也一样能够做到。AOP 和定义良好的 OOP 的接口可以说都是用来解决并且实现需求中的切面问题的方法。但是对于 OOP 中的接口来说,它仍然需要人们在相应的模块中去调用该接口中相关的方法,这是 OOP 所无法避免的,并且一旦接口不得不进行修改的时候,所有事情会变得一团糟;AOP 则不会这样,只需要修改相应的 Aspect,再重新编织(Weaving)即可。当然,AOP 也绝对不会代替 OOP。核心的需求仍然会由 OOP 来加以实现,而 AOP 将会和 OOP 整合起来,以此之长,补彼之短。

AOP 是一个概念,是一个规范,本身并没有设定具体语言的实现,这实际上提供了非常广阔的发展空间。

目前有两种主流的 AOP 实现:静态 AOP 和动态 AOP。静态 AOP,比如 AspectJ,提供了编译器的方法来构建基于 AOP 的逻辑,并把它加入到应用程序中。动态 AOP,比如 Spring AOP,允许在运行时把服务逻辑应用到任意一段代码中。两种不同的 AOP 方法都有其适用面。

目前,AOP 的实现有很多,常见的有 AspectJ、AspectWerkz、JBoss AOP 和 Spring AOP。

1. AspectJ

AspectJ 是一个面向切面的框架,它扩展了 Java 语言。AspectJ 定义了 AOP 语法,所以它有一个专门的编译器用来生成遵守 Java 字节编码规范的 class 文件。

AspectJ(也就是 AOP)的动机是发现那些使用传统的编程方法无法很好处理的问题。考虑一个要在某些应用中实施安全策略的问题。安全性是贯穿于系统所有模块间的问题,每个模块都需要应用安全机制才能保证整个系统的安全性,很明显这里的安全策略的实施问题就是一个横切关注点,使用传统的编程解决此问题非常困难而且容易产生差错,这就正是 AOP 发挥作用的时候了。

AspectJ 框架 2001 年发布,最新版本 1.7。如需进一步了解有关 AspectJ 框架的内容,可以访问其官方网站(www.eclipse.org/aspectj)。

2. AspectWerkz

AspectWerkz 发布于 2002 年,是基于 Java 的简单、动态、轻量级、强大的 AOP 框架。既强大又简单,有助于更容易地集成 AOP 到新的或已存在的项目中。现在,AspectJ 和 AspectWerkz 项目已经合并,以便整合两者的优势打造统一的 AOP 平台。两者整合后的第一个版本是 AspectJ5,并于 2005 年发布。

3. JBoss AOP

2004 年 JBoss 4.0 应用程序服务器中提供支持 AOP 技术。如需了解有关 JBoss AOP 的更多内容,可以访问其官方网站,地址为 www.jboss.org/products/aop。

4. Spring AOP

Spring AOP 使用纯 Java 开发,不需要专用的编译过程和特殊的类装载器,在运行期间通过代理方式向目标织入(Weaving)代码。Spring 并不尝试提供最完整的 AOP 实现,而是侧重提供一种和 Spring IoC 容器整合的 AOP 实现,来解决企业级开发中的常见问题。

10.1.2　Spring3 的 AOP 框架主要术语

Spring3 的 AOP 是在一个服务的流程中,插入与该服务的业务逻辑无关的系统服务逻辑,如日志、安全等。这样的逻辑称为横切关注点(Cross Cutting Concern),将横切关注点独立出来设计为一个对象,这样的特殊对象称为切面(方面)。这样的服务逻辑可应用到应用程序的很多地方。在 AOP 中,它们只需要编写一次,就可以将其自动在整个目标应用程序中实施。AOP 重点关注切面的设计及它们在目标应用程序中的织入(Weaving)。

综上所述,AOP 的许多术语都过于抽象,不容易理解。但是这些术语构成了 AOP 的基础,下面对 AOP 的主要术语进行介绍。

1. 横切关注点(Crosscutting Concern)

像安全检查、事务等系统层面的服务,常被安插到一些程序中各个对象的处理流程中,这样的服务逻辑在 AOP 中称为横切关注点,又称为横切关心。

如果把横切关注点直接编写在负责某业务的对象流程中,会使得维护程序的成本增高。假如某天要从该对象中修改或移除该服务逻辑,需要修改所有与该服务相关的程序代码,然后重新编译;另外,把横切关注点混杂于业务逻辑中,会使得业务对象本身的逻辑或程序的编写更为复杂。

AOP 通过把横切关注点织入到业务逻辑中,比较成功地解决了以上问题。

2. 切面(Aspect)

将横切关注点设计为独立可重用的对象,这些对象称为切面。

3. 连接点(Joinpoint)

切面在应用程序执行时加入目标对象的业务流程中的特定点,称为连接点(Joinpoint)。连接点是 AOP 的核心概念之一。它用来定义在目标程序的哪里通过 AOP 加入新的逻辑。

4. 通知(Advice)

切面在某个具体连接点采取的行为或动作,称为通知。

实际上,通知是切面的具体实现。它是在某一特定的连接点处织入目标业务程序中的服务对象。它又分为在连接点之前执行的前置通知(Before Advice)和在连接点之后执行的后置通知(After Advice)。

5. 切入点(Pointcut)

切入点指定某个通知在哪些连接点被织入到应用程序之中。

切入点是通知要被织入到应用程序中的所有连接点的集合。可以在一个文件中,例如 XML 文件中,定义一个通知的切入点,即它的所有连接点。

6. 织入(Weaving)

将通知加入应用程序的过程,称为织入。对于静态 AOP 而言,织入是在编译时完成的,通常在编译过程中增加一个步骤,而动态 AOP 是在程序运行时动态织入的。

7. 目标（Target）

通知被应用的对象，称为目标。

8. 引入（Introduction）

通过引入，我们可以在一个对象中加入新的方法和属性，而不用修改它的程序。

具体说，可以为某个已编写、编译完成的类，在执行时期动态加入一些方法或行为，而不用修改或新增任何一行程序代码。

9. 代理（Proxy）

代理是由 AOP 框架生成的一个对象，用来执行切面的内容。代理有静态代理和动态代理。

10.2 代 理

10.2.1 静态代理

程序中经常需要为某些动作或者时间保存记录，以便随时检查程序的运行情况，排除错误信息，例 10-1 中代码就是含有日志的动作。

【例 10-1】 HelloSpeaker 类（HelloSpeaker.java）。

```
package proxyexample;
//导入类库中的 Logger 类,日志类
import java.util.logging.Level;
import java.util.logging.Logger;

public class HelloSpeaker {
    //通过 Logger 类的静态方法 Logger.getLogger()来实例化并获取对象
    private Logger logger=Logger.getLogger(this.getClass().getName());
    public void hello(String name)
    {
        logger.log(Level.INFO,"hello()方法开始执行");      //方法开始执行时留下日志
        System.out.println("hello,"+name);                //程序的主要功能
        logger.log(Level.INFO,"hello()方法执行结束");      //方法执行完成时留下日志
    }
}
```

在例 10-1 中，因为方法执行前后都加上了日志动作，所以当执行 hello()方法时，该方法执行开始和执行完成时都会留下日志。然而对于 HelloSpeaker 类来说，日志动作并不是 HelloSpeaker 类的业务逻辑部分，这样就增加了 HelloSpeaker 类的额外负担。

如果程序中的这种日志动作到处都需要，上述实现方法就使得开发人员不得不到处写日志动作，这样就导致维护日志代码比较困难。如果需要的服务不只是日志动作，例如，非本类职责的相关动作也需要加入到类中，如权限控制、事务管理等，会导致类的负担加重，甚至混淆类本身的功能（职责）。另外，如果程序中不再需要日志或者权限控制时，将需要修改所有已编写的相关动作，无法简单地将这些相关服务从现有的程序中删除。

为了解决上述问题，可以使用代理（Proxy）机制。代理方式有两种：静态代理（Static

Proxy)和动态代理(Dynamic Proxy)。

在静态代理的实现中,代理类和被代理的类必须实现同一个接口。在代理类中可以实现记录、权限等服务,并在需要时呼叫(使用)被代理的类。这样被代理的类就可以仅仅保留业务逻辑相关的功能(职责)。

下面通过一个项目介绍静态代理的使用。

1. 项目介绍

项目是一个 Java 应用程序,命名为 ProxyExample,该项目下有一个接口 Hello,代码如例 10-2 所示;实现接口的业务逻辑类为 HelloSpeaker1,在该类中没有插入任何日志代码,如例 10-3 所示,日志服务的实现将被放到代理类中,代理类同样要实现接口;实现接口的代理类为 HelloProxy,在该类的 hello()方法中,实现为业务逻辑添加记录服务,代码如例 10-4 所示;测试该项目的类为 Test,代码如例 10-5 所示。项目的文件结构以及运行结果如图 10-1 所示。

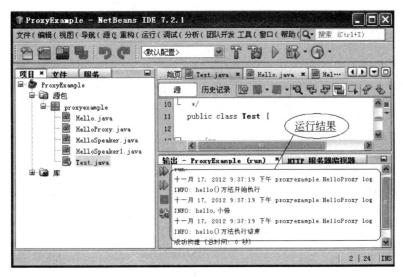

图 10-1　项目的文件结构以及运行结果

2. 接口 Hello 的代码

【例 10-2】　Hello 接口(Hello.java)。

```
package proxyexample;

public interface Hello {
    void hello(String name);
}
```

3. 实现接口类的代码

【例 10-3】　HelloSpeaker1 类(HelloSpeaker1.java)。

```
package proxyexample;

public class HelloSpeaker1 implements Hello{
```

```java
    public void hello(String name){
        System.out.println("hello,"+name);
    }
}
```

请比较例 10-1 与例 10-3。

【例 10-4】 HelloProxy 类（HelloProxy.java）。

```java
package proxyexample;
import java.util.logging.Level;
import java.util.logging.Logger;

public class HelloProxy implements Hello{
    private Hello helloObject;
    private Logger logger=Logger.getLogger(this.getClass().getName());
    public HelloProxy(Hello helloObject)
    {
        this.helloObject=helloObject;
    }
    public void log(String msg)
    {
        logger.log(Level.INFO,msg);
    }
    public void hello(String name)
    {
        log("hello()方法开始执行");            //日志服务
        log("hello,"+name);                    //执行业务逻辑功能
        log("hello()方法执行结束");            //日志服务
    }
}
```

4．测试类

【例 10-5】 Test 类（Test.java）。

```java
package proxyexample;

public class Test {
    public static void main(String[]args) {
        Hello proxy=new HelloProxy(new HelloSpeaker1());
        proxy.hello("小强");
    }
}
```

5．测试结果

运行 Test 类,运行结果如图 10-1 所示。

从图 10-1 所示的运行结果可见,静态代理的一个接口只能服务于一种类型的类,如果要代理的方法很多,需要为每个方法进行代理。很明显,静态代理无法满足大规模的程序,

因此还需要动态代理。

10.2.2 动态代理

从 JDK 1.3 起就提供了开发动态代理的类库。使用动态代理不需要为特定的类和方法编写特定的代理类,可以使用一个处理者(Handler)为各个类服务。

要实现动态代理,也不需要定义所要代理的接口。

下面通过一个项目介绍动态代理的使用。

1. 项目介绍

本项目是一个 Java 应用程序,命名为 ProxyExample1,该项目下有一个接口 Hello;实现接口的业务逻辑类为 HelloSpeaker1,代码与 10.2.1 节中的 Hello 和 HelloSpeaker1 类似;处理者类为 LogHandler,不是代理类,代码如例 10-6 所示;测试该项目的类为 Test,代码如例 10-7 所示。项目的文件结构以及运行结果如图 10-2 所示。

图 10-2 项目的文件结构以及运行结果

2. 接口 Hello 的代码

Hello 接口的代码与 10.2.1 节的例 10-2 类似,只需将包名改为"package proxyexample1;"。

3. 实现接口类的代码

HelloSpeaker1 类的代码与 10.2.1 节的例 10-3 类似,只需将包名改为"package proxyexample1;"。

4. 处理者类 LogHandler

【例 10-6】 处理者类 LogHandler(LogHandler.java)。

```
package proxyexample1;
/*
导入 InvocationHandler,该接口是代理实例的调用处理程序实现的接口。每个代理实例都具有一个关联的调用处理程序。对代理实例调用方法时,将对方法调用进行编码并将其指派到它的调用处理程序的 invoke()方法。
*/
import java.lang.reflect.InvocationHandler;
```

```java
import java.lang.reflect.Method;

public class LogHandler implements InvocationHandler{
    private Object object;
    public LogHandler(){
    }
    public LogHandler(Object obj){
        object=obj;
    }
    /*
    实现接口 InvocationHandler 中的方法,在代理实例上处理方法调用并返回结果。在与方法关联的代理实例上调用方法时,将在调用处理程序上调用此方法。参数 proxy:在其上调用方法的代理实例。参数 method:对应于在代理实例上调用的接口方法的 Method 实例。Method 对象的声明类将是在其中声明方法的接口,该接口可以是代理类赖以继承方法的代理接口的超口。参数 args:包含传入代理实例上方法调用的参数值的对象数组,如果接口方法不使用参数,则为 null。基本类型的参数被包装在适当基本包装器类(如 java.lang.Integer 或 java.lang.Boolean)的实例中。
    */
    public Object invoke(Object proxy,Method method,Object[]args) throws Throwable{
        System.out.println("invoke()方法开始执行...");
        method.invoke(object, args);
        System.out.println("invoke()方法执行完成...");
        return null;
    }
}
```

5. 测试类

【例 10-7】 Test 类(Test.java)。

```java
package proxyexample1;
/*
导入 Proxy 类,该类提供用于创建动态代理类和实例的静态方法,它还是由这些方法创建的所有动态代理类的超类。
*/
import java.lang.reflect.Proxy;

public class Test {
    public static void main(String[]args) {
        HelloSpeaker1 helloSpeaker=new HelloSpeaker1();
        LogHandler logHandler=new LogHandler(helloSpeaker);
        Class cla=helloSpeaker.getClass();
        /*
        newProxyInstance()方法返回一个指定接口的代理类实例,该接口可以将方法调用指派到指定的调用处理程序。
        */
        Hello hello=(Hello)Proxy.newProxyInstance(
                cla.getClassLoader(),cla.getInterfaces(),logHandler);
```

```
        hello.hello("小强");
    }
}
```

6. 测试结果

运行 Test 类,运行结果如图 10-2 所示。

从该项目中可以看出,HelloSpeaker1 类本身的职责是显示文字,日志的程序代码横切(Crosscutting)到 HelloSpeaker1 的程序执行流程中,日志这样的动作被称为横切关注点(Crosscutting Concern)。HelloProxy 类和 LogHandler 类称为切面(Aspect)。

10.3 创建通知

Spring3 提供了 5 种通知(Advice)类型:前置通知(Before Advice)、后置通知(After Advice)、环绕通知(Around Advice)、异常通知(Throws Advice)和引入通知(Introduction Advice)。如果要为目标对象提供通知,则必须为它们建立代理(Proxy)对象。在 Spring3 中,为了创建代理,需要使用代理工厂(Proxy Factory)。

常用的代理工厂是 org.springframework.aop.framework.ProxyFactoryBean。ProxyFactoryBean 是 Sprmg3 提供的一个类,它是在 Spring IoC 环境中创建代理的最底层和最灵活的方法。它有 2 个最重要的属性:proxyInterfaces 和 interceptorNames。proxyInterfaces 属性指定被代理的接口;interceptorNames 属性指定共同建立拦截器链的通知或通知器的名字列表。

下面以实例来介绍通知的使用。在一家公司里,经常有客户来来往往,有接待人员对其接待。接待人员负责服务客户,接待人员的接口为 Reception,代码如例 10-8 所示;实现接口的类为 ConcreteReception,代码如例 10-9 所示。

【例 10-8】 Reception 接口(Reception.java)。

```
package beforeadviceexample;

public interface Reception {
    void serveCustomer(String customerName);           //为客户提供的服务方法
}
```

【例 10-9】 ConcreteReception 类(ConcreteReception.java)。

```
package beforeadviceexample;

public class ConcreteReception implements Reception{
    public void serveCustomer(String customerName){//为客户服务
        System.out.println("我正在为客户服务:"+ customerName+ "。(接待中)");
    }
}
```

10.3.1 前置通知及应用实例

前置通知在连接点(或目标对象的方法调用)之前执行。例如,客户来公司时,接待人员见到客户首先要做见面问候，一般不会直接上来与客户谈生意，但是有的接待人员未必能做到一定见面首先相互问候，因此公司有可能会损失一部分客户(客户会认为太不重视他啦)，所以这里需要实现的是在见到客户时，首先必须礼貌问候，简单地说就是在调用 serveCustomer() 方法之前执行一些动作，即问候客户。

Spring3 提供了一般的 org.spring.framework.aop.BeforeAdvice 接口，可以与任何类型的连接点一起使用。但实际应用中一般通过实现了 BeforeAdvice 接口的子接口 org.spring.framework.aop.MethodBeforeAdvice 来实现前置通知的逻辑，因为该子接口总是用于方法拦截的。

MethodBeforeAdvice 接口的声明如下：

```
public interface MethodBeforeAdvice extends BeforeAdvice{
  void before(Method method,Object[ ] args,Object target) throws Throwable;
}
```

MethodBeforeAdvice 接口可以获取目标方法、参数以及目标值，但是不能改变这些值。
before() 方法会在目标对象(Target)所指定的方法执行之前被执行。该方法返回 void，意味着它不传回任何结果。在 before() 方法执行完毕之后，除非抛出异常，否则目标对象的方法就会被执行。

因此实现 MethodBeforeAdvice 接口就可以在调用 ConcreteReception 类的 serveCustomer() 方法之前执行一些动作，即问候客户，实现该接口的类为 GettingBeforeAdvice，代码如例 10-10 所示。为了降低耦合度，在 Spring3 的配置文件中声明一个代理，配置文件为 applicationContext.xml，代码如例 10-11 所示。为了测试前置通知的实现，写一个测试类 Test，代码如例 10-12 所示。前置通知实例是一个 Java 应用程序项目，项目名为 BeforeAdviceExample，如果使用的 NetBeans7 集成了 Spring 3.0 或者 Spring 3.1 或者使用 Spring 3.1.3，都需要使用第三方插件，即 aopalliance-1.0.jar 文件，否则将报异常信息为 "java.lang.NoClassDefFoundError：org/aopalliance/aop/Advice" 的异常。aopalliance-1.0.jar 下载地址为 http://mirrors.ibiblio.org/pub/mirrors/maven2/aopalliance/aopalliance/1.0/。另外，还需要 Apache 的 LogFactory 类库，该类库需要在 www.apache.org 的网站下载，否则将抛出 "java.lang.NoClassDefFoundError：org/apache/commons/logging/LogFactory" 异常，请参考 9.2.3 节。

项目的文件结构、使用的 JAR 文件以及运行结果如图 10-3 所示。

【例 10-10】 GettingBeforeAdvice 类(GettingBeforeAdvice.java)。

```
package beforeadviceexample;
//Spring3 中的类,需要先加载 Spring3 类库
import java.lang.reflect.Method;
import org.springframework.aop.MethodBeforeAdvice;

public class GettingBeforeAdvice implements MethodBeforeAdvice{
    public void before(Method method,Object[] args,Object target) throws Throwable
{
        String customerName= (String)args[0];
```

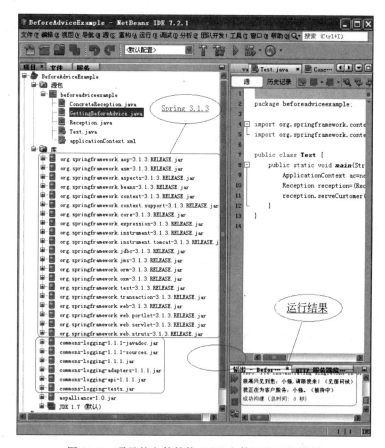

图 10-3 项目的文件结构、JAR 文件以及运行结果

```
        System.out.println("很高兴见到您:"+customerName+",请跟我来!(见面问候)");
    }
}
```

【例 10-11】 配置文件(applicationContext.xml)。

```xml
<?xml version="1.0" encoding="UTF-8"?>
<!DOCTYPE beans PUBLIC "-//SPRING//DTD BEAN 2.0//EN"
http://www.springframework.org/dtd/spring-beans-2.0.dtd
[<!ENTITY contextInclude SYSTEM
   "org/springframework/web/context/WEB-INF/contextInclude.xml">]>
<beans>
    <bean id="gettingBeforeAdvice"
        class="beforeadviceexample.GettingBeforeAdvice">
    </bean>
    <!--使用 Spring 代理工厂配置一个代理-->
    <bean id="reception"
        class="org.springframework.aop.framework.ProxyFactoryBean">
        <!--指定代理接口,如果有多个接口可以使用 list 元素指定-->
        <property name="proxyInterfaces"
```

```xml
            value="beforeadviceexample.Reception"></property>
        <!--指定通知-->
        <property name="interceptorNames"
            value="gettingBeforeAdvice"></property>
        <!--指定目标对象-->
        <property name="target" ref="target"></property>
    </bean>
    <bean id="target" class="beforeadviceexample.ConcreteReception"></bean>
</beans>
```

【例 10-12】 Test 类(Test.java)。

```java
package beforeadviceexample;
import org.springframework.context.ApplicationContext;
import org.springframework.context.support.FileSystemXmlApplicationContext;

public class Test {
    public static void main(String[]args) {
        ApplicationContext ac=new FileSystemXmlApplicationContext ( " src/beforeadviceexample/applicationContext.xml");
        Reception reception= (Reception)ac.getBean("reception");     //生成代理对象
        reception.serveCustomer("小强");                              //接待人员接到客户,提供服务
    }
}
```

运行 Test.java 后的结果如图 10-3 所示。

10.3.2 后置通知及应用实例

与前置通知相反,后置通知是在目标方法执行之后被调用。例如,在 10.3.1 节中讲到客户接待,当客户离开公司时,要求接待人员必须送别,这就可以使用后置通知。可以实现 AfterReturningAdvice 接口来实现后置通知的逻辑。

该接口的定义声明如下:

```java
public interface org.spring.framework.aop.AfterReturningAdvice extends Advice {
    void afterReturning (Object teturnValue, Method method, Object [ ] args, Object target) throws Throwable;
}
```

AfterReturningAdvice 接口继承 Advice 接口。后置通知使程序员有机会得到调用的方法、传入自变量以及目标对象。afterReturning()方法的返回值是 void,当抛出异常时,程序就中止执行。

可以在 10.3.1 节实例中加入后置通知,将实现后置通知的类命名为 GettingAfter-Advice,代码如例 10-13 所示。修改后的配置文件代码如例 10-14 所示,项目的文件结构以及运行结果如图 10-4 所示。

【例 10-13】 GettingAfterAdvice 类(GettingAfterAdvice.java)。

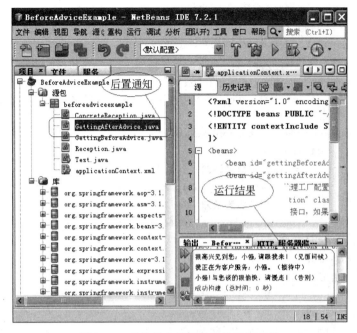

图 10-4 项目的文件结构以及运行结果

```
package beforeadviceexample;
import java.lang.reflect.Method;
import org.springframework.aop.AfterReturningAdvice;

public class GettingAfterAdvice implements AfterReturningAdvice{
    public void afterReturning(Object teturnValue, Method method, Object[] args,
Object target) throws Throwable{
        String customerName=(String)args[0];
        System.out.println(customerName+"!与您谈得很愉快,请慢走!(告别)");
    }
}
```

【例 10-14】 修改后的配置文件(applicationContext.xml)。

```
<?xml version="1.0" encoding="UTF-8"?>
<!DOCTYPE beans PUBLIC "-//SPRING//DTD BEAN 2.0//EN"
http://www.springframework.org/dtd/spring-beans-2.0.dtd
[<!ENTITY contextInclude SYSTEM
"org/springframework/web/context/WEB-INF/contextInclude.xml">]>
<beans>
    <bean id="gettingBeforeAdvice"
        class="beforeadviceexample.GettingBeforeAdvice">
    </bean>
    <bean id="gettingAfterAdvice"
        class="beforeadviceexample.GettingAfterAdvice">
    </bean>
```

```xml
<!--使用Spring代理工厂配置一个代理-->
<bean id="reception"
    class="org.springframework.aop.framework.ProxyFactoryBean">
    <!--指定代理接口,如果有多个接口可以使用list元素指定-->
    <property name="proxyInterfaces"
        value="beforeadviceexample.Reception"></property>
    <!--指定通知-->
    <property name="interceptorNames">
        <list>
            <!--指定前置通知-->
            <value>gettingBeforeAdvice</value>
            <!--指定后置通知-->
            <value>gettingAfterAdvice</value>
        </list>
    </property>
    <!--指定目标对象-->
    <property name="target" ref="target"></property>
</bean>
<bean id="target" class="beforeadviceexample.ConcreteReception"></bean>
</beans>
```

10.3.3 环绕通知及应用实例

环绕通知是最常用的通知类型。前置通知和后置通知分别在目标类的前后织入通知，但是如果需要同时使用这两种类型的通知，不必同时实现这两种类型通知，可以使用环绕通知实现同样效果。

通过实现 MethodInterceptor 接口可以实现环绕通知。环绕通知与前置通知和后置通知的区别如下。

(1) MethodInterceptor 能够控制目标方法是否真的被调用，通过调用方法 MethodInvocation.proceed()来调用目标方法。这一点不同于接口 MethodBeforeAdvice，该接口目标方法总是被调用，除非抛出异常。

(2) MethodInterceptor 可以控制返回的对象，可以返回一个与 proceed()方法返回的是同一个对象。

环绕通知应用实例：项目名称为 AroundAdviceExample。接待人员接口为 Reception，代码同前，实现该接待人员接口的代码类为 ConcreteReception，代码同前。实现环绕通知的实现类为 GettingAroundAdvice，代码如例 10-15 所示。配置文件 applicationContext.xml 的代码如例 10-16 所示。测试类 Test 代码同前，项目的文件结构、JAR 文件以及运行结果如图 10-5 所示。

【例 10-15】 GettingAroundAdvice 类（GettingAroundAdvice.java）。

```
package aroundadviceexample;
import org.aopalliance.intercept.MethodInterceptor;
import org.aopalliance.intercept.MethodInvocation;
```

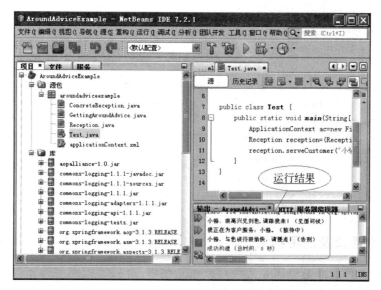

图 10-5　项目的文件结构、JAR 文件以及运行结果

```
public class GettingAroundAdvice implements MethodInterceptor{
    public Object invoke(MethodInvocation invocation) throws Throwable{
        Object args[]=invocation.getArguments();
        String customerName= (String)args[0];
        //在目标方法执行前调用
        System.out.println(customerName+",很高兴见到您,请跟我来!(见面问候)");
        Object object=invocation.proceed();          //通过反射调用执行方法
        //在目标方法执行后调用
        System.out.println(customerName+",与您谈得很愉快,请慢走!(告别)");
        return object;
    }
}
```

【例 10-16】 配置文件(applicationContext.xml)。

```xml
<?xml version="1.0" encoding="UTF-8"?>
<!DOCTYPE beans PUBLIC "-//SPRING//DTD BEAN 2.0//EN"
http://www.springframework.org/dtd/spring-beans-2.0.dtd
[<!ENTITY contextInclude SYSTEM
"org/springframework/web/context/WEB-INF/contextInclude.xml">]>
<beans>
    <bean id="gettingAroundAdvice"
        class="aroundadviceexample.GettingAroundAdvice">
    </bean>
    <bean id="reception"
        class="org.springframework.aop.framework.ProxyFactoryBean">
        <property name="proxyInterfaces"
            value="aroundadviceexample.Reception"></property>
```

```xml
            <!--指定通知-->
            <property name="interceptorNames">
                <list>
                    <!--指定环绕通知-->
                    <value>gettingAroundAdvice</value>
                </list>
            </property>
            <!--指定目标对象-->
            <property name="target" ref="target"></property>
        </bean>
        <bean id="target" class="aroundadviceexample.ConcreteReception"></bean>
</beans>
```

10.3.4 异常通知及应用实例

当异常发生时，如果希望通知某些对象做某些事，可以使用异常通知。可以通过实现ThrowsAdvice接口来实现异常通知。但该接口没有定义任何方法，只是一个标签接口。可以在其中定义任意方法，只要它是以下的形式：

```
methodName([Method],[args],[target],subclassOfthrowable);
```

方括号[ ]中的参数，例如，Method、args、target 等，是可以省略的，但一定要有subclassOfthrowable，这个参数必须是 Throwable 的子类，在异常发生时，它会检验所设定的异常通知中是否有符合异常类型的方法，如果有就通知它执行。

下面是2个方法声明的例子：

```
void beforeThrowing(Throwable throwable);
void beforeThrowing(Method method,Object[ ]obj,Object target,Throwable throwable);
```

在方法中如果声明了不同的 Throwable 类型，则发生的异常类型不同，会通知不同的方法。例如，WalkerException 会通知声明有 WalkerException 参数的方法，而 HikerException 会通知声明有 HikerException 的方法。

当异常发生时，Throws 通知的任务只是执行对应的方法，并不能在 Throws 通知中将异常处理掉。

在 Throws 通知执行完毕后，原先的异常仍被传播到应用程序之中，Throws 通知并不介入应用程序的异常处理。异常处理仍旧是应用程序本身所要负责的。

如果想要在 Throws 通知处理时终止应用程序的处理流程，做法是抛出其他的异常。

异常通知应用实例：项目名称为 ThrowsAdviceExample。有一个接口为 Business，代码如例 10-17 所示，实现该接口的代码类为 BusinessCode，代码如例 10-18 所示。实现异常通知的类为 GettingThrowsAdvice，代码如例 10-19 所示。配置文件为 applicationContext.xml，代码如例 10-20 所示。测试类为 Test，代码如例 10-21 所示，项目的文件结构、JAR 文件以及运行结果如图 10-6 所示。

【例 10-17】 Business 接口（Business.java）。

```
package throwsadviceexample;
```

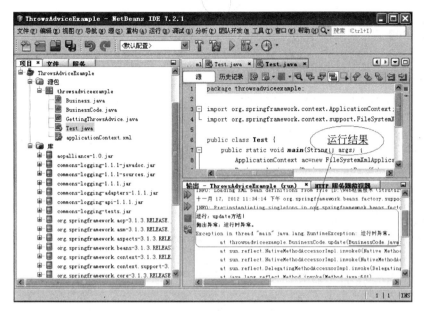

图 10-6　项目的文件结构、JAR 文件以及运行结果

```
public interface Business {
    void update();                              //提供数据更新方法
}
```

【例 10-18】 实现 Business 接口的类 BusinessCode(BusinessCode.java)。

```
package throwsadviceexample;

public class BusinessCode implements Business{
    public void update(){                       //数据更新方法
        //为了简化程序,业务逻辑代码省略
        ⋮
        throw new RuntimeException("运行时异常。");
    }
}
```

【例 10-19】 实现异常通知的类 GettingThrowsAdvice(GettingThrowsAdvice.java)。

```
package throwsadviceexample;
import java.lang.reflect.Method;
import org.springframework.aop.ThrowsAdvice;

public class GettingThrowsAdvice implements ThrowsAdvice{
    public void afterThrowing(Method method,Object[]obj,Object target, Exception re){
        System.out.println("运行:"+method.getName()+"方法!");
        System.out.println("抛出异常:"+re.getMessage());
    }
```

}

【例 10-20】 配置文件(applicationContext.xml)。

```xml
<?xml version="1.0" encoding="UTF-8"?>
<!DOCTYPE beans PUBLIC "-//SPRING//DTD BEAN 2.0//EN"
http://www.springframework.org/dtd/spring-beans-2.0.dtd
[<!ENTITY contextInclude SYSTEM
   "org/springframework/web/context/WEB-INF/contextInclude.xml">]>
<beans>
    <bean id="gettingThrowsAdvice"
        class="throwsadviceexample.GettingThrowsAdvice">
    </bean>
    <bean id="throws"
        class="org.springframework.aop.framework.ProxyFactoryBean">
        <property name="proxyInterfaces"
            value="throwsadviceexample.Business"></property>
        <!--指定通知-->
        <property name="interceptorNames">
            <list>
                <!--指定异常通知-->
                <idref local="gettingThrowsAdvice"></idref>
            </list>
        </property>
        <!--指定目标对象-->
        <property name="target" ref="target"></property>
    </bean>
    <bean id="target" class="throwsadviceexample.BusinessCode"></bean>
</beans>
```

【例 10-21】 Test 类(Test.java)。

```java
package throwsadviceexample;
import org.springframework.context.ApplicationContext;
import org.springframework.context.support.FileSystemXmlApplicationContext;

public class Test {
    public static void main(String[]args) {
        ApplicationContext ac=new
FileSystemXmlApplicationContext ( " src/throwsadviceexample/applicationContext.
xml");
        Business business=(Business)ac.getBean("throws");        //生成代理对象
        business.update();                                        //提供业务功能服务
    }
}
```

10.3.5 引入通知

引入通知与前面几种通知类型不同。其他类型的通知是在目标对象的方法被调用的周围织入。引入通知将给目标对象添加新的方法以及属性，这是比较难理解的通知。为了理解引入通知，需要同时理解切入点，因此这里把引入通知的介绍放在切入点后面讲解。

10.4 定义切入点

切入点决定通知应该作用于哪个连接点，也就是说通过切入点来定义需要增强方法的集合，这些集合的选取可以按照一定的规则来完成。在这种情况下，切入点通常意味着标识方法，例如，这些需要增强的地方可以是被某个正则表达式标识，或根据某个方法名进行匹配等。

切入点确定应该在目标对象的哪里应用通知。AOP 的魅力更多地在于应该在哪里使用动作（切入点），而不在于应该使用哪些动作（通知）。

10.4.1 静态切入点和动态切入点

切入点分为静态切入点和动态切入点。静态切入点是指在目标对象的方法执行之前就已经确定该方法是否织入通知，而动态切入点是指在目标对象的方法执行期间才确定该方法是否织入通知。

按照 Spring 的术语，动态切入点不仅可能有静态标准，而且还依赖于只有在做出调用时才能知道的信息，例如，参数值或者调用者。

动态切入点的运行性能比静态切入点慢。尽管在不可能匹配特定的方法的时候，可以预先排除某些动态切入点，但是在它们慢每一次调用时总是需要评估它们。

Spring AOP 中提供的常用静态切入点实现类有 NameMatchMethodPointcut 和 RegexpMethodPointcut 等。Spring AOP 中提供的常用动态切入点实现类是 ControlFlowPointCut。

Spring 的切入点是通过实现 org.springframework.aop.Pointeut 接口来实现的，其声明如下：

```
Public interface Pointcut {
    ClassFilter getClassFilter();
    MethodMatcher getMethodMatcher();
}
```

ClassFilter 接口决定一个类是否要应用通知（Advice），其声明如下：

```
Public interface ClassFilter {
    boolean matches (Class clazz);
    ClassFilter.TRUE =TrueClassFilter.INSTANCE;
}
```

matches()方法中要决定传入的类是不是符合切入点的定义，ClassFilter.TRUE 是 ClassFilter 接口的简单实现。它的 matches()方法总是传回 true。

在 Spring 中，使用切入点通知器（PointeutAdvisor）将切入点与通知结合成为一个对象。

PointeutAdvisor 为 Advisor 的子接口。

Advisor 接口在 Spring 中的定义如下：

```
import org.aopalliance.aop.Advice;
public interface Advisor {
    boolean isPerInstance();
    Advice getAdvice();
}
```

PointeutAdvisor 接口在 Spring 中的声明如下：

```
public interface PointcutAdvisor extends Advisor {
    Pointcut getPointcut();
}
```

Spring 内建的切入点都有对应的切入点通知器：NameMatchMethodPointcutAdvisor 和 RegexpMethodPointcutAdvisor。

控制流切入点是 Spring 所提供的一个动态切入点。它的作用是判断在某个指定类的某方法的执行过程中，该方法是否曾经要求目标对象执行某个动作。由于在方法执行期间才会确定是否介入通知，因此控制流切入点是一个动态切入点。但是动态切入点系统开销比较大，一般建议不使用动态切入点。

10.4.2 切入点的应用实例

1. NameMatchMethodPointcutAdvisor

NameMatchMethodPointcutAdvisor 是静态切入点 NameMatchMethodPointcut 的通知器。可以指定通知所要应用的目标上的方法名称，或者调用"＊"来指定。例如，hello＊表示调用代理对象上以 hello 作为开头的方法时，都会应用指定的通知（Advice）。

静态切入点 NameMatchMethodPointcut 应用实例：项目名称为 PointcutExample。有一个接口 Hello，代码如例 10-22 所示，实现该接口的代码类为 HelloSpeaker，代码如例 10-23 所示。实现前置通知的类为 LogBeforeAdvice，代码如例 10-24 所示。配置文件 applicationContext.xml 的代码如例 10-25 所示。测试类 Test 的代码如例 10-26 所示，项目的文件结构、JAR 文件以及运行结果如图 10-7 所示。

【例 10-22】 Hello 接口（Hello.java）。

```
package pointcutexample;

public interface Hello {
    void helloQiang(String name);
    void helloXiang(String name);
}
```

【例 10-23】 实现 Hello 接口的类 HelloSpeaker（HelloSpeaker.java）。

```
package pointcutexample;
```

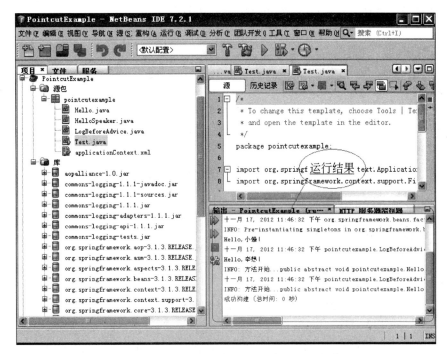

图 10-7 项目的文件结构、JAR 文件以及运行结果

```
public class HelloSpeaker implements Hello{
    public void helloQiang(String name){
        System.out.println("Hello,"+name+"强!");
    }
    public void helloXiang(String name){
        System.out.println("Hello,"+name+"想!");
    }
}
```

【例 10-24】 实现前置通知的日志类 LogBeforeAdvice(LogBeforeAdvice.java)。

```
package pointcutexample;
import java.lang.reflect.Method;
import java.util.logging.Level;
import java.util.logging.Logger;
import org.springframework.aop.MethodBeforeAdvice;

public class LogBeforeAdvice implements MethodBeforeAdvice{
    private Logger logger=Logger.getLogger(this.getClass().getName());
    public void before(Method method,Object[]args,Object target) throws Throwable
    {
        logger.log(Level.INFO,"方法开始..."+method);
    }
}
```

【例 10-25】 配置文件(applicationContext.xml)。

```xml
<?xml version="1.0" encoding="UTF-8"?>
<!DOCTYPE beans PUBLIC "-//SPRING//DTD BEAN 2.0//EN"
http://www.springframework.org/dtd/spring-beans-2.0.dtd
[<!ENTITY contextInclude SYSTEM
   "org/springframework/web/context/WEB-INF/contextInclude.xml">]>
<beans>
    <bean id="logBeforeAdvice" class="pointcutexample.LogBeforeAdvice"/>
    <!--定义切面-->
    <bean id="helloAdvisor"
        class="org.springframework.aop.support.NameMatchMethodPointcutAdvisor">
        <!--通过 mappedNames 属性指定目标方法,这里指定了 hello*。当调用 helloQiang
        () 和 helloXiang () 时,由于这两个方法名称的开头为 hello,就会应用
        logBeforeAdvice 的业务逻辑-->
        <property name="mappedNames">
            <list>
                <!--通知要应用的目标上的方法名称-->
                <value>hello*</value>
            </list>
        </property>
        <!--注入前置通知-->
        <property name="advice">
            <ref bean="logBeforeAdvice"/>
        </property>
    </bean>
    <bean id="helloSpeaker" class="pointcutexample.HelloSpeaker"/>
    <bean id="helloProxy"
        class="org.springframework.aop.framework.ProxyFactoryBean">
        <property name="proxyInterfaces" value="pointcutexample.Hello"/>
        <!--指定目标对象-->
        <property name="target">
            <ref bean="helloSpeaker"/>
        </property>
        <!--指定通知-->
        <property name="interceptorNames">
            <list>
                <!--指定切面-->
                <value>helloAdvisor</value>
            </list>
        </property>
    </bean>
</beans>
```

【例 10-26】 Test 类(Test.java)。

```
package pointcutexample;
```

```java
import org.springframework.context.ApplicationContext;
import org.springframework.context.support.FileSystemXmlApplicationContext;

public class Test {
    public static void main(String[]args) {
        ApplicationContext ac=new FileSystemXmlApplicationContext("src/pointcutexample/applicationContext.xml");
        Hello helloProxy= (Hello)ac.getBean("helloProxy");        //生成代理对象
        helloProxy.helloQiang("小");
        helloProxy.helloXiang("李");
    }
}
```

从上述程序可以看出,在 Spring3 中,使用切入点适配器能够把切入点(Pointcut)和通知(Advice)结合为一个对象。

2. RegexpMethodPointcutAdvisor

RegexpMethodPointcutAdvisor 是静态切入点 RegexpMethodPointcut 的通知器。在 Spring 中,RegexpMethodPointcutAdvisor 是以正则表达式方式匹配切面的实现类。常用格式如例 10-27 中所示。

【例 10-27】 RegexpMethodPointcutAdvisor 的配置。

```xml
<?xml version="1.0" encoding="UTF-8"?>
<!DOCTYPE beans PUBLIC "-//SPRING//DTD BEAN 2.0//EN"
http://www.springframework.org/dtd/spring-beans-2.0.dtd
[<!ENTITY contextInclude SYSTEM
"org/springframework/web/context/WEB-INF/contextInclude.xml">]>
<beans>
...
<bean id="helloAdvisor"
    class="org.springframework.aop.support.RegexpMethodPointcutAdvisor">
    <property name="patterns">
        <list>
            <!--通知要应用的目标上的方法名称-->
            <value>hello*</value>
        </list>
    </property>
    <!--注入前置通知-->
    <property name="advice">
        <ref bean="logBeforeAdvice"/>
    </property>
</bean>
...
</beans>
```

10.5 创建引入

引入(Introduction)是一种特殊的通知(Advice)。引入影响了目标对象的行为定义，直接增加了目标对象的职责。

与 BeforeAdvice、AfterAdvice 等在目标对象的方法前后介入服务不同，引入直接介入整个目标对象的行为，就好像对象凭空多了一些可操作的行为，为对象动态加入原先所没有的方法和属性。使用引入能够动态地建立复合对象。

一般通过实现 IntroductionInterceptor 接口来实现引入。

IntroductionInterceptor 接口包含 2 个主要方法：

```
boolean implementsInterface();
Object invoke();
```

其中，implementsInterface()方法的返回值如果是 true 的话，表示目前的 IntroductionInterceptor 实现了规定的接口（也就是要额外增加行为的接口）。使用 invoke()方法执行接口上的方法，调用引入的方法，让目标对象执行额外行为。

使用接口比较麻烦，为了简化开发，Spring3 提供了已实现接口 IntroductionInterceptor 的类 DelegateingIntroductionInterceptor。使用 DelegateingIntroductionInterceptor 开发程序可以使代码简洁，效率提高。

引入通知后还需要通知器(Advisor)，可以使用 IntroductionAdvisor 接口实现，Spring 提供了一个该接口的实现类 DefaultIntroductionAdvisor，该类有 3 个构造方法。

(1) DefaultIntroductionAdvisor(Advice advice)：通过一个通知创建一个引入切面，引入切面将为目标对象新增通知对象中所有接口的实现。

(2) DefaultIntroductionAdvisor(DynamicIntroductionAdvice advice, Class clazz)：通过一个通知和一个指定的接口类创建引入切面，仅为目标对象新增 clazz 接口的实现。

(3) DefaultIntroductionAdvisor(Advice advice, IntroductionInfo introductionInfo)：通过一个通知和一个 IntroductionInfo 创建一个引入切面，目标对象需要实现哪些接口，由 introductionInfo 对象的 getInterfaces()指定。

10.6 本章小结

本章主要介绍了 Spring3 框架中的 AOP 框架的基础知识以及应用实例，通过本章的学习应了解和掌握如下内容。

(1) AOP 基础知识以及 Spring AOP 的主要术语。
(2) 通知(Advice)及应用实例。
(3) 切点(Pointcut)及应用实例。
(4) 引入(Introduction)。

10.7 习　　题

10.7.1 选择题

1. Spring3 框架中用于切面处理的是(　　)。
   A. IoC　　　　　B. AOP　　　　　C. MVC 框架　　　　D. ORM 框架
2. 通知被应用的对象,称为(　　)。
   A. 切面　　　　B. 切入点　　　　C. 织入　　　　　　D. 目标
3. 影响了目标对象的行为定义,直接增加了目标对象的职责的是(　　)。
   A. 通知　　　　B. 切入点　　　　C. 引入　　　　　　D. 切面

10.7.2 填空题

1. 目前有两种主流的 AOP 实现：_____和_____。
2. 切面在某个具体连接点采取的行为或动作,称为_____。
3. 代理分为_____和_____。
4. Spring3 提供了 5 种通知(Advice)类型：_____、_____、环绕通知(Around Advice)、异常通知(Throws Advice)和引入通知(Introduction Advice)。

10.7.3 简答题

简述 Spring3 的 AOP 框架主要术语。

10.7.4 实训题

1. 使用静态切入点适配器 RegexpMethodPointcutAdvisor 编写一个简单的应用程序。
2. 使用 Spring AOP 框架编写一个简单的应用程序。

第 11 章　基于 Struts2＋Hibernate＋Spring3 的校园论坛 BBS 项目实训

本章详细介绍基于 Struts2＋Hibernate＋Spring3 的 BBS 系统的设计与实现。通过本项目的整合训练，培养熟练运用 Struts2、Hibernate 和 Spring3 框架知识开发 Java Web 项目的能力。

11.1　项目需求分析

公告牌服务(Bulletin Board Service,BBS)是 Internet 上的一种电子信息服务系统，它提供一块公共电子白板，每个用户都可以在上面书写，并且可发布信息或提出看法。

大部分 BBS 由教育机构、研究机构或商业机构管理，像日常生活中的黑板报一样，电子公告牌按不同的主题分成很多个布告栏。布告栏设立的依据是大多数 BBS 使用者的要求和喜好，使用者可以阅读他人关于某个主题的最新看法，也可以将自己的想法毫无保留地贴到公告栏中。

同样，别人对你的观点的回应也是很快的。如果需要私下交流，也可以将想说的话直接发到某个人的电子信箱中，如果想与正在使用的某个人聊天，可以启动聊天程序加入闲谈者的行列。虽然谈话的双方素不相识，却可以亲近地交谈。在 BBS 里，人们之间的交流打破了空间、时间的限制。在与别人进行交往时，无须考虑自身的年龄、学历、知识、社会地位、财富、外貌、健康状况，而这些条件往往是人们在其他交流形式中无可回避的。

同样地，也无从知道交谈的对方的真实社会身份。这样，参与 BBS 的人可以处于一个平等的位置与其他人进行任何问题的探讨。这对于现有的所有其他交流方式来说是不可能的。

目前国内的 BBS 已经十分普遍，可以说不计其数，其中 BBS 大致可以分为 5 类。

(1) 校园 BBS。自建立以来，校园 BBS 很快地发展了起来。目前很多大学都有了 BBS，几乎遍及全国上下，像清华大学、北京大学等都建立了自己的 BBS 系统。清华大学的水木清华很受学生和网民的喜爱。

(2) 商业 BBS 站。这些 BBS 站主要进行有关商业的宣传、产品推广等。目前手机的商业站、计算机的商业站、房地产的商业站比比皆是。

(3) 专业 BBS 站。这里所说的专业 BBS 主要用于建立地域性的文件传输和信息发布系统。

(4) 情感 BBS。主要用于交流情感，是许多娱乐网站的首选。

(5) 个人 BBS。有些个人主页的制作者们在自己的个人主页上建设了 BBS，用于接收别人的想法，更有利于与朋友进行沟通。

本系统开发的目的主要是为学生提供一个用于技术交流、生活娱乐的 BBS，该 BBS 可满足以下用户。

1. 普通（游客）用户

普通用户即游客是该论坛的一种用户，普通用户以游客身份访问系统，只具有查看帖子的权限，不能留言和发帖。

2. 会员用户

会员用户拥有普通用户所有的权限。此外会员用户可以登录系统、修改个人信息、上传个人头像、修改密码、发帖、查看自己所有发帖、回复别人帖子等。

3. 管理员

管理员登录以后可以对论坛系统进行管理，包括发布管理员帖子、管理员个人信息修改、对帖子进行管理、板块管理等功能。

11.2 项目分析与设计

根据前述需求分析，BBS系统功能描述如下。

1. 普通（游客）用户

只能查看BBS系统的论坛，不能发帖和留言。

2. 会员用户

1）会员登录功能

提供会员登录功能。只有登录后方可实施修改个人信息、上传个人头像、修改密码、发帖、查看自己所有发帖、回复别人帖子等操作。

2）网站搜索帖子功能

可以通过搜索功能查找BBS中的帖子。

3）我的帖子功能

可以查看用户本人发过的所有帖子。

4）修改个人信息功能

可以修改用户的昵称、QQ号码、邮箱以及上传头像。

5）修改密码功能

可以修改用户的登录密码。

6）发帖和回帖功能

用户登录后可以在各个板块发帖和回帖。

3. 管理员

1）登录功能

管理员登录后可以实现发布管理员帖子、管理员个人信息修改、对帖子进行管理、板块管理等功能。

2）我的帖子功能

可以发布管理员帖子。

3）修改个人信息功能

可以修改管理员的昵称、QQ号码、邮箱以及上传头像。

4）帖子管理功能

可以修改帖子和删除帖子。

5）板块管理功能

可以增加各级板块功能。

系统功能模块结构如图 11-1 所示。

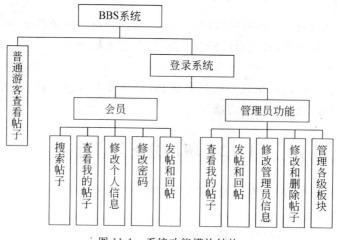

图 11-1　系统功能模块结构

11.3　项目的数据库设计

如果已经掌握 DBMS 相关知识，读者可按照数据库优化的思想自行选择相应 DBMS 并设计项目的数据库及表结构。本章提供的数据库设计仅供参考，读者可根据需要进一步优化。

本项目使用的数据库是 MySQL。项目中用到的数据库（bbs）和表（admin、board、post、reply、student）如图 11-2 所示。

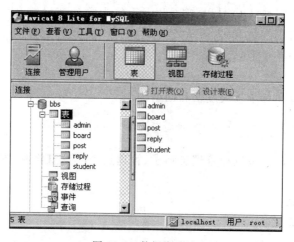

图 11-2　数据库和表

管理员表（admin）用于保存管理员信息，如表 11-1 所示。

表 11-1 管理员表(admin)

admin 管理表

字 段 名	类 型	长度	允许空	键	说 明
Id	Int	4		主键	id 自增
account	Char	10			账号
password	Char	20			密码
qx	Int	4			权限(0 超级,1 普通)
nickName	Char	10	√		昵称
name	VarChar	10	√		姓名
photoPath	VarChar	100	√		头像路径

板块表(board)用于保存板块信息,如表 11-2 所示。

表 11-2 板块表(board)

Board 板块表

字 段 名	类 型	长度	允许空	键	说 明
Id	Int	4		主键	id 自增
name	Char	20			板块名
description	VarChar	50	√		描述
parentId	Int	4	√		上级板块 id
aid	Int	4		外键	管理员 id
boardImg	VarChar	100	√		板块图片

帖子表(post)用于保存帖子信息,如表 11-3 所示。

表 11-3 帖子表(post)

post 帖子表

字 段 名	类 型	长度	允许空	键	说 明
Id	Int	4		主键	id 自增
name	Char	40			帖子名
content	Longtext	0	√		内容
publishTime	Datetime	0	√		发帖时间
sid	Int	4	√	外键	学生 id
bid	Int	4		外键	板块 id
Aid	Int	4	√	外键	管理员 id
Count	Int	11	√		点击量

回帖表(reply)用于保存回复帖子信息,如表 11-4 所示。

表 11-4 回帖表(reply)

reply 回帖表

字段名	类型	长度	允许空	键	说明
Id	Int	4		主键	id 自增
content	Text	0			回复内容
publishTime	Date	0	√		回复时间
pid	Int	4		外键	帖子 id
aid	Int	4	√	外键	管理员 id
sid	Int	4	√	外键	学生 id

会员表,即学生表(student)用于保存会员即学生信息,如表 11-5 所示。

表 11-5 学生表(student)

student 学生表

字段名	类型	长度	允许空	键	说明
Id	Int	4		主键	id 自增
stuNum	Int	15			学号
realName	Char	10	√		真实姓名
nickName	Char	20	√		昵称
password	Char	20			密码
qq	Char	20	√		QQ
email	Char	30	√		邮箱
major	Char	20	√		专业
className	Char	20	√		班级名
photoPath	VarChar	100	√		头像路径

从以上表中可以看出,本项目使用了表的关联关系。

11.4 项目实现

11.4.1 项目的文件结构和主页面

该项目命名为 BBS,在项目 Web 页文件夹中创建一个页面(redirect.jsp),运行该文件后将请求提交到业务控制器 IndexAction 中,用于加载并初始化 BBS 页面,然后把页面跳转到 BBS 论坛主页面(index.jsp),如图 11-3 所示。项目的页面文件结构如图 11-4 所示。项目的库包文件结构如图 11-5 所示。

图 11-3 项目主页面

图 11-4 项目页面文件结构

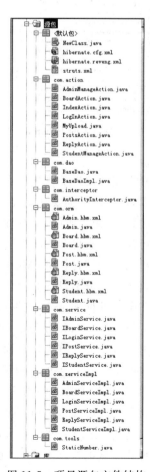

图 11-5 项目源包文件结构

项目使用的 JAR 文件如图 11-6 和图 11-7 所示。该项目使用 Struts 2.3.4、Spring 3.1.3 和 Hibernate 3.6 以及 MySQL 驱动和 SiteMesh 的 JAR 文件。

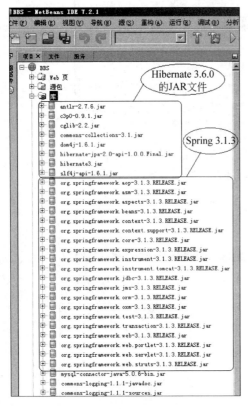
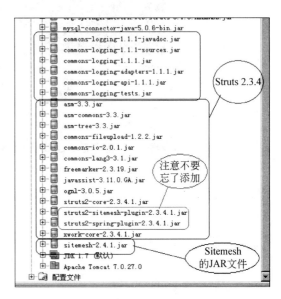

图 11-6　项目所需的 Hibernate 3.6 和 Spring 3.1.3 中的 JAR

图 11-7　项目所需的 Struts 2.3.4 和 Sitemesh 中的 JAR

为了开发视觉效果美观的页面，本项目使用了 Ckeditor 和 SiteMesh 技术。如果不熟悉这些技术，也可以根据实际情况自行选择其他页面设计技术。Ckeditor 是一个专门使用在网页上的、开放源代码的、所见即所得的文字编辑器。它志于轻量化，不需要太复杂的安装步骤即可使用。它可和 PHP、JavaScript、ASP、ASP. NET、ColdFusion、Java 以及 ABAP 等不同编程语言相结合。SiteMesh 是一个用于网页布局和修饰的框架，利用它可以将网页的内容和页面结构分离，以达到页面结构共享的目的。SiteMesh 是一个用于在 JSP 中实现页面布局和装饰的框架组件，能够帮助网站开发人员较容易地实现页面中动态内容和静态装饰外观的分离。Sitemesh 是由一个基于 Web 页面布局、装饰的、与现存 Web 应用整合的框架。它能帮助人们在有大量页面工程的项目中创建一致的页面布局和外观，如一致的导航条、一致的 banner、一致的版权等。它不仅能处理动态的内容，如 JSP、PHP、ASP、CGI 等产生的内容，还能处理静态的内容，比如 HTML 的内容，使得它的内容也符合你的页面结构的要求。甚至它能像 include 那样将 HTML 文件以一个面板的形式嵌入到别的文件中去。所有这些，都是 GOF 的 Decorator 模式的最生动的实现。装饰模式可以在不必改变原类文件和使用集成的情况下，动态地扩展一个对象的功能。它能通过创建一个包装对象，来装饰要包裹的对象。尽管它是由 Java 语言实现的，但是它也能与其他 Web 应用很好地集成。

图 11-4 所示的文件夹 BoardManage 中存放板块管理相关页面,包括板块主页面、增加板块、修改板块和上传板块图片;文件夹 MessageInfo 中存放信息提示页面,包括修改个人信息成功提示页面、增加板块成功提示页面、错误提示信息页面、登录成功提示页面、完善信息提示页面、上传成功提示页面;文件夹 PersonalInfo 中存放修改个人信息的相关页面,包括管理员个人信息页面、会员即学生信息页面和修改密码页面;文件夹 PostManagement 中存放帖子管理的相关页面,包括帖子管理、增加、修改、查询、查看、我的帖子等页面;文件夹 WEB-INF 中存放有 web.xml、applicationContext 和 decorators 文件;文件夹 ReplyManagement 中存放回复帖子管理的相关页面,包括帖子回复和回复成功页面;文件夹 ckeditor_3.6 中存放使用到的 js 文件以及需要的其他模板;文件夹 css 中存放项目用到的 css 文件;文件夹 decorators 中存放 SiteMesh 用到的装饰器页面;文件夹 images 中存放项目用到的图片;文件夹 js 中存放 Ckeditor 使用的 js 文件;文件夹 upload 中存放上传头像或者图片。

项目所需源包和配置文件的存放如图 11-5 所示。

运行图 11-4 中的 redirect.jsp 页面后,出现 BBS 项目的主页面,如图 11-3 所示。

redirect.jsp 的代码如下:

```
<%@page contentType="text/html" pageEncoding="UTF-8"%>
<%response.sendRedirect("index");%>
```

从代码"response.sendRedirect("index")"中可以看出,请求将跳转到名为 index 的 Action 中,index 的配置在 struts.xml 文件中,struts.xml 代码以及名为 index 的 Action 实现类 IndexAction 的代码如下所示。

struts.xml 代码如下:

```
<?xml version="1.0" encoding="UTF-8"?>
<!--指定Struts2配置文件的DTD信息-->
<!DOCTYPE struts PUBLIC
"-//Apache Software Foundation//DTD Struts Configuration 2.1.7//EN"
"http://struts.apache.org/dtds/struts-2.1.7.dtd">
<!--Struts2配置文件的根元素-->
<struts>
    <constant name =" struts.ognl.allowStaticMethodAccess " value =" true ">
    </constant>
    <!--配置了系列常量-->
    <package name="default" extends="struts-default">
        <interceptors>
            <!--定义一个名为authority的拦截器-->
            <interceptor class="com.interceptor.AuthorityInterceptor"
                name="authority"/>
            <!--定义一个包含权限检查的拦截器栈-->
            <interceptor-stack name="mydefault">
                <!--配置自定义的拦截器-->
```

```xml
            <interceptor-ref name="authority">
                <!--配置拦截器拦截的方法-->
                <param name="includeMethods">
                    preparePost,stuReply,viewPostsByUser,personalStuInfo
                </param>
            </interceptor-ref>
            <!--配置内建默认拦截器-->
            <interceptor-ref name="defaultStack"/>
        </interceptor-stack>
    </interceptors>
    <default-interceptor-ref name="mydefault"></default-interceptor-ref>
    <!--定义全局 Result-->
    <global-results>
        <result name="login">/login.jsp</result>
        <result name="stuPersonalInfo">
            /MessageInfo/pleaseComplete.jsp</result>
        <result name="adminPersonalInfo">/</result>
        <result name="error">/MessageInfo/error.jsp</result>
    </global-results>
    <action name="index" class="com.action.IndexAction" >
        <result name="success">/index.jsp</result>
    </action>
    <!--登录配置-->
    <action name="login" class="com.action.LogInAction">
        <result name="loginSuccess">/MessageInfo/loginSuccess.jsp</result>
        <result name="success">/Mainbbs.jsp</result>
        <result name="input">/login.jsp</result>
        <result name="exit">/redirect.jsp</result>

    </action>
     <!--板块管理-->
    <action name="board" class="com.action.BoardAction">
        <result name="success">/BoardManage/MainBoard.jsp</result>
        <result name="addBoard">/BoardManage/AddBoard.jsp</result>
        <result name="addSuccess">/MessageInfo/addBoardSuccess.jsp</result>
        <result name="prepareBoard" type="redirectAction">
            board!prepareAddBoard.action</result>
        <result name="prepareModifyBoard">
            /BoardManage/modifyBoardInfo.jsp</result>
    </action>
     <!--学生管理-->
    <action name="student" class="com.action.StudentManageAction">
        <result name="modifySuccess">
            /MessageInfo/ModifyStuInfoSuccess.jsp</result>
        <result name="success">/PersonalInfo/StuPersonalInfo.jsp</result>
```

```xml
        <result name="modifyPswSuccess">
            /PersonalInfo/StuPswModify.jsp</result>
</action>
<!--管理员-->
<action name="admin" class="com.action.AdminManageAction">
        <result name="success">/PersonalInfo/AdminPersonalInfo.jsp</result>
</action>
<!--发帖配置-->
<action name="post" class="com.action.PostsAction">
        <!--直接跳转到首页-->
        <result name="success">/PostManagement/PostManage.jsp</result>
        <result name="postSuccess">/PostManagement/postSuccess.jsp</result>
        <result name="viewDetail">
            /PostManagement/ViewPostDetail.jsp</result>
        <result name="prepareSuccess">
            /PostManagement/ModifyPost.jsp</result>
        <result name="modifySuccess" type="redirectAction">
                <param name="actionName">post</param>
                <param name="namespace">/</param>
                <param name="result">${result}</param>
        </result>
        <result name="deleteSuccess" type="redirectAction">
                <param name="actionName">post</param>
                <param name="namespace">/</param>
                <param name="result">${result}</param>
        </result>
        <result name="myposts">/PostManagement/myPosts.jsp</result>
        <result name="preparePost">/PostManagement/AddPost.jsp</result>
        <result name="searchSuccess">
            /PostManagement/PostSearchResult.jsp</result>
</action>
<!--回帖配置-->
<action name="reply" class="com.action.ReplyAction">
        <!--跳转返回到帖子详细信息页面-->
        <result name="success" type="redirectAction">
                post!viewDetail.action?pid=%{#request.pid}</result>
        <result name="modifySuccess" type="redirectAction">
            <param name="actionName"></param>
            <param name="namespace">/</param>
            <param name="result">${result}</param>
        </result>
        <result name="deleteSuccess" type="redirectAction">
            <param name="actionName"></param>
            <param name="namespace">/</param>
            <param name="result">${result}</param>
```

```xml
            </result>
        </action>
        <!--上传头像-->
        <action name="myUpload" class="com.action.MyUpload">
            <interceptor-ref name="fileUpload">
                <!--    <param name="allowedTypes">
            image/png,image/gif,image/bmp,image/jpeg,image/jpg,image/pjpg
                </param>-->
                <param name="maximumSize">3000000</param>
            </interceptor-ref>
            <interceptor-ref name="defaultStack"></interceptor-ref>
            <param name="path">/upload</param>
            <result name="success">/MessageInfo/uploadSuccess.jsp</result>
            <result name="input">/upload.jsp</result>
        </action>
    </package>
</struts>
```

IndexAction.java 代码如下:

```java
package com.action;
import com.opensymphony.xwork2.ActionContext;
import com.opensymphony.xwork2.ActionSupport;
import com.orm.Board;                         //导入 Bord 类
import com.orm.Post;                          //导入 Post 类
import com.orm.Student;                       //导入 Student 类
import com.service.IBoardService;             //导入 IboardService 接口
import com.service.IPostService;              //导入 IPostService 接口
import com.service.IStudentService;           //导入 IStudentService 接口
import java.util.List;
import javax.annotation.Resource;

public class IndexAction extends ActionSupport {
    @Resource(name="studentService")
    IStudentService studentService;
    @Resource(name="boardService")
    IBoardService boardService;
    @Resource(name="postService")
    IPostService postService;
    private List<Board> rootBoard;
    private int todayNum;
    private int yestNum;
    private int highestNum;
    private int total;
    private Student student;
    private List<Post> hotPosts;
```

```java
public int getTodayNum() {
    return todayNum;
}
public void setTodayNum(int todayNum) {
    this.todayNum=todayNum;
}
public int getYestNum() {
    return yestNum;
}
public void setYestNum(int yestNum) {
    this.yestNum=yestNum;
}
public int getTotal() {
    return total;
}
public void setTotal(int total) {
    this.total=total;
}
public List<Post>getHotPosts() {
    return hotPosts;
}
public void setHotPosts(List<Post>hotPosts) {
    this.hotPosts=hotPosts;
}
public Student getStudent() {
    return student;
}
public void setStudent(Student student) {
    this.student=student;
}

//初始化BBS主页面
@Override
public String execute() throws Exception {
    try {
        // Student 为 PO 对象
        Student sessionStudent=
            (Student) ActionContext.getContext().getSession().get("student");
        if (sessionStudent!=null) {
            setStudent(studentService.getStudentByStuNum(
                sessionStudent.getStuNum()));
        }
        //加载板块
        setRootBoard(boardService.loadRootBoards());
        setHotPosts(postService.rankPosts(10));
```

```java
            setTotal(postService.countTotalPost());
            setYestNum(postService.countYesteradyPost());
            setTodayNum(postService.countTodayPost());
            return SUCCESS;
        }catch (Exception e) {
            e.printStackTrace();
            return ERROR;
        }
    }
    //显示板块
    public List<Board>getRootBoard() {
        return rootBoard;
    }
    //火热帖子排行榜
    public void setRootBoard(List<Board> rootBoard) {
        this.rootBoard=rootBoard;
    }
}
```

IndexAction.java 中使用到 Bord 类、Post 类、Student 类以及接口 IBoardService、IPostService、IStudentService。下面分别给出这些类和接口的代码。

Bord.java 代码如下：

```java
package com.orm;
import java.util.HashSet;
import java.util.Set;

public class Board  implements java.io.Serializable {
    private Integer id;
    private Board board;
    private Admin admin;
    private String name;
    private String description;
    private String boardImg;
    private Set posts=new HashSet(0);
    private Set boards=new HashSet(0);
    public Board() {
    }
    public Board(Admin admin, String name) {
        this.admin=admin;
        this.name=name;
    }
    public Board(Board board, Admin admin, String name, String description, String boardImg, Set posts, Set boards) {
        this.board=board;
        this.admin=admin;
```

```java
        this.name=name;
        this.description=description;
        this.boardImg=boardImg;
        this.posts=posts;
        this.boards=boards;
    }
    public Integer getId() {
        return this.id;
    }
    public void setId(Integer id) {
        this.id=id;
    }
    public Board getBoard() {
        return this.board;
    }

    public void setBoard(Board board) {
        this.board=board;
    }
    public Admin getAdmin() {
        return this.admin;
    }
    public void setAdmin(Admin admin) {
        this.admin=admin;
    }
    public String getName() {
        return this.name;
    }
    public void setName(String name) {
        this.name=name;
    }
    public String getDescription() {
        return this.description;
    }
    public void setDescription(String description) {
        this.description=description;
    }
    public String getBoardImg() {
        return this.boardImg;
    }
    public void setBoardImg(String boardImg) {
        this.boardImg=boardImg;
    }
    public Set getPosts() {
        return this.posts;
```

```
    }
    public void setPosts(Set posts) {
        this.posts=posts;
    }
    public Set getBoards() {
        return this.boards;
    }
    public void setBoards(Set boards) {
        this.boards=boards;
    }
}
```

该 PO 对象(Bord)对应的映射文件为 Board.hbm.xml,代码如下:

```xml
<?xml version="1.0"?>
<!DOCTYPE hibernate-mapping PUBLIC
"-//Hibernate/Hibernate Mapping DTD 3.0//EN"
"http://hibernate.sourceforge.net/hibernate-mapping-3.0.dtd">
<hibernate-mapping>
    <class name="com.orm.Board" table="board" catalog="bbs" lazy="false">
        <id name="id" type="java.lang.Integer">
            <column name="id" />
            <generator class="identity" />
        </id>
        <many-to-one name="board" class="com.orm.Board" fetch="select">
            <column name="parentId" />
        </many-to-one>
        <many-to-one name="admin" class="com.orm.Admin" fetch="select">
            <column name="aid" not-null="true" />
        </many-to-one>
        <property name="name" type="string">
            <column name="name" length="20" not-null="true" />
        </property>
        <property name="description" type="string">
            <column name="description" length="50" />
        </property>
        <property name="boardImg" type="string">
            <column name="boardImg" length="100" />
        </property>
        <set name="posts" inverse="true">
            <key>
                <column name="bid" not-null="true" />
            </key>
            <one-to-many class="com.orm.Post" />
        </set>
        <set name="boards" inverse="true" lazy="false" order-by="id asc">
```

```xml
                <key>
                    <column name="parentId" />
                </key>
                <one-to-many class="com.orm.Board" />
            </set>
        </class>
</hibernate-mapping>
```

Post.java 的代码如下：

```java
package com.orm;
import java.util.Date;
import java.util.HashSet;
import java.util.Set;

public class Post   implements java.io.Serializable {
        private Integer id;
        private Admin admin;
        private Student student;
        private Board board;
        private String name;
        private String content;
        private Date publishTime;
        private Integer count;
        private Set replies=new HashSet(0);
    public Post() {
    }
    public Post(Board board, String name) {
        this.board=board;
        this.name=name;
    }
    public Post (Admin admin, Student student, Board board, String name, String
    content, Date publishTime, Integer count, Set replies) {
        this.admin=admin;
        this.student=student;
        this.board=board;
        this.name=name;
        this.content=content;
        this.publishTime=publishTime;
        this.count=count;
        this.replies=replies;
    }
    public Integer getId() {
        return this.id;
    }
    public void setId(Integer id) {
```

```java
        this.id=id;
    }
    public Admin getAdmin() {
        return this.admin;
    }
    public void setAdmin(Admin admin) {
        this.admin=admin;
    }
    public Student getStudent() {
        return this.student;
    }
    public void setStudent(Student student) {
        this.student=student;
    }
    public Board getBoard() {
        return this.board;
    }
    public void setBoard(Board board) {
        this.board=board;
    }
    public String getName() {
        return this.name;
    }
    public void setName(String name) {
        this.name=name;
    }
    public String getContent() {
        return this.content;
    }
    public void setContent(String content) {
        this.content=content;
    }
    public Date getPublishTime() {
        return this.publishTime;
    }
    public void setPublishTime(Date publishTime) {
        this.publishTime=publishTime;
    }
    public Integer getCount() {
        return this.count;
    }
    public void setCount(Integer count) {
        this.count=count;
    }
    public Set getReplies() {
```

```
        return this.replies;
    }
    public void setReplies(Set replies) {
        this.replies=replies;
    }
}
```

PO 对象(Post)对应的映射文件为 Post.hbm.xml,代码如下:

```xml
<?xml version="1.0"?>
<!DOCTYPE hibernate-mapping PUBLIC
"-//Hibernate/Hibernate Mapping DTD 3.0//EN"
"http://hibernate.sourceforge.net/hibernate-mapping-3.0.dtd">
<hibernate-mapping>
    <class name="com.orm.Post" table="post" catalog="bbs" lazy="false">
        <id name="id" type="java.lang.Integer">
            <column name="id" />
            <generator class="identity" />
        </id>
        <many-to-one name="admin" class="com.orm.Admin" fetch="select">
            <column name="aid" />
        </many-to-one>
        <many-to-one name="student" class="com.orm.Student" fetch="select" lazy="false">
            <column name="sid" />
        </many-to-one>
        <many-to-one name="board" class="com.orm.Board" fetch="select">
            <column name="bid" not-null="true" />
        </many-to-one>
        <property name="name" type="string">
            <column name="name" length="40" not-null="true" />
        </property>
        <property name="content" type="string">
            <column name="content" />
        </property>
        <property name="publishTime" type="timestamp">
            <column name="publishTime" length="19" />
        </property>
        <property name="count" type="java.lang.Integer">
            <column name="count" />
        </property>
        <set name="replies" inverse="true" cascade="all-delete-orphan" lazy="false" order-by="id asc">
            <key>
                <column name="pid" not-null="true" />
            </key>
```

```
            <one-to-many class="com.orm.Reply" />
        </set>
    </class>
</hibernate-mapping>
```

Student.java 代码如下：

```
package com.orm;
import java.util.HashSet;
import java.util.Set;

public class Student   implements java.io.Serializable {
    private Integer id;
    private String stuNum;
    private String realName;
    private String nickName;
    private String password;
    private String qq;
    private String email;
    private String major;
    private String className;
    private String photoPath;
    private Set replies=new HashSet(0);
    private Set posts=new HashSet(0);
    public Student() {
    }
    public Student(String stuNum, String password) {
        this.stuNum=stuNum;
        this.password=password;
    }
    public Student (String stuNum, String realName, String nickName, String password, String qq, String email, String major, String className, String photoPath, Set replies, Set posts) {
        this.stuNum=stuNum;
        this.realName=realName;
        this.nickName=nickName;
        this.password=password;
        this.qq=qq;
        this.email=email;
        this.major=major;
        this.className=className;
        this.photoPath=photoPath;
        this.replies=replies;
        this.posts=posts;
    }
    public Integer getId() {
```

```java
        return this.id;
    }
    public void setId(Integer id) {
        this.id=id;
    }
    public String getStuNum() {
        return this.stuNum;
    }
    public void setStuNum(String stuNum) {
        this.stuNum=stuNum;
    }
    public String getRealName() {
        return this.realName;
    }
    public void setRealName(String realName) {
        this.realName=realName;
    }
    public String getNickName() {
        return this.nickName;
    }
    public void setNickName(String nickName) {
        this.nickName=nickName;
    }
    public String getPassword() {
        return this.password;
    }
    public void setPassword(String password) {
        this.password=password;
    }
    public String getQq() {
        return this.qq;
    }
    public void setQq(String qq) {
        this.qq=qq;
    }
    public String getEmail() {
        return this.email;
    }
    public void setEmail(String email) {
        this.email=email;
    }
    public String getMajor() {
        return this.major;
    }
    public void setMajor(String major) {
```

```
            this.major=major;
        }
        public String getClassName() {
            return this.className;
        }
        public void setClassName(String className) {
            this.className=className;
        }
        public String getPhotoPath() {
            return this.photoPath;
        }

        public void setPhotoPath(String photoPath) {
            this.photoPath=photoPath;
        }
        public Set getReplies() {
            return this.replies;
        }
        public void setReplies(Set replies) {
            this.replies=replies;
        }
        public Set getPosts() {
            return this.posts;
        }
        public void setPosts(Set posts) {
            this.posts=posts;
        }
}
```

PO 对象(Student)对应的映射文件为 Student.hbm.xml,代码如下:

```
<?xml version="1.0"?>
<!DOCTYPE hibernate-mapping PUBLIC
"-//Hibernate/Hibernate Mapping DTD 3.0//EN"
"http://hibernate.sourceforge.net/hibernate-mapping-3.0.dtd">
<hibernate-mapping>
    <class name="com.orm.Student" table="student" catalog="bbs">
        <id name="id" type="java.lang.Integer">
            <column name="id" />
            <generator class="identity" />
        </id>
        <property name="stuNum" type="string">
            <column name="stuNum" length="15" not-null="true" />
        </property>
        <property name="realName" type="string">
            <column name="realName" length="10" />
```

```xml
        </property>
        <property name="nickName" type="string">
            <column name="nickName" length="20" />
        </property>
        <property name="password" type="string">
            <column name="password" length="20" not-null="true" />
        </property>
        <property name="qq" type="string">
            <column name="qq" length="20" />
        </property>
        <property name="email" type="string">
            <column name="email" length="30" />
        </property>
        <property name="major" type="string">
            <column name="major" length="20" />
        </property>
        <property name="className" type="string">
            <column name="className" length="20" />
        </property>
        <property name="photoPath" type="string">
            <column name="photoPath" length="100" />
        </property>
        <set name="replies" inverse="true">
            <key>
                <column name="sid" />
            </key>
            <one-to-many class="com.orm.Reply" />
        </set>
        <set name="posts" inverse="true" lazy="false">
            <key>
                <column name="sid" />
            </key>
            <one-to-many class="com.orm.Post" />
        </set>
    </class>
</hibernate-mapping>
```

IBoardService(IBoardService.java)接口声明了有关板块的操作方法，代码如下：

```java
package com.service;
import com.orm.Board;
import java.util.List;

public interface IBoardService {
    //加载 Board
    public Board loadBoard(int id);
```

```java
//加载子板块
public List<Board> loadChildBoards(int parentId);
//加载板块
public List<Board> loadAllBoards();
//加载根栏目
public List<Board> loadRootBoards();
//保存更新板块
public boolean saveOrUpdateBoard(Board board);
}
```

该接口的实现类为BoardServiceImpl(BoardServiceImpl.java),代码如下:

```java
package com.serviceImpl;
import com.dao.BaseDao;                    //导入BaseDao接口
import com.orm.Board;
import com.service.IBoardService;
import java.util.List;
import javax.annotation.Resource;

public class BoardServiceImpl implements IBoardService{
    @Resource(name="dao")
     BaseDao dao;
    public Board loadBoard(int id) {
        return (Board) dao.loadById(Board.class,new Integer(id));
    }
    public List<Board> loadChildBoards(int parentId) {
        final String queryChilds="from Board as b where b.board = '"+parentId+"' order by b.id asc";
        return dao.query(queryChilds);
    }
    public List<Board> loadAllBoards() {
        return dao.listAll("Board");
    }
    public List<Board> loadRootBoards() {
        return dao.query("from Board as b where b.board is  null order by b.id asc");
    }
    public boolean saveOrUpdateBoard(Board board) {
        try
        {
            dao.saveOrUpdate(board);
            return true;
        }catch (Exception e) {
            e.printStackTrace();
            return false;
        }
    }
}
```

}

BoardServiceImpl 类中导入了 BaseDao 接口(BaseDao.java),该接口封装了对 PO 对象的常用操作方法,该接口的实现类是 BaseDaoImpl(BaseDaoImpl.java),它们的代码分别如下:

```java
// BaseDao.java
package com.dao;
import com.mysql.jdbc.Connection;
import java.io.Serializable;
import java.util.List;

public interface BaseDao {
    //加载指定 ID 的持久化对象
    public Object loadById(Class clazz, Serializable id);
    // 加载满足条件的持久化对象
    public Object loadObject( String hql);
    // 删除指定 ID 的持久化对象
    public void delById(Class clazz, Serializable id);
    //保存或更新指定持久化对象
    public void saveOrUpdate(Object obj);
    // 装载指定类的所有持久化对象
    public List listAll(String clazz);
    //分页装载指定类的所有持久化对象
    public List listAll(String clazz, int pageNo, int pageSize);
    // 统计指定类的所有持久化对象
    public int countAll(String clazz);
    //查询指定类的满足条件的持久化对象
    public List query(String hql);
    //分页查询指定类的满足条件的持久化对象
    public List query(String hql, int pageNo, int pageSize);
    //统计指定类的查询结果
    public int countQuery(String hql);
    //条件更新数据库
    public int update(String hql);
    //从连接池获取 JDBC 连接
    public Connection getConnection();
}
// BaseDaoImpl.java
package com.dao;
import com.mysql.jdbc.Connection;
import java.io.Serializable;
import java.sql.SQLException;
import java.util.List;
import javax.annotation.Resource;
import org.hibernate.Hibernate;
```

```java
import org.hibernate.HibernateException;
import org.hibernate.Query;
import org.hibernate.Session;
import org.springframework.orm.hibernate3.HibernateCallback;
import org.springframework.orm.hibernate3.HibernateTemplate;

public class BaseDaoImpl implements BaseDao {
    @Resource(name="hibernateTemplate")
    HibernateTemplate hibernateTemplate;
    //加载指定 id 的对象
    public Object loadById(Class clazz, Serializable id) {
        return hibernateTemplate.get(clazz, id);
    }
//根据 HQL 查找指定对象
    public Object loadObject(String hql) {
        final String hql1=hql;
        Object obj=null;
        List list=hibernateTemplate.executeFind(new HibernateCallback() {
            //重写查询方法
            public Object doInHibernate(Session session) throws HibernateException {
                Query query=session.createQuery(hql1);
                return query.list();
            }
        });
        if (list.size() >0) {
            obj=list.get(0);
        }
        return obj;
    }
    //删除
    public void delById(Class clazz, Serializable id) {
        hibernateTemplate.delete(hibernateTemplate.load(clazz, id));
    }
    //保存或更新
    public void saveOrUpdate(Object obj) {
        hibernateTemplate.saveOrUpdate(obj);
    }
    //获取全部
    public List listAll(String clazz) {
        return hibernateTemplate.find("from "+clazz+" as c ");
    }
    public List listAll(String clazz, int pageNo, int pageSize) {
        final int pNo=pageNo;
        final int pSize=pageSize;
        final String hqlString="from "+clazz+" as c order by c.id desc";
```

```java
        List list=hibernateTemplate.executeFind(
            new HibernateCallback<Object>() {
                public Object doInHibernate(Session sn) throws
                HibernateException, SQLException {
                    Query query=sn.createQuery(hqlString);
                    query.setMaxResults(pSize);
                    query.setFirstResult((pNo-1) * pSize);
                    List result=query.list();
                    if (!Hibernate.isInitialized(result)) {
                        Hibernate.initialize(result);
                    }
                    return result;
                }
            });
        return list;
    }
    public int countAll(String clazz) {
        final String hql="select count(*) from  "+clazz+" asc";
        Long count=
        (Long) hibernateTemplate.execute(new HibernateCallback<Object>() {
            public Object doInHibernate (Session sn) throws HibernateException,
            SQLException {
                Query query=sn.createQuery(hql);
                query.setMaxResults(1);
                return query.uniqueResult();
            }
        });
        return count.intValue();
    }
    public List query(String hql) {
        final String hql1=hql;
        return hibernateTemplate.executeFind(new HibernateCallback<Object>() {
            public Object doInHibernate (Session sn) throws HibernateException,
            SQLException {
                Query query=sn.createQuery(hql1);
                return query.list();
            }
        });
    }
    public List query(String hql, int pageNo, int pageSize) {
        final int pNo=pageNo;
        final int pSize=pageSize;
        final String hqlString=hql ;
        List list=hibernateTemplate.executeFind(
            new HibernateCallback<Object>() {
```

```java
            public Object doInHibernate(Session sn)throws HibernateException,
            SQLException {
                Query query=sn.createQuery(hqlString);
                query.setMaxResults(pSize);
                query.setFirstResult((pNo-1) * pSize);
                List result=query.list();
                if (!Hibernate.isInitialized(result)) {
                    Hibernate.initialize(result);
                }
                return result;
            }
        });
        return list;
    }
    //根据HQL查询数量
    public int countQuery(String hql) {
        final String hql1=hql;
        Long count= (Long) hibernateTemplate.execute(new HibernateCallback(){
            public Object doInHibernate (Session sn) throws HibernateException,
            SQLException {
                Query query=sn.createQuery(hql1);
                query.setMaxResults(1);
                return query.uniqueResult();
            }
        });
        return  count.intValue();
    }
    public int update(String hql) {
        throw new UnsupportedOperationException("Not supported yet.");
    }
    public Connection getConnection() {
        throw new UnsupportedOperationException("Not supported yet.");
    }
}
```

接口IPostService(IpostService.java)声明了操作帖子的常用方法,代码如下:

```java
package com.service;
import com.orm.Post;
import java.util.List;

public interface IPostService {
    public boolean saveOrUpdate(Post post);
    public List<Post>allPost();
    public List<Post>pageAllPost(int bid,int pageNo, int pageSize);
    public int getPostsCount();
```

```java
    public Post loadPost(int id);
    public List<Post>allPostsByUser(Object user);
    public boolean deletePost(int id);
    public List<Post>searchPosts(String searchKey);
    public List<Post>rankPosts(int size);
    public int countTotalPost();
    public int countTodayPost();
    public int countYesteradyPost();
    public int countDayLargestPost();
}
```

该接口的实现类为 PostServiceImpl(PostServiceImpl.java),代码如下:

```java
package com.serviceImpl;

import com.dao.BaseDao;
import com.orm.Admin;
import com.orm.Post;
import com.orm.Student;
import com.service.IPostService;
import java.text.SimpleDateFormat;
import java.util.ArrayList;
import java.util.Date;
import java.util.List;
import javax.annotation.Resource;

public class PostServiceImpl implements IPostService {
    @Resource(name="dao")
    BaseDao dao;
    public boolean saveOrUpdate(Post post) {
        try {
            dao.saveOrUpdate(post);
            return   true;
        } catch (Exception e) {
            return false;
        }
    }
    public List<Post>allPost() {
        List<Post>list=dao.listAll("Post");
        return list;
    }
    public Post loadPost(int id) {
        return   (Post) dao.loadById(Post.class, new Integer(id));
    }
    public List<Post>allPostsByUser(Object user) {
        if (user instanceof Student ) {
```

```java
            Student s=(Student)user;
            List<Post>list= (List<Post>) dao.query("from Post as p where p.student
            ="+s.getId()+" order by p.publishTime desc  ");
            return list;
        }
        if (user instanceof Admin) {
            Admin a=(Admin)user;
            List<Post>list=dao.query("from Post as p where p.admin.id ='"+a.getId
            ()+"' ");
            return list;
        }
        return null;
    }
    public List<Post>pageAllPost(int bid,int pageNo, int pageSize) {
        return dao.query("from Post as p where p.board ='"+bid+"' order by p.
        publishTime desc  ", pageNo, pageSize);
    }
    public int getPostsCount() {
        return  dao.countAll("Post");
    }
    public boolean deletePost(int id) {
        try
        {
            dao.delById(Post.class, id);
            return true;
        } catch (Exception e) {
            return false;
        }
    }
    public List<Post>searchPosts(String searchKey) {
        return dao.query("from Post as p where p.name like '%"+searchKey+"%' ");
    }
    public List<Post>rankPosts(int size) {
        List<Post>list=dao.query("from Post as p order by p.count desc ");
        List<Post>result=new ArrayList<Post>();
        for (int i=0; i<size; i++) {
            result.add(list.get(i));
        }
        return result;
    }
    public int countTotalPost() {
       return dao.countAll("Post");
    }
    public int countYesteradyPost() {
        Date todayDate=new Date();
```

```java
            Date ysterDate=new Date(System.currentTimeMillis()-1000*60*60*24);
            SimpleDateFormat sf=new SimpleDateFormat("yyyy-MM-dd");
            String today=sf.format(todayDate)+" 00:00:00";
            String yesterday=sf.format(ysterDate)+" 00:00:00";
            return dao.countQuery("select count(*) from Post as p where p.publishTime between '"+yesterday+"' and '"+today+"'    ");
            //return dao.countQuery(" select count ( * ) from Post as p where p.publishTime >'"+today+"' ");
    }
    public int countDayLargestPost() {
        throw new UnsupportedOperationException("Not supported yet.");
    }
    public int countTodayPost() {
        Date todayDate=new Date();
        Date tomorrowDate=new Date(System.currentTimeMillis()+1000*60*60*24);
        SimpleDateFormat sf=new SimpleDateFormat("yyyy-MM-dd");
        String today=sf.format(todayDate)+" 00:00:00";
        String tomorrow=sf.format(tomorrowDate)+" 00:00:00";
        return dao.countQuery("select count(*) from Post as p where p.publishTime between '"+today+"' and '"+tomorrow+"'    ");
    }
}
```

PostServiceImpl 类中使用了 Admin 类，Admin 类的代码如下：

```java
package com.orm;
import java.util.HashSet;
import java.util.Set;

public class Admin   implements java.io.Serializable {
    private Integer id;
    private String account;
    private String password;
    private int qx;
    private String nickName;
    private String name;
    private String photoPath;
    private Set posts=new HashSet(0);
    private Set replies=new HashSet(0);
    private Set boards=new HashSet(0);
    public Admin() {
    }
    public Admin(String account, String password, int qx) {
        this.account=account;
        this.password=password;
        this.qx=qx;
```

```java
    }
    public Admin(String account, String password, int qx, String nickName, String
    name, String photoPath, Set posts, Set replies, Set boards) {
        this.account=account;
        this.password=password;
        this.qx=qx;
        this.nickName=nickName;
        this.name=name;
        this.photoPath=photoPath;
        this.posts=posts;
        this.replies=replies;
        this.boards=boards;
    }
    public Integer getId() {
        return this.id;
    }
    public void setId(Integer id) {
        this.id=id;
    }
    public String getAccount() {
        return this.account;
    }
    public void setAccount(String account) {
        this.account=account;
    }
    public String getPassword() {
        return this.password;
    }
    public void setPassword(String password) {
        this.password=password;
    }
    public int getQx() {
        return this.qx;
    }
    public void setQx(int qx) {
        this.qx=qx;
    }
    public String getNickName() {
        return this.nickName;
    }

    public void setNickName(String nickName) {
        this.nickName=nickName;
    }
    public String getName() {
```

```
        return this.name;
    }
    public void setName(String name) {
        this.name=name;
    }
    public String getPhotoPath() {
        return this.photoPath;
    }
    public void setPhotoPath(String photoPath) {
        this.photoPath=photoPath;
    }
    public Set getPosts() {
        return this.posts;
    }
    public void setPosts(Set posts) {
        this.posts=posts;
    }
    public Set getReplies() {
        return this.replies;
    }
    public void setReplies(Set replies) {
        this.replies=replies;
    }
    public Set getBoards() {
        return this.boards;
    }
    public void setBoards(Set boards) {
        this.boards=boards;
    }
}
```

PO 对象(Admin)对应的映射文件是 Admin.hbm.xml,代码如下:

```xml
<?xml version="1.0"?>
<!DOCTYPE hibernate-mapping PUBLIC
"-//Hibernate/Hibernate Mapping DTD 3.0//EN"
"http://hibernate.sourceforge.net/hibernate-mapping-3.0.dtd">
<!--Generated 2012-11-22 22:11:50 by Hibernate Tools 3.2.1.GA-->
<hibernate-mapping>
    <class name="com.orm.Admin" table="admin" catalog="bbs">
        <id name="id" type="java.lang.Integer">
            <column name="id" />
            <generator class="identity" />
        </id>
        <property name="account" type="string">
            <column name="account" length="10" not-null="true" />
```

```xml
        </property>
        <property name="password" type="string">
            <column name="password" length="20" not-null="true" />
        </property>
        <property name="qx" type="int">
            <column name="qx" not-null="true" />
        </property>
        <property name="nickName" type="string">
            <column name="nickName" length="10" />
        </property>
        <property name="name" type="string">
            <column name="name" length="10" />
        </property>
        <property name="photoPath" type="string">
            <column name="photoPath" length="100" />
        </property>
        <set name="posts" inverse="true">
            <key>
                <column name="aid" />
            </key>
            <one-to-many class="com.orm.Post" />
        </set>
        <set name="replies" inverse="true">
            <key>
                <column name="aid" />
            </key>
            <one-to-many class="com.orm.Reply" />
        </set>
        <set name="boards" inverse="true">
            <key>
                <column name="aid" not-null="true" />
            </key>
            <one-to-many class="com.orm.Board" />
        </set>
    </class>
</hibernate-mapping>
```

接口 IStudentService(IStudentService.java)声明了对会员即学生的操作方法，代码如下：

```java
package com.service;
import com.orm.Student;

public interface IStudentService {
    public Student getStudentByStuNum(String StuNum);
    public boolean modifyStudent(Student student);
}
```

该接口的实现类为 StudentServiceImpl(StudentServiceImpl.java)，代码如下：

```java
package com.serviceImpl;
import com.dao.BaseDao;
import com.orm.Student;
import com.service.IStudentService;
import javax.annotation.Resource;

public class StudentServiceImpl implements IStudentService {
    @Resource(name="dao")
    BaseDao dao;
    public Student getStudentByStuNum(String StuNum) {
        Student s=(Student) dao.loadObject("from Student as s where s.stuNum='"+
        StuNum+"' ");
        if (s !=null) {
            return s;
        }
        return null;
    }
    public boolean modifyStudent(Student student) {
        try {
            dao.saveOrUpdate(student);
            return true;
        } catch (Exception e) {
            return false;
        }
    }
}
```

另外，在调用 BaseDaoImpl 类时需要加载 Hibernate 配置文件。在 hibernate.xml 和 applicationContext.xml 中对 Hibernate 均有配置，另外需要在 web.xml 中配置 Struts2、Spring 3.1.3 以及 SiteMesh。

web.xml 的代码如下：

```xml
<?xml version="1.0" encoding="UTF-8"?>
<web-app version="3.0"
xmlns="http://java.sun.com/xml/ns/javaee"
xmlns:xsi=http://www.w3.org/2001/XMLSchema-instance
xsi:schemaLocation="http://java.sun.com/xml/ns/javaee
http://java.sun.com/xml/ns/javaee/web-app_3_0.xsd">
    <filter>
        <filter-name>struts-cleanup</filter-name>
        <filter-class>
            org.apache.struts2.dispatcher.ActionContextCleanUp
        </filter-class>
    </filter>
```

```xml
<filter>
    <filter-name>sitemesh</filter-name>
    <filter-class>
        com.opensymphony.module.sitemesh.filter.PageFilter
    </filter-class>
</filter>
<context-param>
    <param-name>contextConfigLocation</param-name>
    <param-value>/WEB-INF/applicationContext.xml</param-value>
</context-param>
<filter>
    <filter-name>struts2</filter-name>
    <filter-class>org.apache.struts2.dispatcher.FilterDispatcher</filter-class>
</filter>
<filter>
    <filter-name>openSessionInViewFilter</filter-name>
    <filter-class>
        org.springframework.orm.hibernate3.support.OpenSessionInViewFilter
    </filter-class>
    <init-param>
        <param-name>openSessionInViewFilter</param-name>
        <param-value>/*</param-value>
    </init-param>
</filter>
<filter-mapping>
    <filter-name>struts-cleanup</filter-name>
    <url-pattern>/*</url-pattern>
</filter-mapping>
<filter-mapping>
    <filter-name>sitemesh</filter-name>
    <url-pattern>/*</url-pattern>
</filter-mapping>
<filter-mapping>
    <filter-name>struts2</filter-name>
    <url-pattern>/*</url-pattern>
</filter-mapping>
<listener>
    <listener-class>
        org.springframework.web.context.ContextLoaderListener
    </listener-class>
</listener>
<session-config>
    <session-timeout>
        30
    </session-timeout>
```

```xml
        </session-config>
        <welcome-file-list>
            <welcome-file>redirect.jsp</welcome-file>
        </welcome-file-list>
</web-app>
```

hibernate.xml 的代码如下：

```xml
<?xml version="1.0" encoding="UTF-8"?>
<!DOCTYPE hibernate-configuration PUBLIC
"-//Hibernate/Hibernate Configuration DTD 3.0//EN"
"http://hibernate.sourceforge.net/hibernate-configuration-3.0.dtd">
<hibernate-configuration>
    <session-factory>
        <property name="hibernate.dialect">
            org.hibernate.dialect.MySQLDialect</property>
        <property name="hibernate.connection.driver_class">
            com.mysql.jdbc.Driver</property>
        <property name="hibernate.connection.url">
            jdbc:mysql://localhost:3307/bbs</property>
        <property name="hibernate.connection.username">root</property>
        <property name="hibernate.connection.password">root</property>
        <mapping resource="com/orm/Admin.hbm.xml"/>
        <mapping resource="com/orm/Student.hbm.xml"/>
        <mapping resource="com/orm/Reply.hbm.xml"/>
        <mapping resource="com/orm/Post.hbm.xml"/>
        <mapping resource="com/orm/Board.hbm.xml"/>
    </session-factory>
</hibernate-configuration>
```

applicationContext.xml 的代码如下：

```xml
<?xml version="1.0" encoding="GBK"?>
<!--指定 Spring 配置文件的 Schema 信息-->
<beans xmlns="http://www.springframework.org/schema/beans"
    xmlns:xsi="http://www.w3.org/2001/XMLSchema-instance"
        xmlns:context="http://www.springframework.org/schema/context"
    xmlns:aop="http://www.springframework.org/schema/aop"
    xmlns:p="http://www.springframework.org/schema/p"
    xmlns:tx="http://www.springframework.org/schema/tx"
    xsi:schemaLocation="http://www.springframework.org/schema/beans
    http://www.springframework.org/schema/beans/spring-beans-3.0.xsd
    http://www.springframework.org/schema/tx
    http://www.springframework.org/schema/tx/spring-tx-3.0.xsd
        http://www.springframework.org/schema/context
    http://www.springframework.org/schema/context/spring-context-3.0.xsd
    http://www.springframework.org/schema/aop
```

```xml
         http://www.springframework.org/schema/aop/spring-aop-3.0.xsd">

    <!--定义数据源Bean,使用C3P0数据源实现-->
    <bean id="dataSource" class="com.mchange.v2.c3p0.ComboPooledDataSource"
         destroy-method="close">
        <!--指定连接数据库的驱动-->
        <property name="driverClass" value="com.mysql.jdbc.Driver"/>
        <!--指定连接数据库的URL-->
        <property name="jdbcUrl" value="jdbc:mysql://localhost/bbs"/>
        <!--指定连接数据库的用户名-->
        <property name="user" value="root"/>
        <!--指定连接数据库的密码-->
        <property name="password" value="root"/>
        <!--指定数据库连接池的最大连接数-->
        <property name="maxPoolSize" value="40"/>
        <!--指定数据库连接池的最小连接数-->
        <property name="minPoolSize" value="1"/>
        <!--指定数据库连接池的初始化连接数-->
        <property name="initialPoolSize" value="1"/>
        <!--指定数据库连接池的连接的最大空闲时间-->
        <property name="maxIdleTime" value="20"/>
    </bean>
    <!--定义Hibernate的SessionFactory-->
    <bean id="sessionFactory"
         class="org.springframework.orm.hibernate3.LocalSessionFactoryBean">
        <!--依赖注入数据源,注入的正是上面定义的dataSource-->
        <property name="dataSource" ref="dataSource"/>
        <!--mappingResouces属性用来列出全部映射文件-->
        <property name="mappingResources">
            <list>
                <!--列出Hibernate映射文件-->
                <value>com/orm/Admin.hbm.xml</value>
                <value>com/orm/Board.hbm.xml</value>
                <value>com/orm/Post.hbm.xml</value>
                <value>com/orm/Student.hbm.xml</value>
                <value>com/orm/Reply.hbm.xml</value>
            </list>
        </property>
        <!--定义Hibernate的SessionFactory的属性-->
        <property name="hibernateProperties">
            <!--配置Hibernate属性-->
            <props>
                <prop key="hibernate.dialect">
                    org.hibernate.dialect.MySQLInnoDBDialect</prop>
                <prop key="hibernate.show_sql">true</prop>
```

```xml
            <prop key="hibernate.format_sql">true</prop>
            <prop key="hibernate.hbm2ddl.auto">update</prop>
            <prop key="hibernate.jdbc.fetch_size">50</prop>
            <prop key="hibernate.jdbc.batch_size">50</prop>
        </props>
    </property>
</bean>
<bean id="hibernateTemplate"
    class="org.springframework.orm.hibernate3.HibernateTemplate">
    <property name="sessionFactory">
        <ref bean="sessionFactory"/>
    </property>
    <property name="allowCreate">
        <value>true</value>
    </property>
</bean>
    <!--依赖注入,通过注解注入-->
    <!--即将被注入 dao-->
<bean id="dao" class="com.dao.BaseDaoImpl" ></bean>
    <!--业务层注入 即将注入 action 控制层-->
<bean id="loginService" class="com.serviceImpl.LoginServiceImpl" ></bean>
<bean id="boardService" class="com.serviceImpl.BoardServiceImpl" ></bean>
<bean id="postService" class="com.serviceImpl.PostServiceImpl" ></bean>
<bean id="replyService" class="com.serviceImpl.ReplyServiceImpl" ></bean>
<bean id="studentService" class="com.serviceImpl.StudentServiceImpl" ></bean>
<bean id="adminService" class="com.serviceImpl.AdminServiceImpl" ></bean>
<!--开启 Spring 的 Annotation 注解处理器-->
<context:annotation-config />
<!--开启 Spring 的 Bean 自动扫描机制来查找和管理 Bean 实例-->
<context:component-scan base-package="com" />
</beans>
```

根据 struts.xml 中的配置,业务控制器 IndexAction 执行后,页面将跳转到 index.jsp,运行效果如图 11-3 所示。在执行 index.jsp 时将使用 Sitemesh 的装饰器页面(myDecorator.jsp)加载 index.jsp,另外需要在 decorators.xml 中进行配置。

index.jsp 的代码如下:

```jsp
<%@page contentType="text/html" pageEncoding="UTF-8"%>
<%@taglib  prefix="s" uri="/struts-tags" %>
<!DOCTYPE html>
<html>
    <head>
        <meta http-equiv="Content-Type" content="text/html; charset=UTF-8">
        <title>郑州轻工业学院 BBS</title>
        <script type="text/javascript" src="js/jquery-1.2.6.pack.js"></script>
        <style type="text/css">
```

```
#Lboard{    }
.subBoard{ min-height:100px;    }
.btitle{ padding-left:10px; text-align:left; line-height:30px;
    color:#1B72AF; border:1px solid #D6E8F4;
    background: url('<%=request.getContextPath()%>
                /images/h.png') repeat-x;}
#Lboard ul li { float:left; width:250px; margin:5px; padding:5px; }
#Lboard ul li a{ text-decoration:none;}
#Lboard ul li a:hover{ text-decoration:underline;}
#Lboard .bimg{ float:left;}
#Lboard strong { margin-left:15px; height:24px; font-size:18px;}
#Lboard p { padding-left:60px;}
.clear{clear:both;}
#rank{float:left;display:block; margin-left:30px;
    position:relative; bottom:45px; top:5px; }
#rank li{ margin:7px;}
/**/
#banner {position:relative; float:left; width:378px; height:256px;
    margin-left:20px; margin-bottom:15px;
    border:1px solid #666; overflow:hidden;}
#banner_list img {border:0px;}
#banner_bg {position:absolute; bottom:0;background-color:#000;
    height:30px;filter: Alpha(Opacity=30);opacity:0.3;
    z-index:1000;cursor:pointer; width:478px; }
#banner_info{position:absolute; bottom:0; left:5px;
    height:22px;color:#fff;z-index:1001;cursor:pointer}
#banner_text {position:absolute;width:120px;z-index:1002;
    right:3px; bottom:3px;}
#banner ul {position:absolute;list-style-type:none;
    filter: Alpha(Opacity=80);opacity:0.8; border:1px solid
    #fff;z-index:1002;
    margin:0; padding:0; bottom:3px; right:5px;}
#banner ul li {padding:0px 8px;
    float:left;display:block;color:#FFF;
    border:#e5eaff1px solid;
    background:#6f4f67;cursor:pointer}
#banner ul li.on { background:#900}
#banner_list a{position:absolute;}            /*让四张图片重叠在一起*/
</style>
<script type="text/javascript">
    var t=n=0, count;
    $(document).ready(function(){
        count=$("#banner_list a").length;
        $("#banner_list a:not(:first-child)").hide();
        $("#banner_info").html($("#banner_list a:first-child").find(
```

```
                "img").attr('alt'));
            $("#banner_info").click(function(){window.open($(
               "#banner_list a:first-child").attr('href'),"_blank")});
            $("#banner li").click(function() {
                var i=$(this).text()-1;//获取 Li 元素内的值,即 1、2、3、4
                n=i;
                if (i >=count) return;
                $("#banner_info").html($("#banner_list a").eq(
                    i).find("img").attr('alt'));
                $("#banner_info").unbind().click(function(){window.open($(
                    "#banner_list a").eq(i).attr('href'),"_blank")})
                $("#banner_list a").filter(":visible").fadeOut(
                    2000).parent().children().eq(i).fadeIn(2000);
                document.getElementById("banner").style.background="";
                $(this).toggleClass("on");
                $(this).siblings().removeAttr("class");
            });
            t=setInterval("showAuto()", 3000);
            $("#banner").hover(function(){clearInterval(t)}, function(){t=
            setInterval("showAuto()", 3000);});
        })
        function showAuto()
        {
            n=n >= (count-1) ? 0 : ++n;
            $("#banner li").eq(n).trigger('click');
        }
    </script>
</head>
<body>
    <s:set name="total" value="total" scope="application"></s:set>
    <s:set name="yestNum" value="yestNum" scope="application"></s:set>
    <s:set name="todayNum" value="todayNum" scope="application" ></s:set>
    <s:set name="student" value="student" scope="session" ></s:set>
        <div id="banner">
            <div id="banner_bg"></div>   <!--标题背景-->
            <div id="banner_info"></div><!--标题-->
            <ul>
                <li class="on">1</li>
                <li>2</li>
                <li>3</li>
                <li>4</li>
            </ul>
            <div id="banner_list">
                <a href="#" target="_blank">
                    <img src="images/imgs/10794203.jpg" title="郑州轻工业学院" alt
```

```
                ="郑州轻工业学院" /></a>
                <a href="#" target="_blank"><img src="images/imgs/p5.jpg"
                    title="橡树小屋的blog" alt="橡树小屋的blog" /></a>
                <a href="#" target="_blank"><img src="images/imgs/p3.jpg"
                    title="橡树小屋的blog" alt="橡树小屋的blog" /></a>
                <a href="#" target="_blank"><img src="images/imgs/p4.jpg"
                    title="橡树小屋的blog" alt="橡树小屋的blog" /></a>
            </div>
        </div>
        <div id="rank">
            <h3 style="color:red;">帖子排行榜:</h3>
            <ul>
            <s:iterator value="hotPosts" id="row" status="st">
                <li><a href="post!viewDetail.action?pid=<s:property value="
                    id" />"><s:property value="#row.name" />  </a>【点
                    击量<s:property value="#row.count" />】</li>
            </s:iterator>
            </ul>
</div>
<div class="clear" ></div>
<div id="Lboard">
    <s:iterator value="rootBoard" id="row">
        <div class="btitle"><s:property value="#row.name" /></div>
        <div class="subBoard">
            <ul>
                <s:iterator value="#row.boards" id="sub">
                    <li>
                        <a href="login!showAll.action?bid=
                            <s:property value="#sub.id" />">
                            <s:if test="#sub.boardImg!=null">
                                <img width="60" height="60" src="<%=request.
                                    getContextPath()%>/upload/<s:property value
                                    ="#sub.boardImg"/>" class="bimg" ></img>
                            </s:if>
                            <s:else>
                                <img width="60" height="60" src="<%=
                                    request.getContextPath()%>/images/bimg.
                                    gif" class="bimg" ></img>
                            </s:else>
                            <h4><s:property
                                value="#sub.name" /></h4>
                        </a>
                    </li>
                </s:iterator>
            </ul>
```

```
            </div>
            <div class="clear" ></div>
        </s:iterator>      </div>
    <div class="btitle">友情链接</div>
    <div class="subBoard">
        <ul>
            <li><a href="http://www.zzuli.edu.cn/">郑州轻工业学院</a></li>
            <li><a href="http://www.tup.tsinghua.edu.cn/">清华大学出版社
            </a></li>
        </ul>
    </div>
</body>
</html>
```

装饰器页面(myDecorator.jsp)代码如下：

```
<%@page language="java" import="java.util.*" pageEncoding="UTF-8"%>
<%@taglib prefix="decorator"
    uri="http://www.opensymphony.com/sitemesh/decorator" %>
<%@taglib prefix="page" uri="http://www.opensymphony.com/sitemesh/page" %>
<%@taglib prefix="s" uri="/struts-tags" %>
<!DOCTYPE HTML PUBLIC "-//W3C//DTD HTML 4.01 Transitional//EN">
<html>
<head>
    <title><decorator:title default="装饰器页面"/></title>
    <decorator:head/>
    <link rel="stylesheet" type="text/css" href="../Style.css">
    <style type="text/css">
        * { margin: 0; padding: 0; font-size: 13px;}
        #outer{ width: 980px; margin: 0 auto;  margin-left: 15%);}
        #main{ border: 1px solid #A6CBE7; border-top: none; min-height: 500px;
             height: auto; overflow:visible; padding-top: 15px;}
        #searchBar {height:50px; margin-top: 5px; border:1px solid #D0E3F4;
            border-top:none; background-color: #F4F9FF;   }
        #searchKey {border:1px solid sliver; height:25px; line-height:30px;
            width:200px;}
        #searchBar ul{ float: left; margin-top: 15px; margin-left: 5%;}
        #searchBar ul li{float: left; width: 310px;}
    </style>
      <script type="text/javascript">
        function checkForm()
        {
            if (document.getElementById("searchKey").value=="") {
                return false;
            }
            else{return true;}
```

```jsp
            }
        </script>
</head>
<body>
    <div id="outer">
    <%@ include file="../top.jsp" %>
    <div id="searchBar"  >
        <form action="post!searchPost.action" id="searchForm" style="float: left;" >
            <span>
                <img style="position: relative; top:10px;" src="<%=request.getContextPath()%>/images/search_1.png" />    <input id="searchKey" type="text" name="post.name"  placeholder="搜索其实很简单" ></input><input type="submit" value="搜索" style="border: none; background: url(images/search_btn.png) no-repeat; width: 45px; text-align: left; text-indent: 10px;  height: 25px; margin-left: 15px; line-height: 30px; position: relative; top: 0px; cursor: pointer; "  onclick="return checkForm()" ></input>

            </span>
        </form>
        <!--<s:property value="@com.tools.StaticNumber@give()" />a-->
        <ul>
            <li>昨日帖子：<%=application.getAttribute("yestNum")%>    今日帖子：<%=application.getAttribute("todayNum")%>   共：<s:property value="#application.total" />    </li>
            <li id="time"></li>
            <script>
                function show()
                {
                    now=new Date();
                    year=now.getFullYear();
                    month=now.getMonth()+1;
                    date=now.getDate();
                    hours=now.getHours();
                    minutes=now.getMinutes();
                    seconds=now.getSeconds();
                    if(minutes<=9)
                        minutes="0"+minutes
                    if(seconds<=9)
                        seconds="0"+seconds
                        time.innerHTML=year+"年"+month+"月"+date+"日"+"   现在时间:"+hours+":"+minutes+":"+seconds;
                    setTimeout("show()",1000);
                }
```

```
                show();
            </script>
        </ul>
    </div>
    <div id="main" >
        <decorator:body/>
    </div>
    <hr />
    <div>
        <center>
        <%@include file="../bottom.html" %>
        </center>
    </div>
</div>
</body>
</html>
```

myDecorator.jsp 页面中使用 JSP 指令导入 top.jsp 和 bottom.html。top.jsp 页面的运行效果如图 11-8 所示。

图 11-8 top.jsp 页面运行效果

top.jsp 代码如下：

```
<%@page import="com.orm.Student"%>
<%@page contentType="text/html" pageEncoding="UTF-8"%>
<%@taglib prefix="s" uri="/struts-tags" %>
<!DOCTYPE html>
<html>
    <head>
        <meta http-equiv="Content-Type" content="text/html; charset=UTF-8">
        <title>JSP Page</title>
```

```html
<style type="text/css">
    .ul { list-style-type: none;}
    #po{}
</style>
</head>
<body>
    <div id="wrap">
        <div id="Top">
            <div id="logo">
                <img src="<%=request.getContextPath()%>/images/media.jpg"
                    alt="郑州轻工业学院 BBS" border="0" />
                <label style="font-size:30px; font-weight: bold;">
                    郑州轻工业学院论坛</label>
                <div id="logRegist">
                    <s:if test="#session.student!=null">
                        <span id="po"><img style=" float: right; width:50px;
                        height:50px;" src="<%= request.getContextPath()%>/
                        upload/
                        <s:property value="#session.student.photoPath"/>"
                        /></span>
                        <h4><a href="student!personalStuInfo.action">
                            <s:property value="#session.student.nickName" />
                            </a>欢迎你!</h4>
                    </s:if>
                    <s:else>
                        <span id="po"><img style="float: right;width:50px;
                        height:50px;" src="<%= request.getContextPath()%>/
                        images/bbsPhoto.jpg" /></span>
                         <h4><a href="student!personalStuInfo.action"><s:
                         property value="#session.student.nickName" /></a>
                         </h4>
                    </s:else><%if (request.getSession().getAttribute(
                    "admin") !=null) {%>
                    <h4><a href="#">管理员</a>,欢迎您!</h4>
                    <%}
                        if (request.getSession().getAttribute("student")
                            ==null && request.getSession().getAttribute(
                            "admin") ==null) {%>
                    <h4>您 好,请<a href="<%=request.getContextPath()%>
                        /login.jsp" style="color:orange;">登 录</a></h4>
                    <%}
                    %>
                </div>
            </div>
        </div>
```

```html
<div id="s_head">
    <div id="menu">
        <ul id="menu_left">
            <li id="m_01"><a id="a_01"
            href="<%=request.getContextPath()%>/index.action">首
             页</a></li>
            <li class="menu_ge"></li>
            <li id="m_03"><a href="post!viewPostsByUser.action">我的
            帖子</a></li>
            <s:if test="#session.student!=null">
                <li class="menu_ge"></li>
            <li id="m_04"><a href="student!personalStuInfo.action">
            个人资料</a></li>
            <li class="menu_ge"></li>
            <li id="m_07"><a href="<%=
            request.getContextPath()%>/PersonalInfo/StuPswModify.
            jsp">修改密码</a></li>
            </s:if>
            <s:elseif test="#session.admin!=null">
                <li class="menu_ge"></li>
            <li id="m_04"><a href=
            "admin!personalAdminInfo.action">个人资料</a></li>
            </s:elseif>
            <li class="menu_ge"></li>
            <%if (request.getSession().getAttribute("admin") !=null) {
             %>
            <li id="m_07"><a href="post">帖子管理</a></li>
            <li class="menu_ge"></li>
            <li id="m_07"><a href="board!loadRootBoards.action">
                板块管理</a></li>
             <li class="menu_ge"></li>
             <li id="m_07"><a href="#">系统维护</a></li>
             <li class="menu_ge"></li>
            <%}%>
            <li id="m_08"><a href="login!exit.action">退出</a></li>
        </ul>
    </div>
</div>
        </body>
</html>
```

bottom.html 代码如下：

```
<!DOCTYPE HTML PUBLIC "-//W3C//DTD HTML 4.01 Transitional//EN">
<html>
```

```
<head>
    <meta http-equiv="content-type" content="text/html; charset=utf-8">
    <link rel="stylesheet" type="text/css" href="Style.css">
</head>
<body>
    <center>
        <div>
            <br/>
            郑州轻工业学院版权所有 电话:86-(0)371-63556666
            地址:郑州市科学大道166号
            <br/>
            豫 ICP 备 050023456 号    网络导航
        </div>
    </center>
</body>
</html>
```

myDecorator.jsp 使用 Sitemesh 转载了 index.jsp 页面,还需要对该页面进行配置,配置文件 decorators.xml 的代码如下:

```
<?xml version="1.0" encoding="UTF-8"?>
<decorators defaultdir="/decorators">
    <decorator name="myDecorator" page="myDecorator.jsp">
        <pattern>/*</pattern>
        <pattern>/*.action</pattern>
        <pattern>/*.jsp</pattern>
    </decorator>
</decorators>
```

11.4.2 BBS 登录功能的实现

单击图 11-3 所示页面中的超链接"登录"(参考 top.jsp 中的代码),出现如图 11-9 所示的登录页面(login.jsp)。可用学生或管理员两种身份登录,未登录用户只能浏览板块和帖子,不能发帖和回帖。

login.jsp 的代码如下:

```
<%@page contentType="text/html" pageEncoding="UTF-8"%>
<%@taglib prefix="s" uri="/struts-tags" %>
<html>
    <head>
        <meta http-equiv="Content-Type" content="text/html; charset=UTF-8">
        <title>郑州轻工业学院论坛</title>
        <script type="text/javascript">
            window.onload=function(){
                var tabS=document.getElementById("tabL");
                var tabA=document.getElementById("tabA");
```

图 11-9 登录页面

```
        var stuL=document.getElementById("stuLoginF");
        var adminL=document.getElementById("adminLoginF");

        tabS.onclick=function(){
            if (adminL.style.display =="block") {
                adminL.style.display ="none"
            }
         if (tabA.style.backgroundColor!="") {
        tabA.style.backgroundColor="";
        }
            tabS.style.backgroundColor="#2FB4D6";
            stuL.style.display="block";
        }
        tabA.onclick=function(){
            if (stuL.style.display=="block") {
            stuL.style.display="none";
            }
            if ( tabS.style.backgroundColor!="") {
              tabS.style.backgroundColor="";
            }
                tabA.style.backgroundColor  ="#2FB4D6";
                adminL.style.display ="block";
            }
        }
    </script>
    <style type="text/css">
        #tabs{ height: 20px; width: 300px; margin: 0 auto; }
```

```
            #tabContent{ width: 400px; height: 200px;  border: 1px solid #CACACA;
            margin-top: 10px;   border-radius:15px 15px 0px 0px; box-shadow:15px
            2px 4px #000;    }
            ul{ float: left; list-style-type: none;   line-height: 25px; }
            li{ width: 120px; height: 30px; text-align: center; float: left; display:
            block; cursor: pointer;   }
            #tabs{position:relative; left:-30px;}
            #tabs li { display:block;   border-radius:5px 35px 0px 0px;   }
            li:hover{  }
            .loginBtn{ border: none; width: 128px; height: 40px;
            background:url('<%=request.getContextPath() %>/images/
            btn-submit.png') no-repeat;}
            .loginBtn:hover{cursor: pointer;}
            .outerBorder{ width: 200px; height:25px; background-color: #F1FAFF;
            border: #C1C1C1 1px solid; }
            .tdLabel{ font-family:宋体; font-size: 14px;}
            td { padding-top: 15; }
            #login{margin-left:-30px;}
        </style>
    </head>
    <body>

    <center>
        <h3>Sorry,您还未登录,请登录:</h3>
        <div id="tabs">
            <ul>
                <li id="tabL" style=" background-color:#2FB4D6;"
                    margin-right: 5px;" ><b>学生登录</b></li>
                <li id="tabA"><b>管理员登录</b></li>
            </ul>
        </div>
        <div id="tabContent">
        <div id="stuLoginF" style="display: block;" >
            <s:form action="login">
                <s:textfield name="student.stuNum" label="学号"
                    cssClass="outerBorder"></s:textfield>
                <s:password name="student.password" label="密码"
                    cssClass="outerBorder"></s:password>
                <s:submit value=""   cssClass="loginBtn"></s:submit>
            </s:form>
            <s:actionmessage/>
        </div>
            <div id="adminLoginF" style="display: none;" >

                <s:form action="login!adminLogin.action">
```

```
            <s:textfield name="admin.account" label="账号"
                cssClass="outerBorder" ></s:textfield>
            <s:password name="admin.password" label="密码"
                cssClass="outerBorder"></s:password>
            <s:submit value="" cssClass="loginBtn" ></s:submit>
        </s:form>
        <s:actionmessage/>
      </div>
    </div>
  </center>
  </body>
</html>
```

登录页面(login.jsp)对应的业务控制器为 LogInAction(LogInAction.java),该业务控制器是动态 Action,可分别处理两类登录用户的请求,提供 execute()和 adminLogin()方法,也实现了分页、注销登录功能,对应的方法分别为 showAll()和 exit(),代码如下:

```
package com.action;
import com.opensymphony.xwork2.ActionContext;
import com.opensymphony.xwork2.ActionSupport;
import com.orm.Admin;
import com.orm.Board;
import com.orm.Post;
import com.orm.Student;
import com.service.IBoardService;
import com.service.ILoginService;            //导入 ILoginService 接口
import com.service.IPostService;
import com.service.IStudentService;
import java.util.Iterator;
import java.util.List;
import javax.annotation.Resource;
import javax.servlet.http.HttpServletRequest;
import org.apache.struts2.ServletActionContext;

public class LogInAction extends ActionSupport {
    private Student student;
    private Admin admin;
    private String tip;
    private int pageNo;                       //当前页
    private int pageSize;                     //每页条数
    private int count;
    private List<Post>list;
    private int bid;
    private Board board;
    @Resource(name="loginService")
    ILoginService loginService;
```

```java
@Resource(name="postService")
IPostService postService;
@Resource(name="studentService")
IStudentService studentService;
@Resource(name="boardService")
IBoardService boardService;
public int getCount() {
    return count;
}
public void setCount(int count) {
    this.count=count;
}
public LogInAction() {
}
public int getPageNo() {
    return pageNo;
}
public void setPageNo(int pageNo) {
    this.pageNo=pageNo;
}
public int getPageSize() {
    return pageSize;
}
public void setPageSize(int pageSize) {
    this.pageSize=pageSize;
}
public List getList() {
    return list;
}
public void setList(List list) {
    this.list=list;
}
public Student getStudent() {
    return student;
}
public void setStudent(Student student) {
    this.student=student;
}
public String getTip() {
    return tip;
}
public void setTip(String tip) {
    this.tip=tip;
}
public Admin getAdmin() {
```

```java
        return admin;
    }
    public void setAdmin(Admin admin) {
        this.admin=admin;
    }
    public int getBid() {
        return bid;
    }
    public void setBid(int bid) {
        this.bid=bid;
    }
    public Board getBoard() {
        return board;
    }
    public void setBoard(Board board) {
        this.board=board;
    }
    public String execute() throws Exception {
        //登录前清空所有 Session
        ActionContext.getContext().getSession().clear();
        Student s=loginService.stuLogin(student);
        if (s !=null) {
            ActionContext.getContext().getSession().put("student", s);
            List<Post> result=postService.allPost();
            setList(result);
            System.out.println("OK");
            return "loginSuccess";
        }
        addActionMessage("用户名或密码错误!");
        System.out.println("BAD");
        return INPUT;
    }
    public String adminLogin() throws Exception {
        //登录前清空所有 Session
        ActionContext.getContext().getSession().clear();
        Admin a=loginService.adminLogin(getAdmin());
        if (a !=null) {
            ActionContext.getContext().getSession().put("admin", a);
            List<Post> result=postService.allPost();
            setList(result);
            return "loginSuccess";
        }
        addActionMessage("用户名或密码错误!");
        return INPUT;
    }
```

```java
//帖子分页列表
public String showAll() {
    //设置每页条数
    setPageSize(15);
    HttpServletRequest request=ServletActionContext.getRequest();
    Student tempStudent = (Student) request.getSession().getAttribute("student");
    setBoard(boardService.loadBoard(getBid()));
    if (request.getParameter("page")!=null) {
        setPageNo(Integer.parseInt(request.getParameter("page")));
        System.out.println(pageNo);
    } else {
        setPageNo(1);
    }
    if (request.getParameter("bid")!=null) {
        setBid(Integer.parseInt(request.getParameter("bid")));
    }
    try {
        setCount(postService.getPostsCount());
        setList(postService.pageAllPost(getBid(),getPageNo(), getPageSize()));
        if(tempStudent!=null) {
            setStudent(studentService.getStudentByStuNum(
                            tempStudent.getStuNum()));
        }
        return SUCCESS;
    } catch (Exception e) {
        return ERROR;
    }
}
//注销登录
public String exit() {
    ActionContext.getContext().getSession().clear();
    return "exit";
}
```

该类导入的 PO 类包括 Admin、Board、Post、Student,这些类的代码如前所述。该类导入的接口包括 IBoardService、IPostService、IStudentService,这三个接口及其实现类的代码前面也已经介绍。此外还导入了接口 ILoginService(ILoginService.java),该接口封装了管理员登录和用户登录要使用的方法,其实现类是 LoginServiceImpl(LoginServiceImpl.java),代码分别如下:

```java
//ILoginService.java
package com.service;
import com.orm.Admin;
```

```java
import com.orm.Student;

public interface ILoginService {
    //学生即会员登录
    public Student stuLogin(Student student);
    //管理员登录
    public Admin adminLogin(Admin admin);
}
//LoginServiceImpl.java
package com.serviceImpl;
import com.dao.BaseDao;
import com.orm.Admin;
import com.orm.Student;
import com.service.ILoginService;
import javax.annotation.Resource;

public class LoginServiceImpl implements ILoginService {
    @Resource(name="dao")
    BaseDao dao ;
    public Student stuLogin(Student student) {
        //验证学号和密码
        Student stu= (Student) dao.loadObject("from Student as s where s.stuNum='"+student.getStuNum()+"' and s.password='"+student.getPassword()+"' ");
        if (stu!=null) {
            return  stu;
        }
        return  null;
    }
    public Admin adminLogin(Admin admin) {
        //管理员登录验证
        Admin a= (Admin) dao.loadObject("from Admin as a where a.account='"+admin.getAccount()+"' and a.password='"+admin.getPassword()+"' ");
        if (a!=null) {
            return  a;
        }
        return  null;
    }
}
```

从业务控制器 LogInAction 中可以看出，如果会员即学生登录成功会返回字符串 "loginSuccess"，如果管理员登录成功返回的也是 "loginSuccess"，该逻辑视图对应的页面为文件夹 MessageInfo 中的 loginSuccess.jsp，页面运行效果如图 11-10 所示。

代码如下：

```jsp
<%@page contentType="text/html" pageEncoding="UTF-8"%>
<html>
```

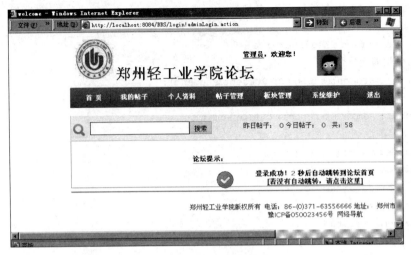

图 11-10 登录成功页面

```html
<head>
    <meta http-equiv="Content-Type" content="text/html; charset=UTF-8">
    <script type="text/javascript">
        function countDown(secs,surl){
            //alert(surl);
            var jumpTo=document.getElementById('jumpTo');
            jumpTo.innerHTML=secs;
            if(--secs>0){
                setTimeout("countDown("+secs+",'"+surl+"')",1000);
            }
            else{
                location.href=surl;
            }
        }
    </script>
    <title>welcome</title>
</head>
<body>
    <center>
        <div style="margin-top:5%;">
        <div id="title" style="border-bottom:1px solid #C8DCEC;
            width: 50%;text-align: left; "><h3>论坛提示:</h3></div>
        <span style="position:relative; left:-180px; top:40px;">
        <img src="<%=request.getContextPath()%>/images/right_big.gif" />
        </span>
        <h3>登录成功!<span id="jumpTo"
        style="color:orange;">5</span> 秒后自动跳转到论坛首页</h3>
        <script type="text/javascript">countDown(5,'index.action');</script>
        <h3><a href="index.action">[若没有自动跳转,请点击这里]</a></h3>
```

```
            </div>
        </center>
    </body>
</html>
```

从图11-10中可以看出,页面提示登录成功并提示转到论坛首页,转到论坛首页时会加载用户管理界面,如图11-11所示。图11-11所示页面是管理员页面,图11-3所示页面是游客访问的页面,比较图11-3和图11-11可以看出,管理员页面比游客访问页面增加了"个人资料"和"帖子管理"、"板块管理"等功能。

图11-11　管理员登录成功后的主页面

11.4.3　BBS板块管理功能的实现

单击图11-11所示页面中的板块管理,出现如图11-12所示的页面,从其地址栏中信息可以看出,该请求被提交到了 http://localhost:8084/BBS/board!loadRootBoards.action,其中参数 board!loadRootBoards.action 表示在名为 board 的业务控制器中的 loadRootBoards()方法,参考 struts.xml 配置文件。名为 board 的业务控制器是 BoardAction(BoardAction.java),该业务控制器是个动态 Action,提供有关板块的多个处理方法。参照 struts.xml 可以看出,该页面对应的 JSP 页面是 MainBoard.jsp,在该页面中加载了已经创建的板块信息。

业务控制器 BoardAction(BoardAction.java)的代码如下:

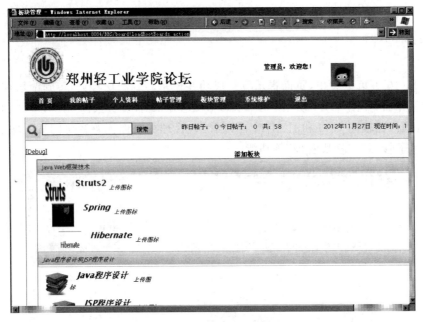

图 11-12 板块管理的主页面

```java
package com.action;
import com.opensymphony.xwork2.ActionSupport;
import com.orm.Admin;
import com.orm.Board;
import com.service.IBoardService;
import java.util.List;
import javax.annotation.Resource;
import javax.servlet.http.HttpServletRequest;
import org.apache.struts2.ServletActionContext;

public class BoardAction extends ActionSupport{
    @Resource(name="boardService")
    IBoardService boardService;

    private List<Board> boardList;
    //子节点
    private List<Board> childBoards;
    //根节点
    private List<Board> rootBoards;
    private Board board;
    public Board getBoard() {
        return board;
    }
    public void setBoard(Board board) {
        this.board=board;
```

```java
    }
    public List<Board> getChildBoards() {
        return childBoards;
    }
    public void setChildBoards(List<Board> childBoards) {
        this.childBoards=childBoards;
    }
    public List<Board> getBoardList() {
        return boardList;
    }
    public void setBoardList(List<Board> boardList) {
        this.boardList=boardList;
    }
    public List<Board> getRootBoards() {
        return rootBoards;
    }
    public void setRootBoards(List<Board> rootBoards) {
        this.rootBoards=rootBoards;
    }
    //加载主板块
    @Override
    public String execute() throws Exception {
        try {
            loadRootBoards();
            return SUCCESS;
        } catch (Exception e) {
            return ERROR;
        }

    }
    //加载所有板块
    public String listBoard() throws Exception
    {
        try {
                setBoardList(boardService.loadAllBoards());
                return SUCCESS;
        } catch (Exception e) {
            e.printStackTrace();
            return ERROR;
        }
    }
    //加载二级板块
    public String listChildBoards() throws Exception
    {
        int parentId=
```

```
        Integer.valueOf(ServletActionContext.getRequest().getParameter("parentId"));
        try {
            loadRootBoards();
            setChildBoards(boardService.loadChildBoards(parentId));
            return SUCCESS;
        } catch (Exception e) {
            e.printStackTrace();
            return ERROR;
        }
    }
    //加载根板块
    public String loadRootBoards() throws Exception
    {
        try {
            setRootBoards(boardService.loadRootBoards());
            return SUCCESS;
        } catch (Exception e) {
            e.printStackTrace();
            return ERROR;
        }
    }
    //跳转添加板块
    public String prepareAddBoard() throws Exception
    {
        try {
            setBoardList(boardService.loadAllBoards());
            return "addBoard";
        } catch (Exception e) {
            e.printStackTrace();
            return ERROR;
        }
    }
    //添加板块
    public String addBoard() throws Exception
    {
        Admin admin=
(Admin)ServletActionContext.getRequest().getSession().getAttribute("admin");
        //获取板块 id
        //判断添加一级板块或者下级板块,首先取出 parentId
        int bid=getBoard().getId();
        int parentId=0;
        Board tempBoard =new Board() ;
        //赋值的时候一定要判断是否为空
        //有些空指针异常是由于 Session 失效导致的
        try
```

```java
        {       //处理添加一级板块
            if (bid==-1)
            {
                tempBoard.setName(getBoard().getName());
                tempBoard.setAdmin(admin);
                tempBoard.setDescription(getBoard().getDescription());
                //提交
                if(boardService.saveOrUpdateBoard(tempBoard)) {
                    return "addSuccess";
                }
                return ERROR;
            }
            //处理添加二级板块
            else
            {
                //逻辑错误
                parentId=
                boardService.loadBoard(bid).getBoard().getId();
                //根据id加载被添加子板块的父板块
                Board board=boardService.loadBoard(bid);
                tempBoard.setName(getBoard().getName());
                tempBoard.setAdmin(admin);
                tempBoard.setBoard(board);
                tempBoard.setDescription(getBoard().getDescription());
                //提交
                if ( boardService.saveOrUpdateBoard(tempBoard)) {
                    return "addSuccess";
                }
                return ERROR;
            }
        }
        catch (Exception e) {
            e.printStackTrace();
            return ERROR;
        }
    }
    public String prepareModifyBoard() throws Exception
    {
        //获取板块id
        HttpServletRequest request=ServletActionContext.getRequest();
        int bid=0;
        if(request.getParameter("bid")!=null) {
            bid=Integer.valueOf(request.getParameter("bid").toString());
        }
        setBoard(boardService.loadBoard(bid));
```

```
            loadRootBoards();
            return "prepareModifyBoard";
    }
}
```

MainBoard.jsp 页面的代码如下：

```
<%@page contentType="text/html" pageEncoding="UTF-8"%>
<%@taglib  prefix="s" uri="/struts-tags" %>
<html>
    <head>
        <meta http-equiv="Content-Type" content="text/html; charset=UTF-8">
        <title>板块管理</title>
        <style type="text/css">
            #Lboard{ margin-left:24px; width:90%; border: 1px solid #A6CBE7;  }
            .subBoard{}
            .btitle{padding-left:10px; text-align:left; line-height:30px;  color: #
                1B72AF; border: 1px solid # D6E8F4; background: url ('<%= request.
                getContextPath()%>/images/h.png') repeat-x;}
            #Lboard  ul li{  height:30px; margin:5px; display:block; width:250px;
                height:60px;  float:left; padding:5px;}
            #Lboard ul li a{ text-decoration:none;}
            #Lboard ul li a:hover{ text-decoration:underline;}
            #Lboard .bimg{ float:left;}
            #Lboard strong { margin-left:15px; height:24px; font-size:18px;}
            #Lboard p { padding-left:60px;}
            .clear{clear:both;}
        </style>
    </head>
    <body>
        <s:debug/>
<center><h2><a href="<%=request.getContextPath() %>
        /board!prepareAddBoard.action">添加板块</a></h2>
</center>
        <div id="Lboard">

        <s:iterator value="rootBoards" id="row">
        <div class="btitle"><s:property value="#row.name" /></div>
        <div class="subBoard">
        <s:iterator value="#row.boards" id="sub">
        <ul>
            <li>
                <a href="login!showAll.action?bid=<s:property value="#sub.id" />">
                    <s:if test="#sub.boardImg!=null">
                    <img width="50" height="50"
                       src="<%=request.getContextPath()%>
```

```
                    /upload/<s:property value="#sub.boardImg" />"
                    class="bimg" ></img>
                </s:if>
                <s:else>
                    <img width="50" height="50"
                        src="<%=request.getContextPath()%>
                        /images/bimg.gif" class="bimg" ></img>
                </s:else>
                <strong><s:property value="#sub.name" /></strong>
                <em><a href="<%=request.getContextPath()%>
                    /BoardManage/modifyBoardImg.jsp?bid=<s:property value="#sub.
                    id" />">上传图标</a></em>
                </a>
            </li>
        </s:iterator>
        </ul>
    </div>
        <div class="clear" ></div>
    </s:iterator>
    </div>
    <div style="clear:both;"></div>
    </body>
</html>
```

单击图 11-12 所示页面中的"添加板块",页面跳转到如图 11-13 所示的页面。从图 11-13 中的 URL 可以看出业务请求提交给了业务控制器 BoardAction(BoardAction.java)的 prepareAddBoard()方法。参照 struts.xml 可以看出,对应的 JSP 页面为 AddBoard.jsp。

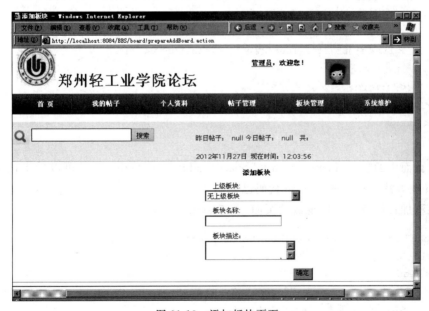

图 11-13 添加板块页面

其中,"无上级板块"表示可以直接创建一个大板块,如图 11-12 所示页面中的"Java Web 框架技术"就是大板块。也可以选择一个已经存在的上级板块后为其添加下级板块。

AddBoard.jsp 代码如下:

```jsp
<%@page contentType="text/html" pageEncoding="UTF-8"%>
<%@taglib prefix="s" uri="/struts-tags" %>
<html>
    <head>
        <meta http-equiv="Content-Type" content="text/html; charset=UTF-8">
        <title>添加板块</title>
        <style type="text/css">
            li{ margin: 10px;}
        </style>
    </head>
<body>
<center>
        <h1>添加板块</h1>
        <s:form action="board!addBoard.action">
            <ul>
            <li>上级板块:<select name="board.id">
            <option value="-1">无上级板块</option>
            <s:iterator value="boardList" id="row">
                <option value="<s:property value="#row.id" />">
                    <s:property value="#row.name" /></option>
            </s:iterator>
            </select></li>
            <li>板块名称:<input type="text" name="board.name"></input></li>
            <li>板块描述:
                <textarea type="text" name="board.description"></textarea></li>
            <li><s:submit value="确定" ></s:submit></li>
            </ul>
        </s:form>
</center>
</body>
</html>
```

从该页面中的"<s:form action="board! addBoard.action">"可以看出该页面对应的 Action 及方法。对应的 Action 代码已给出。

如果添加板块成功将出现成功页面,参考 struts.xml,

```jsp
<%@page contentType="text/html" pageEncoding="UTF-8"%>
<html>
    <head>
        <meta http-equiv="Content-Type" content="text/html; charset=UTF-8">
        <title>添加板块成功</title>
        <script type="text/javascript">
```

```
function countDown(secs,surl){
    var jumpTo=document.getElementById('jumpTo');
    jumpTo.innerHTML=secs;
    if(--secs>0){
        setTimeout("countDown("+secs+",'"+surl+"')",1000);
    }
    else{
        location.href=surl;
    }
}
</script>
</head>
<body>
  <center>
    <div style="margin-top:5%;">
        <div id="title" style="border-bottom:1px solid #C8DCEC; width:
        50%;text-align: left; "><h3>论坛提示:</h3></div>
        <span style="position:relative; left:-180px; top:40px;"><img src="
        <%=request.getContextPath()%>/images/right_big.gif" /></span>
        <h3>板块添加成功!<span id="jumpTo" style="color:orange;">5</span
        > 秒后自动跳转</h3>
        <script t ype="text/javascript">countDown(5,'<%=
        request.getContextPath()%>/board!prepareAddBoard.action');
        </script>
        <h3><a href="<%=request.getContextPath()%>
        /board!prepareAddBoard.action">[若没有自动跳转,请点击这里]</a></h3>
    </div>
  </center>
</body>
</html>
```

单击图11-12所示页面中的"上传图标"可以对某一个板块的图片进行修改,页面如图11-14所示。

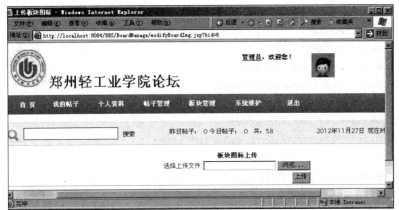

图11-14 上传板块图标

该页面文件 modifyBoardImg.jsp 的代码如下：

```jsp
<%@page contentType="text/html" pageEncoding="UTF-8"%>
<%@taglib prefix="s" uri="/struts-tags" %>
<html>
    <head>
        <meta http-equiv="Content-Type" content="text/html; charset=UTF-8">
        <title>修改板块信息</title>
    </head>
    <body>
        <center>
            <h1>修改板块信息</h1>
            <s:form>
                <s:textfield label="板块标题" value="board.name"></s:textfield>
                <s:textarea label="板块描述"
                    value="board.description"></s:textarea>
                <select name="board.parentId">
                    <s:iterator value="rootBoards.name" id="row">
                        <option></option>
                    </s:iterator>
                </select>
                <s:submit value="提交"></s:submit>
            </s:form>
        </center>
    </body>
</html>
```

上传文件使用了 Struts2 的文件上传拦截器，文件上传后交由 MyUpload(MyUpload.java)控制器进行处理。文件上传拦截器以及 Action 的配置请参考 struts.xml。

MyUpload.java 的代码如下：

```java
package com.action;
import com.opensymphony.xwork2.ActionContext;
import com.opensymphony.xwork2.ActionSupport;
import com.orm.Board;
import com.orm.Student;
import com.service.IBoardService;
import com.service.IStudentService;
import java.io.*;
import javax.annotation.Resource;
import org.apache.struts2.ServletActionContext;

@SuppressWarnings("serial")
public class MyUpload extends ActionSupport {
    @Resource(name="studentService")
    IStudentService studentService;
```

```java
@Resource(name="boardService")
IBoardService boardService;
//封装上传文件的属性
private File doc;
//封装上传文件的类型
private String docContentType;
//封装上传文件名
private String docFileName;
private String path;
public File getDoc() {
    return doc;
}
public void setDoc(File doc) {
    this.doc=doc;
}
public String getDocContentType() {
    return docContentType;
}
public void setDocContentType(String docContentType) {
    this.docContentType=docContentType;
}
public String getDocFileName() {
    return docFileName;
}
public void setDocFileName(String docFileName) {
    this.docFileName=docFileName;
}
public void setPath(String value) {
    this.path=value;
}
@SuppressWarnings("deprecation")
public String getPath() throws Exception {
    return ServletActionContext.getServletContext().getRealPath(path);
}
@Override
public String execute() throws Exception {
    docFileName=getFileName(docFileName);
    FileOutputStream fos=new
        FileOutputStream(getPath()+"\"+docFileName);
    FileInputStream fis=new FileInputStream(doc);
    byte[]b=new byte[1024];
    int length=0;
    while ((length=fis.read(b)) >0) {
        fos.write(b, 0, length);
    }
```

```
        //将头像路径写入数据库
        Student stu   =
            (Student)ActionContext.getContext().getSession().get("student");
        if (stu!=null) {
            Student temp=studentService.getStudentByStuNum(stu.getStuNum());
            temp.setPhotoPath(getDocFileName());
            studentService.modifyStudent(temp);
        }
        return SUCCESS;
    }
    public String boardImgUpload() throws Exception
    {
        docFileName=getFileName(docFileName);
        FileOutputStream fos=
           new FileOutputStream(getPath()+"\"+docFileName);
        FileInputStream fis=new FileInputStream(doc);
        byte[] b=new byte[1024];
        int length=0;
        while ((length=fis.read(b)) >0) {
            fos.write(b, 0, length);
        }
        if (ServletActionContext.getRequest().getParameter("bid")!=null) {
            int bid=
            Integer.valueOf(ServletActionContext.getRequest().getParameter("bid"));
                Board board=boardService.loadBoard(bid);
                board.setBoardImg(getDocFileName());
                if (boardService.saveOrUpdateBoard(board)) {
                return SUCCESS;
            }
                return  ERROR;
        }
        return ERROR;
    }   private String getFileName(String fileName) {
        int position=fileName.lastIndexOf(".");
        String extension=fileName.substring(position);
        return System.currentTimeMillis()+extension;
    }
}
```

文件上传成功后的效果如图 11-15 所示,页面文件是 uploadSuccess.jsp。
uploadSuccess.jsp 代码如下:

```
<%@page language="java" import="java.util.*" pageEncoding="gb2312"%>
<%@taglib prefix="s" uri="/struts-tags"%>
<!DOCTYPE HTML PUBLIC "-//W3C//DTD HTML 4.01 Transitional//EN">
<html>
```

图 11-15　文件上传成功页面

```
<head>
    <title>文件上传成功</title>
    <link rel="stylesheet" type="text/css" href="Style.css">
</head>
<body>
    <center>
        <h3>文件上传成功</h3>
        <hr/>
        <img src="<%=request.getContextPath()%>/upload/
            <s:property value="docFileName"/>"/>
        <br/>
        <s:property value="docFileName"/>
    </center>
</body>
</html>
```

11.4.4　BBS 帖子管理功能的实现

单击图 11-11 所示页面中的"帖子管理",请求提交给名为 post 的业务控制器 PostsAction(PostsAction.java)来处理,该业务控制器是动态 Action,封装了帖子管理的常用操作。处理后把所有帖子返回到帖子管理页面上(PostManage.jsp),请参考 struts.xml。返回的页面效果如图 11-16 所示。

PostsAction.java 的代码如下:

```
package com.action;
import com.opensymphony.xwork2.ActionContext;
import com.opensymphony.xwork2.ActionSupport;
import com.orm.Admin;
import com.orm.Board;
import com.orm.Post;
import com.orm.Reply;
import com.orm.Student;
```

图 11-16 帖子管理页面

```
import com.service.IAdminService;
import com.service.IBoardService;
import com.service.IPostService;
import com.service.IStudentService;
import java.util.Date;
import java.util.List;
import java.util.Set;
import javax.annotation.Resource;
import javax.servlet.http.HttpServletRequest;
import org.apache.struts2.ServletActionContext;

public class PostsAction extends ActionSupport {
    private Post post;
    private List<Post>list;
    private Set<Reply>replies;
    private List<Post>myPosts;
    private Student student;
    private Admin admin;
    private String result;
    public String getResult() {
        return result;
    }
    public void setResult(String result) {
        this.result=result;
    }
    @Resource(name="postService")
    IPostService postService;
    @Resource(name="boardService")
    IBoardService boardService;
```

```java
@Resource(name="studentService")
IStudentService studentService;
@Resource(name="adminService")
IAdminService adminService;
public Admin getAdmin() {
    return admin;
}
public void setAdmin(Admin admin) {
    this.admin=admin;
}
public Student getStudent() {
    return student;
}
public void setStudent(Student student) {
    this.student=student;
}
public List<Post> getMyPosts() {
    return myPosts;
}
public void setMyPosts(List<Post> myPosts) {
    this.myPosts=myPosts;
}
public Set<Reply> getReplies() {
    return replies;
}
public void setReplies(Set<Reply> replies) {
    this.replies=replies;
}
public List<Post> getList() {
    return list;
}
public void setList(List<Post> list) {
    this.list=list;
}
public Post getPost() {
    return post;
}
public void setPost(Post post) {
    this.post=post;
}
//单击"帖子管理"后默认执行该方法,请参考 struts.xml
@Override
public String execute() throws Exception {
    setList(postService.allPost());
    return SUCCESS;
}
//跳转到发帖页面
```

```java
public String preparePost() throws Exception
{
    HttpServletRequest request=ServletActionContext.getRequest();
    Student s=(Student) request.getSession().getAttribute("student");
    if (s!=null) {
    setStudent(studentService.getStudentByStuNum(s.getStuNum()));
    }

    return "preparePost";
}
//发帖
public String addPost() throws Exception
{
    HttpServletRequest request=ServletActionContext.getRequest();
    int bid=0;
    if (request.getParameter("bid")!=null) {
        bid=Integer.valueOf(request.getParameter("bid"));
    }
    Post p=getPost();
    Board b=boardService.loadBoard(bid);
    Student s=(Student) ActionContext.getContext().getSession().get("student");
    Admin  a  =(Admin) ActionContext.getContext().getSession().get("admin");
    if (s!=null)
    {
        p.setStudent(s);
        p.setBoard(b);
        p.setContent(getPost().getContent());
        p.setPublishTime(new Date());
        p.setCount(0);
        if (postService.saveOrUpdate(getPost()))
        {
            post=p;
            return "postSuccess";
        }
        return ERROR;
    }
    if (a!=null) {
        p.setAdmin(a);
        p.setBoard(b);
        p.setContent(getPost().getContent());
        p.setPublishTime(new Date());
        p.setCount(0);
        if (postService.saveOrUpdate(getPost()))
        {
            post=p;
            return "postSuccess";
        }
```

```java
            return ERROR;
        }
        return ERROR;
    }
//查看帖子
    public String viewDetail()
    {
        HttpServletRequest request=ServletActionContext.getRequest();
        //学生登录
        if (request.getSession().getAttribute("student")!=null) {
            Student s= (Student) request.getSession().getAttribute("student");
            setStudent(studentService.getStudentByStuNum(s.getStuNum()));
        }
        //管理员登录
        if (request.getSession().getAttribute("admin")!=null) {
            Admin a= (Admin) request.getSession().getAttribute("admin");
            setAdmin(admin);
        }
        int pid = Integer.parseInt(request.getParameter("pid"));
        Post p=postService.loadPost(pid);
        //令点击量增1
        p.setCount(p.getCount()==null?0:p.getCount()+1);
        postService.saveOrUpdate(p);
        setPost(postService.loadPost(pid));
        if (getPost()!=null) {
            setReplies(postService.loadPost(pid).getReplies());
            return "viewDetail";
        }
        return   ERROR;
    }


//分页
    public void viewAll(int pageNo, int pageSize)
    {
        int bid=
            Integer.valueOf(ServletActionContext.getRequest().getParameter("bid"));
        setList(postService.pageAllPost(bid,pageNo, pageSize));
    }
//查看我的帖子
    public String viewPostsByUser()
    {
        Student s= (Student) ActionContext.getContext().getSession().get("student");
        Admin a= (Admin) ActionContext.getContext().getSession().get("admin");
        if (s!=null) {
            setMyPosts(postService.allPostsByUser(s));
            setStudent(studentService.getStudentByStuNum(s.getStuNum()));
            return "myposts";
```

```java
        }
        if (a!=null) {
                setMyPosts(postService.allPostsByUser(a));
                setAdmin(adminService.loadAdmin(a.getId()));
                return "myposts";
        }
        return ERROR;
}
//准备修改
public String prepareModify() throws Exception
{
    int pid=-1;
    if (ServletActionContext.getRequest().getParameter("pid")!=null) {
        pid= Integer.valueOf(ServletActionContext.getRequest().getParameter
           ("pid"));
    }
    setPost(postService.loadPost(pid));
    return "prepareSuccess";
}
//修改帖子
public String modifyPost() throws Exception
{
    int pid=-1;
    if (ServletActionContext.getRequest().getParameter("pid")!=null) {
        pid= Integer.valueOf(ServletActionContext.getRequest().getParameter
           ("pid"));
        Post tempPost=postService.loadPost(pid);
        tempPost.setName(getPost().getName());
        tempPost.setContent(getPost().getContent());
        if (postService.saveOrUpdate(tempPost)) {
            setResult("Mok");
            return "modifySuccess";
        }
        return ERROR;
    }
    return ERROR;
}
//删除帖子
public String deletePost() throws Exception
{
    int pid=-1;
    if (ServletActionContext.getRequest().getParameter("pid")!=null) {
        pid= Integer.valueOf(ServletActionContext.getRequest().getParameter
           ("pid"));
        if (postService.deletePost(pid)) {
            setResult("Dok");
            return "deleteSuccess";
```

```
            }
            return ERROR;
        }
        return ERROR;
    }
    //按帖子名称搜索帖子
    public  String searchPost() throws Exception
    {
        // 要先输入关键字,才能执行搜索函数以搜索帖子
        try {
        String searchKey=getPost().getName().trim();
            if (searchKey!="") {
            setList(postService.searchPosts(searchKey));
            return "searchSuccess";
            }
            return ERROR;
        } catch (Exception e) {
            e.printStackTrace();
            return ERROR;
        }
    }
}
```

业务控制器 PostsAction 用到了 PO（Reply.java）和接口 IAdminService（IadminService.java）。PO（Reply.java）的映射文件为 Admin.hbm.xml，接口的实现类为 AdminServiceImpl(AdminServiceImpl.java)。

PostManage.jsp 的代码如下：

```
<%@page contentType="text/html" pageEncoding="UTF-8"%>
<%@taglib prefix="s" uri="/struts-tags" %>
<html>
    <head>
        <meta http-equiv="Content-Type" content="text/html; charset=UTF-8">
        <title>帖子管理</title>
        <script type="text/javascript">
            function confirmDel()
            {
                if (!confirm("您确认要删除吗?此删除为级联删除?删除操作不可恢复!")) {
                    return false;
                }
                return true;
            }
        </script>
    </head>
    <body>
        <center>
```

```
<s:if test="#request.result=='Mok'">
    <h4 style="color:red;">
    <em>修改成功,您可以继续操作...</em>
   </h4>
</s:if>
<s:elseif test="#request.result=='Dok' ">
    <h4 style="color:red;">
    <em>删除成功,您可以继续操作...</em>
   </h4>
</s:elseif>
<div>
    <!--帖子列表-->
    <table style="width:98%;">
        <tr style="background-color:#E7EFEF;">
            <th style="width:10px;"></th>
            <th style="text-align:left;">帖子标题</th>
            <th>作者</th>
            <th>回复 / 点击</th>
            <th>发布时间</th>
            <th>可选操作</th>
        </tr>
        <s:iterator value="list" status="st" id="row">
            <tr>
                <td><img src="<%=request.getContextPath()%>/images/folder_new.gif" style="display: inline-block; margin: 4px 5px 0px 0px;" /></td>
                <td style="text-align: left;"><a href="post!viewDetail.action?pid=<s:property value="id" />"><s:property value="name" /></a></td>
                <td><s:property value="#row.getStudent().getNickName()" />
                </td>
                <td><s:property value="#row.getReplies().size()" />/
                    <s:property value="count" /></td>

                <td>
                    <!--<s:property value="publishTime" />-->
                    <s:date format="yyyy-MM-dd hh:mm:ss"
                        name="publishTime" />
                </td>
                <td>
                    <a href="post!prepareModify.action?pid=
                    <s:property value="#row.id"/>">修改</a>
                    <a href="post!deletePost.action?pid=
                        <s:property value="#row.id"/>" onclick="return confirmDel()">删除</a>
```

```html
                </td>
            </tr>
        </s:iterator>
    </table>
</center>
<!--生成分页-->
<div style="padding-left:30px; margin-top: 20px;">
    <img src="<%=request.getContextPath()%>/images/pn_post.png" style="cursor:pointer;" onclick="javascript:location.href='<%=request.getContextPath()%>/post!preparePost.action?bid=<s:property value="%{#request.bid}"/>'"/>
</div>
<center>
    <div id="displayPagination">
        <script type="text/javascript">
            var pg=new showPages('pg');
            var total=<s:property value="count" />;
            var pageSize=<s:property value="pageSize" />
            if (total%pageSize==0) {
                pg.pageCount=total/pageSiz;
            }
            pg.pageCount=total / pageSize+1 ;             //定义总页数(必需)
            pg.printHtml(2);
            pg.printHtml(5);
        </script>
    </div>
</center>
</body>
</html>
```

Reply.java 的代码如下：

```java
package com.orm;
import java.util.Date;

public class Reply    implements java.io.Serializable {
    private Integer id;
    private Admin admin;
    private Student student;
    private Post post;
    private String content;
    private Date publishTime;
    public Reply() {
    }
    public Reply(Admin admin, Student student, Post post, String content, Date publishTime) {
```

```java
        this.admin=admin;
        this.student=student;
        this.post=post;
        this.content=content;
        this.publishTime=publishTime;
    }
    public Integer getId() {
        return this.id;
    }
    public void setId(Integer id) {
        this.id=id;
    }
    public Admin getAdmin() {
        return this.admin;
    }
    public void setAdmin(Admin admin) {
        this.admin=admin;
    }
    public Student getStudent() {
        return this.student;
    }
    public void setStudent(Student student) {
        this.student=student;
    }
    public Post getPost() {
        return this.post;
    }
    public void setPost(Post post) {
        this.post=post;
    }
    public String getContent() {
        return this.content;
    }
    public void setContent(String content) {
        this.content=content;
    }
    public Date getPublishTime() {
        return this.publishTime;
    }
    public void setPublishTime(Date publishTime) {
        this.publishTime=publishTime;
    }
    public Reply(Post post, String content) {
        this.post=post;
        this.content=content;
```

        }
    }

IAdminService.java 的代码如下：

```java
package com.service;
import com.orm.Admin;

public interface  IAdminService {
    //获取管理员
    public Admin loadAdmin(int id);
}
```

Admin.hbm.xml 的代码如下：

```xml
<?xml version="1.0"?>
<!DOCTYPE hibernate-mapping PUBLIC
"-//Hibernate/Hibernate Mapping DTD 3.0//EN"
"http://hibernate.sourceforge.net/hibernate-mapping-3.0.dtd">
<!--Generated 2012-11-22 22:11:50 by Hibernate Tools 3.2.1.GA-->
<hibernate-mapping>
    <class name="com.orm.Admin" table="admin" catalog="bbs">
        <id name="id" type="java.lang.Integer">
            <column name="id" />
            <generator class="identity" />
        </id>
        <property name="account" type="string">
            <column name="account" length="10" not-null="true" />
        </property>
        <property name="password" type="string">
            <column name="password" length="20" not-null="true" />
        </property>
        <property name="qx" type="int">
            <column name="qx" not-null="true" />
        </property>
        <property name="nickName" type="string">
            <column name="nickName" length="10" />
        </property>
        <property name="name" type="string">
            <column name="name" length="10" />
        </property>
        <property name="photoPath" type="string">
            <column name="photoPath" length="100"·/>
        </property>
        <set name="posts" inverse="true">
            <key>
                <column name="aid" />
```

```xml
            </key>
            <one-to-many class="com.orm.Post" />
        </set>
        <set name="replies" inverse="true">
            <key>
                <column name="aid" />
            </key>
            <one-to-many class="com.orm.Reply" />
        </set>
        <set name="boards" inverse="true">
            <key>
                <column name="aid" not-null="true" />
            </key>
            <one-to-many class="com.orm.Board" />
        </set>
    </class>
</hibernate-mapping>
```

AdminServiceImpl.java 的代码如下:

```java
package com.serviceImpl;
import com.dao.BaseDao;
import com.orm.Admin;
import com.service.IAdminService;
import javax.annotation.Resource;

public class AdminServiceImpl implements IAdminService {
    @Resource(name="dao")
    BaseDao dao;
    @Override
    public Admin loadAdmin(int id) {
        return (Admin)dao.loadById(Admin.class, id);
    }
}
```

在图 11-16 所示页面中可以搜索帖子。输入帖子名后单击"搜索"按钮,请求提交到 post！searchPost,即由业务名为 post 的 Action 中的 searchPost() 方法执行,返回结果对应的页面为 PostSearchResult.jsp,搜索的结果如图 11-17 所示。

PostSearchResult.jsp 代码如下:

```jsp
<%@page contentType="text/html" pageEncoding="UTF-8"%>
<%@taglib prefix="s" uri="/struts-tags" %>
<!DOCTYPE html>
<html>
    <head>
        <meta http-equiv="Content-Type" content="text/html; charset=UTF-8">
        <title>搜索结果</title>
```

图 11-17　帖子搜索功能

```html
<style type="text/css">
    table{ border-collapse: collapse; text-align: center; width: 98%; }
    th{ width: 150px;}
    tr,td{  border: 1px solid silver; line-height:20px; }
    th,tr, td{ border-left: none; border-right: none; padding-bottom:5px;}
    th:hover,td:hover,tr:hover{ background-color: #F0F0F0;}
</style>
</head>
<body>
    <s:debug/>
    <s:property value="post.name"/>
    <h1>Hello World!</h1>
    <table>
        <tr style="background-color:#E7EFEF;">
            <th style="text-align:left;">帖子标题</th>
            <th>作者</th>
            <th>回复 / 点击</th>
            <th>发布时间</th>
        </tr>
        <s:iterator value="list" status="st" id="row">
            <tr>
                <td style="text-align: left;"><img src="<%=request.getContextPath()%>/images/folder_new.gif" style="display: inline-block; margin: 4px 5px 0px 0px;" /><a href="post!viewDetail.action?pid=<s:property value="id" />"><s:property value="name" /></a></td>
                <td><s:property value="#row.getStudent().getNickName()" /></td>
                <td><s:property value="#row.getReplies().size()" />/<s:property value="count" /></td>
                <td>
```

```
                    <!--<s:property value="publishTime" />-->
                    <s:date format="yyyy-MM-dd hh:mm:ss"
                    name="publishTime" />
                </td>
            </tr>
        </s:iterator>
        </table>
    </center>
</body>
</html>
```

如果找到了要搜索的帖子，可单击帖子，将出现如图 11-18 所示的查看帖子页面 （ViewPostDetail.jsp）。URL 中的参数为 post! viewDetail。

图 11-18　查看帖子页面

ViewPostDetail.jsp 的代码如下：

```
<%@page contentType="text/html" pageEncoding="UTF-8"%>
<%@taglib prefix="s" uri="/struts-tags" %>
<html>
    <head>
        <meta http-equiv="Content-Type" content="text/html; charset=UTF-8">
        <title>帖子</title>
        <script type="text/javascript" src="<%=request.getContextPath()%>
            /ckeditor_3.6.2/ckeditor/ckeditor.js"></script>
        <script type="text/javascript">
            window.onload=function(){
                var replyBtn=document.getElementById("replyBtn");
```

```
            var replyDiv=document.getElementById("replyDiv");
            replyBtn.onclick =function(){
                if (replyDiv.style.display=="none") {
                    replyDiv.style.display="block";
                }
                else if (replyDiv.style.display=="block") {
                    replyDiv.style.display="none";
                }
            }
        }
    </script>
    <style type="text/css">
        table{ border-collapse: collapse; width: 95%;}
        th, tr,td{ border: 1px solid silver; background-color: #E3F2E1;}
        th{ height: 32px;}
        #nav{ margin:10px 0 10px 30px; font-size: 15px; }
        .left_td{width:15%; text-align: left; padding-left:30px;   }
        .right_td{width:70%; text-align: left; padding-left: 30px; vertical-align: text-top; background-color: white;}
        #replyDiv{display: block; margin-left: 50px;   }
        #myImg{ width: 128px; height:128px; }
        #pmain li{ margin: 8px ;}
        #fastReply{ margin-top: 30px;}
    </style>
</head>
<body>
    <div id="nav">
    当前位置:    <a href="index.action">首页 </a>  >>   
    <s:propertyvalue="%{post.board.name}" />   >>     <s:property
    value="%{post.name}" />
    </div>
    <!--帖子主题-->
    <center>
    <div id="pmain">
    <table>
        <th>作者</th><th>正文</th>
        <tr>
            <td class="left_td" valign="top">
                <br/>
                <s:if test="%{post.student.photoPath!=null}">
                    <img id="myImg"
                      src="<%= request.getContextPath()%>/upload/<s:property
                      value="%{post.student.photoPath}"   />" />
                </s:if>
                <s:else>
```

```jsp
            <img src="<%=request.getContextPath()%>
                /images/bbsPhoto.jpg" />
        </s:else>
        <br/>
        <ul>
        <s:if test="#session.student!=null">
            <li>   <h5>
                <b>昵称：
                <s:property value="%{post.getStudent().getNickName()
                    }" /></b>
            </h5>
            </li>
            <li><h5>
                <b>专业：
                <s:property value="%{post.getStudent().major
                    }" /></b>
            </h5>
            </li>
            <li>
            <h5>
                <b>班级：
                <s:property value="%{post.getStudent().className
                    }" /></b>
            </h5>
            </li>
        </s:if>
        <s:elseif test="#session.admin!=null">
            <li><h5>
                管理员
            </h5>
                </li>
            <li><h5>
                <b>姓名：<s:property value="%{post.admin.name}" />
                </b>
            </h5>
            </li>
        </s:elseif>
        </ul>
</td>
<td class="right_td">
    <h5>发帖时间：<s:date format="yyyy-MM-dd hh:mm:ss"
        name="%{post.publishTime}" /></h5>
    <s:property value="%{post.content}" escape="false" />
    <s:if test="#session.admin!=null">
        <h5><a href="reply!prepareModify.action">修改
```

```
                </a></h5>
                <h5><a href="#">删除</a></h5>
            </s:if>
        </td>
    </tr>
    <!--回帖列表-->
    <s:iterator value="replies" id="row" status="st">
        <tr>
            <td class="left_td" align="top" valign="top">
                <br/>
                <s:if test="#row.getStudent().photoPath!=null">
                    <img id="myImg"
                        src="<%=request.getContextPath()%>
                        /upload/<s:property value="#row.student.
                        photoPath" />" />
                </s:if>
                <s:else>
                    <img src="<%=request.getContextPath()%>
                        /images/bbsPhoto.jpg" />
                </s:else>
                <br/>
                <h5>
                    <b>昵称：
                    <s:property value="#row.getStudent().nickName" />
                    </b>
                </h5>
                <h5>
                    <b>专业：
                    <s:property value="#row.getStudent().major" />
                    </b>
                </h5>
                <h5>
                    <b>班级：
                    <s:property value="#row.getStudent().className" />
                    </b>
                </h5>
            </td>
            <td class="right_td">
                <h5>回复时间：<s:date format="yyyy-MM-dd
                hh:mm:ss" name="#row.getPublishTime()" /></h5>
                <s:property value="#row.getContent()" escape="false"/>
            </td>
        </tr>
    </s:iterator>
</table>
```

```html
<div id="fastReply">
    <s:form action="reply!stuReply.action?pid=%{post.id} ">
        <tr>
            <td style="width:19.5%;" valign="middle" align="center">
                <s:if test="#session.admin==null">
                    <img id="myImg"
                         src="<%=request.getContextPath()%>
                         /upload/<s:property
                         value="%{student.photoPath}" />" />
                </s:if>
                <h2>回复:</h2></td>
            <td>
                <textarea id="context" name="reply.content" ></textarea>
            </td>
        </tr>
        <tr>
            <td></td>
            <td>
                <input type="submit" value="" style=" border: none;
                cursor: pointer; width: 74px; height: 31px; background:
                url('<%=request.getContextPath()%>/images/btn_reply.
                png') no-repeat;" />
            </td>
        </tr>
    </s:form>
    <script type="text/javascript">
    //<![CDATA[
      //Replace the <textarea id="editor1"> with a CKEditor instance
      using default configuration.
    CKEDITOR.replace( 'context',
    {
        filebrowserImageUploadUrl   :'uploadImg.action',
        filebrowserImageBrowseUrl   :'showImage.jsp?type=image',
        toolbar :'Full',
        width:'100%',
        height:'50%',
        filebrowserWindowWidth  : 700,
        filebrowserWindowHeight : 500
    });
    //]]>
    </script>
</div>
</center>
</div>
</body>
```

```
</html>
```

在图 11-18 所示页面中可回复帖子,回复后请求提交到 reply!stuReply,参考 ViewPostDetail.jsp 中代码,即请求提交到业务控制器 ReplyAction(ReplyAction.java)的 stuReply()方法。该 Action 是一个动态 Action,其还封装了修改和删除帖子的方法。

ReplyAction.java 的代码如下:

```
package com.action;
import com.opensymphony.xwork2.ActionContext;
import com.opensymphony.xwork2.ActionSupport;
import com.orm.Post;
import com.orm.Reply;
import com.orm.Student;
import com.service.IPostService;
import com.service.IReplyService;
import java.util.List;
import javax.annotation.Resource;
import javax.servlet.http.HttpServletRequest;
import org.apache.struts2.ServletActionContext;

public class ReplyAction extends ActionSupport {
    private int pid;
    public int getPid() {
        return pid;
    }
    public void setPid(int pid) {
        this.pid=pid;
    }
    private Reply reply;
    public Reply getReply() {
        return reply;
    }
    public void setReply(Reply reply) {
        this.reply=reply;
    }
    private List<Reply> replys;
    @Resource(name="replyService")
    IReplyService replyService;
    @Resource(name="postService")
    IPostService postService;
    @Override
    public String execute() throws Exception {
        return super.execute();
    }
    //回复帖子
    public String stuReply() throws Exception
```

```java
    {
        HttpServletRequest request=ServletActionContext.getRequest();
        int id=Integer.parseInt(request.getParameter("pid"));
        Student student =
            (Student)ActionContext.getContext().getSession().get("student");
        try
        {
            Post post=postService.loadPost(id);
            replyService.stuReplyPost(student, post, reply);
            //获取学生帖子
            setReplys(replyService.getReplysByPid(id));
            //设置 pid 回传给查看帖子方法
            request.setAttribute("pid", id);
            return SUCCESS;
        }
        catch (Exception e) {
            return ERROR;
        }
    }
    //准备修改回复
    public String prepareModifyReply() throws Exception
    {
        //要到达修改页面就要先获取回复 id 号
        int rid=-1;
        if (ServletActionContext.getRequest().getParameter("rid")!=null) {
            rid=Integer.valueOf(ServletActionContext.getRequest().getParameter
                ("rid"));
            setReply(replyService.loadReply(rid));
            return SUCCESS;
        }
        return ERROR;
    }
    public String modifyReply() throws Exception
    {
        //要先到达修改页面再提交修改
        //仅修改内容
        getReply().setContent(getReply().getContent());
        if (replyService.modifyReply(getReply())) {
            return "modifySuccess";
        }
        return ERROR;
        
    }
     public String delReply() throws Exception
     {
```

```
            int rid=-1;
            if (ServletActionContext.getRequest().getParameter("rid")!=null) {
                rid=Integer.valueOf(ServletActionContext.getRequest().getParameter
                    ("rid"));
                replyService.delReply(rid);
                return "deleteSuccess";
            }
            return ERROR;
        }

        public List<Reply>getReplys() {
            return replys;
        }   public void setReplys(List<Reply>replys) {
            this.replys=replys;
        }
    }
```

ReplyAction 类中使用了接口 IReplyService(IReplyService.java)，该接口封装了帖子回复常用方法，实现该接口的类为 ReplyServiceImpl(ReplyServiceImpl.java)，注意只有登录后方可回帖，为了进行权限控制，定义了一个拦截器类 AuthorityInterceptor (AuthorityInterceptor.java)，并在 struts.xml 中配置该拦截器。

IReplyService.java 的代码如下：

```
package com.service;
import com.orm.Post;
import com.orm.Reply;
import com.orm.Student;
import java.util.List;

public interface IReplyService {
    //加载一条回复
    public Reply loadReply(int rid);
    //获取回复
    public List<Reply>getReplysByPid(int pId);
    //学生回帖
    public boolean stuReplyPost(Student student,Post post,Reply reply);
    //修改回复
    public boolean modifyReply(Reply reply);
    //删除回复
    public boolean delReply(int rid);
}
```

ReplyServiceImpl.java 的代码如下：

```
package com.serviceImpl;
```

```java
import com.dao.BaseDao;
import com.orm.Post;
import com.orm.Reply;
import com.orm.Student;
import com.service.IReplyService;
import java.text.SimpleDateFormat;
import java.util.Date;
import java.util.List;
import javax.annotation.Resource;

public class ReplyServiceImpl  implements IReplyService {
    @Resource(name="dao")
    BaseDao dao;
    public boolean stuReplyPost(Student student,Post post,Reply reply) {
        try {
            reply.setStudent(student);
            reply.setPost(post);
            reply.setContent(reply.getContent());
            reply.setPublishTime(new Date());
            dao.saveOrUpdate(reply);
            return true;
        }
        catch (Exception e){
           return false;
        }
    }
    public List<Reply>getReplysByPid(int pId) {
         try {
            List<Reply>replys=(List<Reply>) dao.loadObject("from reply as r
                where r.pid='"+pId+"'    ");
            return replys;
        } catch (Exception e) {
            return null;
        }
    }
    public boolean modifyReply(Reply reply) {
        try {
            dao.saveOrUpdate(reply);
            return true;
        } catch (Exception e) {
            return false;
        }
    }
    public boolean delReply(int rid) {
        try {
```

```java
            dao.delById(Reply.class, rid);
            return true;
        } catch (Exception e) {
            e.printStackTrace();
            return false;
        }
    }
    public Reply loadReply(int rid) {
        return (Reply) dao.loadById(Reply.class, rid);
    }
}
```

AuthorityInterceptor.java 的代码如下：

```java
package com.interceptor;
import com.opensymphony.xwork2.Action;
import com.opensymphony.xwork2.ActionContext;
import com.opensymphony.xwork2.ActionInvocation;
import com.opensymphony.xwork2.interceptor.MethodFilterInterceptor;
import com.orm.Admin;
import com.orm.Student;
import java.util.Map;

public class AuthorityInterceptor extends MethodFilterInterceptor {
    protected String doIntercept(ActionInvocation ai) throws Exception {
        ActionContext ctx=ai.getInvocationContext();
        Map session=ctx.getSession();
        f (session.get("student") ==null && session.get("admin") ==null ) {
            return Action.LOGIN;
        }
        else
        {
            Student stu= (Student)session.get("student");
            Admin admin= (Admin)session.get("admin");
            if (stu!=null) {
                if (stu.getNickName() ==null) {
                    return  "stuPersonalInfo";
                }
                return ai.invoke();
            }
            if (admin!=null) {
                if (admin.getNickName() ==null) {
                    return "adminPersonalInfo";
                }
                return ai.invoke();
            }
```

```
        }
        return Action.LOGIN;
    }
}
```

回复帖子成功页面 replySuccess.jsp 的代码如下:

```jsp
<%@page contentType="text/html" pageEncoding="UTF-8"%>
<%@taglib prefix="s"  uri="/struts-tags"%>
<html>
    <head>
        <meta http-equiv="Content-Type" content="text/html; charset=GBK">
        <title>回复成功</title>
         <script type="text/javascript">
            function countDown(secs,surl){
                //alert(surl);
                var jumpTo=document.getElementById('jumpTo');
                jumpTo.innerHTML=secs;
                if(--secs>0){
                    setTimeout("countDown("+secs+",'"+surl+"')",1000);
                }
                else{
                    location.href=surl;
                }
            }
        </script>
    </head>
    <body>
        <center>
        <div style="margin-top:5%;">
            <div id="title" style="border-bottom:1px solid #C8DCEC; width: 50%;
            text-align: left; "><h3>论坛提示:</h3></div>
             <span style="position:relative; left:-180px; top:40px;"><img src=
            "<%=request.getContextPath()%>/images/right_big.gif" /></span>
            <h3>回复成功!<span id="jumpTo" style="color:orange;">5</span> 
            秒后自动跳转到论坛首页</h3>
             <script type="text/javascript">countDown(5,'login!index.action');
             </script>
            <h3>< a href ="<% = request. getContextPath () % >/post! viewDetail.
            action?pid=%{#request.pid}">[若没有自动跳转,请点击这里]</a></h3>
        </div>
        </center>
    </body>
</html>
```

单击图 11-16 所示页面中的"修改",出现如图 11-19 所示的页面。根据配置文件和

Action可以看出图11-19所示页面地址栏中的参数为post！prepareModify。对应的JSP页面为ModifyPost.jsp。

图11-19 修改帖子

ModifyPost.jsp的代码如下：

```
<%@page import="java.util.Date"%>
<%@page contentType="text/html" pageEncoding="UTF-8"%>
<%@taglib prefix="s"  uri="/struts-tags" %>%>
<html>
    <head>
        <meta http-equiv="Content-Type" content="text/html; charset=UTF-8">
        <title>修改帖子</title>
        <script type="text/javascript" src="<%=request.getContextPath() %>
            /ckeditor_3.6.2/ckeditor/ckeditor.js"></script>
        <style type="text/css">
            th:hover,td:hover,tr:hover{ background-color: #F0F0F0;}
        </style>
    </head>
<body>
    <s:debug/>
<center>
    <s:url id="modifyPost" action="post!modifyPost.action" >
        <s:param name="pid"><%=request.getParameter("pid") %></s:param>
    </s:url>
    <s:form action="%{modifyPost}">
```

```html
<table>
    <tr>
        <td><b>帖子标题:</b></td>
        <td><input type="text" name="post.name"
            value="${post.name}" size="50" /></td>
    </tr>
    <tr>
        <td valign="top"><b>帖子内容:</b></td>
        <td><textarea id="context" name="post.content">
            <s:property value="post.content"/></textarea>  </td>
    </tr>
    <tr>
        <td></td><td><input type="submit"  value="发表" /></td>
    </tr>
</table>
</s:form>
<script type="text/javascript">
    //<![CDATA[
    //Replace the <textarea id="editor1"> with a CKEditor instance using
     default configuration.
    CKEDITOR.replace( 'context',
    {
    filebrowserImageUploadUrl  :'uploadImg.action',
    filebrowserImageBrowseUrl  :'showImage.jsp?type=image',
    toolbar :'Full',
    width:'700',
    height:'500',
    filebrowserWindowWidth  : 700,
    filebrowserWindowHeight : 500
    });
    //]]>
</script>
</center>
</body>
</html>
```

修改后提交到 post!modifyPost。单击"删除"请求提交到 post!deletePost。

单击图 11-16 所示页面中的"发帖"按钮,出现如图 11-20 所示的发帖页面。请求提交到 post!preparePost,页面为 AddPost.jsp。

AddPost.jsp 的代码如下:

```
<%@page import="java.util.Date"%>
<%@page contentType="text/html" pageEncoding="UTF-8"%>
<%@taglib prefix="s"  uri="/struts-tags" %>
<html>
    <head>
```

图 11-20 发帖页面

```
        <meta http-equiv="Content-Type" content="text/html; charset=UTF-8">
        <title>发表帖子</title>
        <script type="text/javascript" src="<%=request.getContextPath() %>
        /ckeditor_3.6.2/ckeditor/ckeditor.js"></script>
</head>
<body>
<center>
    <s:url id="addPost" action="post!addPost.action" >
        <s:param name="bid"><%=request.getParameter("bid") %></s:param>
    </s:url>
    <s:form action="%{addPost}">
    <table>
        <tr>
            <td><b>帖子标题:</b></td>
            <td><input type="text" name="post.name" size="50" /></td>
        </tr>
        <tr>
            <td valign="top"><b>帖子内容:</b></td>
            <td><textarea id="context" name="post.content" ></textarea></td>
        </tr>
        <tr>
            <td></td><td><input type="submit"  value="发表" /></td>
        </tr>
    </table>
    </s:form>
    <script type="text/javascript">
```

```
            //<![CDATA[
            //Replace the <textarea id="editor1"> with a CKEditor instance using
             default configuration.
            CKEDITOR.replace( 'context',
            {
            filebrowserImageUploadUrl  :'uploadImg.action',
            filebrowserImageBrowseUrl  :'showImage.jsp?type=image',
            toolbar :'Full',
            width:'700',
            filebrowserWindowWidth  : 700,
            filebrowserWindowHeight : 500
            });
            //]]>
        </script>
        </center>
    </body>
</html>
```

发帖后请求提交到 post!addPost，并返回到成功页面 postSuccess.jsp。postSuccess.jsp 的代码如下：

```
<%@page contentType="text/html" pageEncoding="UTF-8"%>
<%@taglib prefix="s"  uri="/struts-tags"%>
<!DOCTYPE html>
<html>
    <head>
        <meta http-equiv="Content-Type" content="text/html; charset=UTF-8">
        <title>发表成功</title>
        <script type="text/javascript">
            function countDown(secs,surl){
                //alert(surl);
                var jumpTo=document.getElementById('jumpTo');
                jumpTo.innerHTML=secs;
                if(--secs>0){
                    setTimeout("countDown("+secs+",'"+surl+"')",1000);
                }
                else{
                    location.href=surl;
                }
            }
        </script>
    </head>
    <body>
        <center>
        <div style="margin-top:5%;">
            <div id="title" style="border-bottom:1px solid #C8DCEC;
```

```html
                width: 50%;text-align: left; "><h3>论坛提示:</h3></div>
                <span style="position:relative; left:-180px; top:40px;">
                <img src="<%=request.getContextPath()%>/images/right_big.gif" />
                </span>
                <h3>发表成功!<span id="jumpTo" style="color:orange;">5</span> 
                秒后自动跳转到论坛首页</h3>
                <script type="text/javascript">countDown(5,'index.action');</script>
                <h3><a href="index.action">[若没有自动跳转,请点击这里]</a></h3>
            </div>
        </center>
    </body>
</html>
```

单击图 11-20 所示页面中的"我的帖子",请求提交到 post!viewPostsByUser,页面效果如图 11-21 所示,其 JSP 页面文件为 myPosts.jsp。

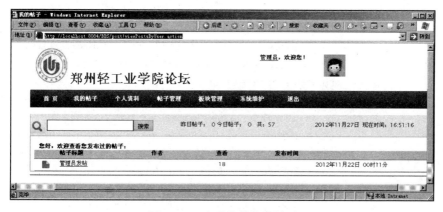

图 11-21 查看我的帖子页面

myPosts.jsp 的代码如下:

```jsp
<%@page contentType="text/html" pageEncoding="UTF-8"%>
<%@taglib prefix="s"  uri="/struts-tags" %>
<html>
    <head>
        <meta http-equiv="Content-Type" content="text/html; charset=UTF-8">
        <title>我的帖子</title>
        <style type="text/css">
            table{ border-collapse: collapse; text-align: center; width: 98%; }
            th{ width: 150px;}
            tr,td{  border: 1px solid silver; }
            th,tr, td{ border-left: none; border-right: none; padding-bottom:5px;}
            th:hover,td:hover,tr:hover{ background-color: #F0F0F0;}
        </style>
    </head>
    <body>
```

```html
            <h3 style="padding-left:20px;">
                <s:property value="#session.student.nickName" />
            您好,欢迎查看您发布过的帖子:</h3>
        <center>

            <!--帖子列表-->
            <table>
                <tr style="background-color:#E7EFEF;">
                    <th style="width:10px;"></th>
                    <th style="text-align:left;">帖子标题</th>
                    <th>作者</th>
                    <th>查看</th>
                    <th>发布时间</th>
                </tr>
                <s:iterator value="myPosts" status="st" id="row">
                <tr>
                    <td><img src="<%=request.getContextPath() %>
                        /images/folder_new.gif" style="display: inline-block;
                        margin:4px 5px 0px 0px;" /></td>
                    <td style="text-align: left;" >
                    <a href="post!viewDetail.action?pid=<s:property value="id" />">
                        <s:property value="name" /></a></td>
                    <td><s:property value="#row.getStudent().getNickName()" />
                    </td>
                    <td><s:property value="count" /></td>
                    <td>
                        <!--<s:property value="publishTime" />-->
                        <s:date format="yyyy年MM月dd日 HH时MM分"
                        name="publishTime" />
                    </td>
                </tr>
                </s:iterator>
            </table>
        </center>
    </body>
</html>
```

11.4.5 个人信息管理功能的实现

单击图 11-21 所示页面中的"个人信息",如图 11-22 所示,可以显示个人信息、上传头像以及修改个人信息(因为当前登录用户为管理员,所以显示的是管理员信息)。请求提交到 admin!personalAdminInfo,对应的 Action 是 AdminManageAction(AdminManageAction.java),修改管理员个人信息的页面为 AdminPersonalInfo.jsp。

AdminManageAction.java 的代码如下:

```
package com.action;
```

图 11-22 修改管理员信息页面

```
import com.opensymphony.xwork2.ActionContext;
import com.opensymphony.xwork2.ActionSupport;
import com.orm.Admin;
import com.service.IAdminService;
import javax.annotation.Resource;

public class AdminManageAction extends ActionSupport {
    @Resource(name="adminService")
    IAdminService adminService;
    private Admin admin;
    public Admin getAdmin() {
        return admin;
    }
    public void setAdmin(Admin admin) {
        this.admin=admin;
    }
    public String execute() throws Exception {
        return super.execute();
    }
    //显示个人信息
    public String personalAdminInfo() throws Exception{
        Admin sessionAdmin=
            (Admin) ActionContext.getContext().getSession().get("admin");
        if (sessionAdmin !=null) {
            setAdmin(adminService.loadAdmin(sessionAdmin.getId()));
            return SUCCESS;
        }
        return ERROR;
```

            }
        }

AdminPersonalInfo.jsp 的代码如下：

```jsp
<%@page contentType="text/html" pageEncoding="UTF-8"%>
<%@taglib prefix="s" uri="/struts-tags" %>
<html>
    <head>
        <meta http-equiv="Content-Type" content="text/html; charset=UTF-8">
        <title>个人信息</title>
        <style type="text/css">
            table{ border-collapse: collapse;}
            th,td{ height: 29px; }
            #profile_act{ margin-left:5%;}
            #profile_act >li { height: 20px;}
            #profile_act >li >a { padding-left: 20px; }
            #silder{float:left;}
            #info{float: left; margin-left: 5%; width: 300px;}
            #modifyInfo{ float: left; padding-left: 3%; width: 500px;}
        </style>
        <script type="text/javascript">
            function checkForm()
            {
                if (document.getElementById("nick").value=="") {
                    document.getElementById("msg").innerHTML=
                        "昵称不能为空！";
                    return false;
                }
                else{return true;}

            }
        </script>
    </head>
    <body>
        <s:debug/>
        <div style="margin-left:2%;">
            <div id="silder">
                <s:if test="admin.photoPath!=null">
                    <p>
                        <img style="width:200px; height:200px; " src="<%
                            = request.getContextPath()%>/upload/<s:property value="
                            admin.photoPath" />" />
                    </p>
                </s:if>
                <s:else>
```

```
                <p>
                    <img style="width:200px; height:200px; "
                    src="<%=request.getContextPath()%>/images/bbsPhoto.jpg" />
                </p>
            </s:else>
            <br/>
            <ul id="profile_act">
                <li><a href="#" style=" background: url('<%=request.
                getContextPath()%>/images/pmto.gif') no-repeat; ">发短消息
                </a></li>
                <li><a href="#" style=" background: url('<%=request.
                getContextPath()%>/images/addbuddy.gif') no-repeat; ">加
                为好友</a></li>
                <li><a href="#" style=" background: url('<%=request.
                getContextPath()%>/images/fastreply.gif') no-repeat; ">搜
                索帖子</a></li>
                <li><a href="#" style=" background: url('<%=request.
                getContextPath()%>/images/home.gif') no-repeat; ">个人空间
                </a></li>
            </ul>
    </div>
    <div id="info">
        <table>
            <tr>
                <th>昵称:</th>
                <td><s:property value="admin.name" /></td>
            </tr>
            <tr>
                <th>姓名:</th>
                <td><s:property value="admin.nickName" /></td>
            </tr>
        </table>
    </div>
    <div id="modifyInfo">
        信息修改
        <s:form action="student!modifyStuInfo.action"
            onsubmit="return checkForm()" >
                <s:textfield id="nick" name="student.nickName" value="%
                {student.nickName}" label="昵称" required="true"></s:
                textfield><br/>
                <s:textfield name="student.qq" value="%{student.qq}"
                label="QQ"></s:textfield><br/>
                <s:textfield name="student.email" value="%{student.
                email}" label="邮箱"></s:textfield><br/>
            <em id="msg" style="color:red;"></em>
```

```
            <s:submit value="提交" onsubmit="return checkModify()" />
        </s:form>
    </div>
        <h3><a href="<%=request.getContextPath()%>/upload.jsp">
        上传头像</a></h3>
    </div>
    <div style="clear: both;"></div>
</body>
</html>
```

如果登录的是会员(学生),单击会员的"个人资料",会出现如图 11-23 所示的页面。请求提交到 student! personalStuInfo,即由业务控制器类 StudentManageAction (StudentManageAction.java)处理,JSP 页面为 StuPersonalInfo.jsp。

图 11-23　会员个人信息页面

StudentManageAction.java 的代码如下:

```
package com.action;
import com.opensymphony.xwork2.ActionContext;
import com.opensymphony.xwork2.ActionSupport;
import com.orm.Student;
import com.service.IStudentService;
import javax.annotation.Resource;

public class StudentManageAction extends ActionSupport {
    @Resource(name="studentService")
    IStudentService studentService;
    private Student student;
    private String psw;
    private String newPsw;
```

```java
private String confirmPsw;
public Student getStudent() {
    return student;
}
public void setStudent(Student student) {
    this.student=student;
}
public String execute() throws Exception {
    return super.execute();
}
public String getPsw() {
    return psw;
}
public void setPsw(String psw) {
    this.psw=psw;
}
public String getNewPsw() {
    return newPsw;
}
public void setNewPsw(String newPsw) {
    this.newPsw=newPsw;
}
public String getConfirmPsw() {
    return confirmPsw;
}
public void setConfirmPsw(String confirmPsw) {
    this.confirmPsw=confirmPsw;
}
//显示个人信息
public String personalStuInfo() throws Exception
{
    Student sessionStudent=
        (Student)ActionContext.getContext().getSession().get("student");
    if (sessionStudent !=null) {
        setStudent(studentService.getStudentByStuNum(
            sessionStudent.getStuNum()));
        return SUCCESS;
    }
    return ERROR;
}
//修改个人信息
public String modifyStuInfo() throws  Exception
{
    Student snStudent=
        (Student) ActionContext.getContext().getSession().get("student");
```

```java
        Student s=studentService.getStudentByStuNum(snStudent.getStuNum());
        Student temp=getStudent();
        s.setNickName(temp.getNickName());
        s.setQq(temp.getQq());
        s.setEmail(temp.getEmail());

        if (studentService.modifyStudent(s)) {
            return "modifySuccess";
        }
        return ERROR;
    }
    //修改密码
    public String ModifyPsw() throws Exception
    {
        //获取当前登录学生个人信息
        Student sessionStudent=
            (Student) ActionContext.getContext().getSession().get("student");
        Student s=
          studentService.getStudentByStuNum(sessionStudent.getStuNum());
        if (getPsw().equals(s.getPassword()) &&
          getNewPsw().equals(getConfirmPsw())  ) {
            s.setPassword(getNewPsw());
            studentService.modifyStudent(s);
            addActionMessage("密码修改成功!");
            return "modifyPswSuccess";
        }
        return ERROR;
    }
}
```

StuPersonalInfo.jsp 的代码如下：

```jsp
<%@page contentType="text/html" pageEncoding="UTF-8"%>
<%@taglib prefix="s" uri="/struts-tags" %>
<!DOCTYPE html>
<html>
    <head>
        <meta http-equiv="Content-Type" content="text/html; charset=UTF-8">
        <title>个人信息</title>
        <style type="text/css">
            table{ border-collapse: collapse;}
            th,td{ height: 29px; }
            #profile_act{ margin-left:5%;}
            #profile_act >li { height: 20px;}
            #profile_act >li >a { padding-left: 20px; }
            #silder{float:left;}
```

```
            #info{float: left; margin-left: 5%; width: 200px;}
            #modifyInfo{ float: left; padding-left: 3%; }
        </style>
        <script type="text/javascript">
            function checkStu()
            {
                if (document.getElementById("nick").value=="") {
                    document.getElementById("msg").innerHTML=
                        "昵称不能为空!";
                    return false;
                }
                else{return true;}

            }
        </script>
</head>
<body>
    <div style="margin-left:2%;">
        <div id="silder">
            <s:if test="student.photoPath!=null">
            <p>
                < img style="width:200px; height:200px; " src="<%=request.
                getContextPath()%>/upload/<s:property value="student.
                photoPath" />" />
            </p>
            </s:if>
            <s:else>
            <p>
                < img style="width:200px; height:200px; " src="<%=request.
                getContextPath()%>/images/bbsPhoto.jpg" />
            </p>
            </s:else>
            <br/>
            <ul id="profile_act">
                <li><a href="#" style=" background: url('<%=request.
                getContextPath()%>/images/pmto.gif') no-repeat; ">发短
                消息</a></li>
                <li><a href="#" style=" background: url('<%=request.
                getContextPath()%>/images/addbuddy.gif') no-repeat; ">
                加为好友</a></li>
                <li><a href="#" style=" background: url('<%=request.
                getContextPath()%>/images/fastreply.gif') no-repeat; ">
                搜索帖子</a></li>
                <li><a href="#" style=" background: url('<%=request.
                getContextPath()%>/images/home.gif') no-repeat ; ">个人
```

```html
                    空间</a></li>
            </ul>
        </div>
        <div id="info">
            <table>
                <tr>
                    <th>昵称:</th>
                    <td><s:property value="student.nickName" /></td>
                </tr>
                <tr>
                    <th>姓名:</th>
                    <td><s:property value="student.realName" /></td>
                </tr>
                <tr>
                    <th>学号:</th>
                    <td>
                        <s:property value="student.stuNum" />
                    </td>
                </tr>
                <tr>
                    <th>QQ:</th>
                    <td style="word-break:all">
                        <s:property value="student.qq" />
                    </td>
                </tr>
                <tr>
                    <th>邮箱:</th>
                    <td style="word-break:all">
                      <s:property value="student.email" />
                    </td>
                </tr>
                <tr>
                    <th>班级:</th>
                    <td style="word-break:all">
                        <s:property value="student.className" />
                    </td>
                </tr>
                <tr>
                    <th>专业:</th>
                    <td style="word-break:all">
                        <s:property value="student.major" />
                    </td>
                </tr>
            </table>
        </div>
```

```
<div id="modifyInfo">
    信息修改
    <s:form action="student!modifyStuInfo.action"
        onsubmit="return checkStu()" >
        <s:textfield id="nick" name="student.nickName" value="%
        {student.nickName}" label="昵称" required="true"></s:
        textfield><br/>
        <s:textfield name="student.qq" value="%{student.qq}" label
        ="QQ"></s:textfield><br/>
        <s:textfield name="student.email" value="%{student.
        email}" label="邮箱"></s:textfield><br/>
        <em id="msg" style="color:red;"></em>
        <s:submit value="提交" onsubmit="return checkModify()" />
    </s:form>
</div>
<h3><a href="<%=request.getContextPath()%>/upload.jsp">上传头像
    </a>
</h3>
</div>
<div style="clear: both;"></div>
</body>
</html>
```

处理完后请求提交到 student! modifyStuInfo, 也可以上传头像。

单击图 11-23 所示页面中的"修改密码", 可以对当前登录用户的密码进行修改, 修改页面(StuPswModify.jsp)如图 11-24 所示。

图 11-24 会员修改密码

StuPswModify.jsp 的代码如下:

```
<%@page contentType="text/html" pageEncoding="UTF-8"%>
<%@taglib prefix="s" uri="/struts-tags" %>
<!DOCTYPE html>
```

```html
<html>
    <head>
        <meta http-equiv="Content-Type" content="text/html; charset=UTF-8">
        <title>修改密码</title>
    </head>
    <body>
    <center>
        <h1>温馨提示:请定期维护您的账户信息</h1>
        <s:debug/>
        <br/>
        <h4><s:actionmessage/></h4>
        <s:form action="student!ModifyPsw.action">
            <s:password name="psw" label="原始密码"></s:password>
            <s:textfield name="newPsw" label="新密码"></s:textfield>
            <s:textfield name="confirmPsw" label="确认密码"></s:textfield>
            <s:submit value="确认"></s:submit>
        </s:form>
    </center>
    </body>
</html>
```

修改后请求提交到"<s:form action="student!ModifyPsw.action">"。

11.5 本章小结

本章实现了一个校园 BBS 系统的基本功能,综合运用了 Java Web 的多种技术,包括 JSP、JavaBean、Struts2、Spring3、Hibernate 以及其他页面开发技术。通过本章项目的训练有助于充分理解、有效整合 Java Web 知识体系中的多项相关技术,培养和积累项目开发经验。

11.6 习题

实训题

1. 请根据所熟悉的 BBS 系统改进本章项目的功能。
2. 请为本章 BBS 系统添加会员管理功能。

参 考 文 献

[1] 张志锋.Struts2+Hibernate框架技术教程[M].北京:清华大学出版社,2012.
[2] 王建国.Struts+Spring+Hibernate框架及应用开发[M].北京:清华大学出版社,2011.
[3] 王颖玲.基于Struts和Hibernate技术的Web开发应用[M].北京:清华大学出版社,2011.
[4] 王伟平.Struts2完全学习手册[M].北京:清华大学出版社,2010.
[5] 蒲子明.Struts2+Hibernate+Spring整合开发技术详解[M].北京:清华大学出版社,2010.
[6] 李刚.Struts 2.1权威指南[M].北京:电子工业出版社,2009.
[7] 邹继成.Struts与Hibernate实用教程[M].北京:电子工业出版社,2008.